WITHDRAWN

SCIENCE
IN
CIVILIZATION

Cover design and line illustrations by
Frank Reising and William Chrysler

SCIENCE IN CIVILIZATION

Theodore W. Jeffries

Lorain County Community College

KENDALL/HUNT PUBLISHING COMPANY
DUBUQUE, IOWA

Preface

For some years I have felt the need for a humanistic directed text in Physical Science written primarily for community college students. One of the purposes of this book is to break down the artificial walls which have been erected among various fields of knowledge by showing some of the ties, links, and interconnections among disciplines. Toward this end the ability of science and technology to stimulate responses in areas such as literature, philosophy, economics, and political science has been explored as well as the effects that other disciplines and areas have had on science and technology.

This book teaches about science rather than teaching science. In a broad sense, it may be considered science appreciation just as the liberal arts' or general studies' student has an exposure to art appreciation and music appreciation.

Many institutions have survey courses in physical science which are designed to acquaint the general college student not majoring in science with some of the methods of, and an appreciation of the various fields—chemistry, physics, astronomy, and geology. Since these attempts are surveys of rather broad fields, they tend to lack in unity and cohesion and degenerate into a hodge-podge scattering of unrelated facts, theories, and experiments at about a high school level.

The more successful survey courses of physical science have a central theme and restrict the areas with which they deal. The central theme in this vehicle is the ideas and discoveries from science and technology which have had the greatest impact on the totality of mankind and an exploration of some of the facets of that impact. Normally such a course would not include the steam engine or automation as topics; even though there would be a discussion of the laws of thermodynamics and some of the theory of electronics. Although the scientist would admit the importance of such devices to our society, he would not allow that the discussion of their impact is germane to a course in the natural sciences. But if the student is not exposed here to the consequences of scientific

discoveries and technological innovations, where will he get it? Perhaps in history? But the historian, although he would agree to some discussion of consequences, would exclaim that he lacked sufficient scientific and technical training for an adequate presentation.

In some respects this course deals with a sort of never-never-land. It includes areas that are important to the educated man of today but that are not normally treated in undergraduate courses. In order to treat these areas in the traditional manner would proliferate additional courses in an already overcrowded curriculum. As a general rule, Einstein's general theory of relativity is not included in the course work of the physics major until late in his graduate studies, nor is the mathematics major exposed to non-Euclidian geometry. It is true that the scientist or the mathematician require considerable training and mathematic background before tackling such advanced topics but for the non-science major the same mathematical training is not needed for an understanding of the pertinent ideas and their implications.

Once every educated man had a knowledge of medicine, law and theology as well as the ability to read Latin and Greek. To know the latest theories and discoveries of science and to have read the latest novel was fashionable and stylish. Today a man is orientated early towards his vocational goals and he counts himself well off to know his field and the latest developments in it. Knowledge has become specialized fields and each field has developed its own brand of terminology which sets up a barrier between the person who would learn and the field that he wishes to learn about. Many of the proofs in science are very indirect ones and are not necessarily apparent. In addition, each field has grown to the extent that no man can know even one field, but rather selects only a portion for detailed study.

Under a totalitarian government, there is little reason for a general education other than the cultural development of the individual. In most European countries, a general education is restricted to the "best" students. The other students receive a vocational training.

In a representative form of government, such as we have, the premise is made that not only should the citizen have vocational training for his livelihood, cultural development for his appreciation and enjoyment, but also enough of a general education to be able to judge the value of government actions. This judgment the citizens give at the polls.

Many of the decisions made by government today involve problems arising from science and technology—the "atomic" bomb, automation, space travel—and you are, or will be, paying the bill. You might ask, "Why not let the expert solve the problem?"

Unfortunately, the social problems and consequences that arise from a scientific discovery or a technical innovation may have a number of solutions and the scientist or expert cannot use his methods to determine which one should be used. In any situation that requires a value judgment, your solution may be as valid as that of the expert.

As a part of this course, we shall examine some of the methods of the scientist to allow the student to see the types of problems that the scientist can handle, as well as to determine those areas that the scientist cannot solve by his methods. We will see that some problems have had different explanations and that one set of data does not give a unique solution to the problem.

One problem that looms over all is the attitude that many persons hold about scientists and science. To some, science is the religion of the twentieth century and the scientist is the high priest. Science is the miracle worker, the ultimate authority. The latest development from the laboratory is more important than the second coming. The public has bought science—hook, line, and sinker. This is not to say that science is not important but it must be placed in proper perspective. Science is not the answer to everything.

Some of the ideas of science which have had the least practical application have had the greatest impact on the way that man thinks and the way in which he views his world. Does it really matter if the sun goes around the earth or if the earth goes around the sun? If fossils were created instantly, came from the Flood, or are a product of millions of years—what practical difference does it make? Is there order in the universe? Regardless of whether gravity is to be explained by Newton or Einstein—objects still fall. The ways we think in the twentieth century come, in part, from ideas in science.

The content of the major theme lies in four broad areas. These include: the revolutions of motion—heavenly, earthly, and the minuscule; the technological revolutions—agricultural, commercial, printing, industrial (steam, mass production, automation), transport, electronics, and synthesis; the nature of matter-energy; and environmental—the interaction of man with his environment.

The major themes are cast against the milieu of the times and are sometimes arranged chronologically. The heavenly revolution of Copernicus could be, and sometimes is, considered outside the framework of the Renaissance and Reformation but to study the revolution with understanding requires the context. It could be asked, why study the Copernican revolution since few would argue that the earth is the center of the solar system? We study it not only for the terrific impact that the idea had at that time, but also because it gives us a paradigm of idea revolutions. In addition, such a study may help us in our understanding of the idea revolution that we are in today. Today's revolution rose from a consideration of the motion of the minuscule and the nature of matter-energy. Just as Descartes tried to reorient man to an order from the chaos caused by Copernicus we will consider Sartre's attempt to do the same for the twentieth century.

In the earthly motion, Newton (and the industrial revolution) accomplishes what Descartes cannot. Perhaps we of the twentieth century are on the verge of another breakthrough that will permit the confidence and optimism of the eighteenth and nineteenth centuries to be resumed. At any rate, the chaos of the twentieth century will be examined from the standpoint of science as a contributory factor.

Although an attempt will be made to present several points of view on the various questions, I am sure that my biases and prejudices will prevail. I am addicted to punning and tongue in cheek remarks to the extent that the reader needs to be warned to take comments with a grain of salt although others may find the material too spicy.

While I cannot trace all those who have contributed to my thoughts in the preparation of this book, I would like to identify those that I remember best or those who have contributed identifiable strands of development. I wish to thank my students of the past five years in whose forge of criticism fragments of the book have been used and tested for their criticism and tolerance. Professors Losh and Tonser of the University of Michigan who gave me the basic idea for the chapter dealing with the Rise of Nineteenth Century Astronomy. My thanks are also due to Professor Zumberger, now president of Grand Valley State College, for some of the material on geology and to W. H. Cowley of Stanford University for some of his ideas. Professors Brooks and Gilmore of our institution have been kind enough to read portions of the material and offer their suggestions as well as Professor Sandrew, formerly of our institution. While a number of typists have labored in the preparation of the manuscript, special thanks goes to Mrs. Patsy J. Swegard who has been through several revisions of the work as well as most of the final draft.

Elyria, Ohio   Theodore W. Jeffries
Christmas, 1970

ACKNOWLEDGMENTS

The author wishes to acknowledge the following as special sources of material.

The council of the Early English Text Society for their kind permission to reproduce portions of Oliver H. Prior (ed.) Caxton's MINOAR OF THE WORLD.

The Horizon Press for permission to use "Newton, The Man" from ESSAYS IN BIOGRAPHY by Maynard Keynes, copyright 1951, reprinted by permission of the publisher, Horizon Press, New York.

Encyclopaedia Britannica for their permission to quote from the Books of Ptolemy, Gilbert, and Milton in their series GREAT BOOKS OF THE WESTERN WORLD.

For a definition of mathematics, the ENCYCLOPEDIA AMERICANA © Americana Corporation 1971 gave its permission.

Special thanks to Ray Osrin of the Cleveland PLAIN DEALER for permission to use his cartoon THE THINKER.

The author also wishes to acknowledge the following sources from which various quotes were used by permission.

From KEPIER by Max Caspar. © Copyright 1959 by Abelard-Schuman Limited. By permission of Abelard-Schuman Limited.

Reprinted by permission of DODD, MEAD & COMPANY, INC. from THE AUTOMOBILE by Frank Ernest Hill. Copyright © 1967 by Frank Ernest Hill.

Reprinted with permission from SCIENCE NEWS, the weekly news magazine of science, copyright 1970 by Science Service, Inc.

Reprinted by permission of Grove Press, Inc. Translated from the French by Donald M. Allen. Copyright © 1958 by Grove Press, Inc.

Originally published by the University of California Press; reprinted by permission of The Regents of the University of California.

© 1970 by the New York Times Company. Reprinted by permission.

From GENERAL CHEMISTRY, First Edition, by Linus Pauling. W.H. Freeman and Company. Copyright © 1970.

From GENERAL COLLEGE CHEMISTRY by M. Snead and Lewis J. Maynard, Copyright © 1944 by Litton Educational Publishing, Inc.

From INTRODUCTORY CHEMISTRY by O.W. Nitz, Copyright © 1956, 1961 by Litton Educational Publishing, Inc.

From MODERN SCIENCE IN THE CHRISTIAN LIFE, by John W. Klotz, copyright 1961. Used by permission of Concordia Publishing House.

Reprinted by permission of Charles Scribner's Sons from AXEL'S CASTLE by Edmund Wilson. Copyright 1931 Charles Scribner's Sons; renewal copyright © 1959 Edmund Wilson.

From MODERN CHEMISTRY by Metcalfe, Dull, and Williams, copyright © 1958 by Holt, Rinehart and Winston, Inc. Reprinted by permission of the publishers.

Contents

Introduction to Science and Scientists

The way in which science is defined is dependent on who is defining it, for what purpose, at what time, and under what conditions. Looking to a dictionary is not a great help for a number of definitions are given including a very general one that science is knowledge. If this definition is used then everything known or knowable must be included within science leaving outside only the class of the emotions or that which is felt rather than known. "Tell me where is fancy bred . . ."—if in the heart then it is not science but if in the head then under the general definition it must be.

If the definition is narrowed to systematized knowledge or any knowledge for which rules can be manufactured and followed then this only excludes art or skill, or any field of creativity. Under this definition there are parts of science which would be excluded from science—any creative endeavor, since creativity is not subject to rules—and some areas of art are sciences—the laws of perspective, and pigment manufacture for example. Sometimes science has been defined as that which can be taught while art or skill has been defined as that which cannot be taught. This statement may be considered only if the definition is worded to fit. The science of boxing can be taught while the art of boxing cannot. This particular separation of knowledge probably began with Francis Bacon who saw science as something that anyone could be taught to do and continually emphasized this idea.

Considering a different definition: science is a branch of study in which facts are observed and classified to establish verifiable general laws by the methods of induction and hypotheses. If the verifiable part of the definition is emphasized then those parts of sciences or those sciences that are primarily concerned with classification only—as parts of botany and a good deal of astronomy would need to be excluded since they cannot be verified. Almost every area of biological science would need to be excluded since biological entities are one time events and cannot be repeated for verification. Dictionaries sometimes use mathematical sciences as an example but

they are deductive in method rather than inductive and most mathematicians would object vehemently to being classed under science. Many theoretical physicists would object that since their method is more deductive than inductive and in some instances not subject to verification that they would also have to be excluded. Verification here meaning reproducibility but if another meaning is substituted for verification then other areas may fit the definition.

Frequently the meaning of science today is natural science or knowledge about or relating to the physical world. Before the nineteenth century it was more apt to be called "natural philosophy," and the persons working in this area were called natural philosophers or sometimes just philosophers. Christian Science and such areas as parapsychology would be excluded by restricting science to the physical world. A question raised by some scientists is whether man himself is a legitimate object of study by science. "Many physical scientists refuse to recognize sociology and anthropology as sciences and at best consider psychology as an inferior science."[1] This question revolves not only from a consideration as to the nature of man but also as to whether, firstly, if man can be objective in dealing with himself and, secondly, if the individual is unique then can anything be said about the behavior of man as a whole or as an entity?

In this country, particularly, there is the tendency to confuse science and technology. Often there is the belief that scientists, engineers and inventors can all be lumped together because it is from them that our present material world has arisen. Most of the Nobel awards in pure science have been won by persons born or educated in Europe while most of the awards in applied science have been won by persons from the United States or England. This difference in attitude comes from the dissimilar historical developments in Europe and England and from the belief that the practical is related to work with the hands. Working with the hands is related to the belief that the craftsman is lower socially than the intellectual.

Little in the way of progress can be achieved where such an attitude prevails. What are the differences between the, so called, "pure" sciences and "applied?"

Here too we must return to the origins and note that it is at that point that "pure" science elects not to deal with the "real" world that this difference arises. To use a simple case as an illustration, when Galileo chooses to ignore air, when talking of falling bodies, he not only differs with Aristotle but he has left the "real" world.[2] The scientist is bound to the "real" world in that his self and senses are planted within a body. His senses can only record subjective impressions of the "real" world but what he describes is not that "real" world. On the other hand engineers and inventors are very much concerned with the "real" world. They want to get to particular cases while the scientist wants to get away from them.

Both the scientist and the engineer use mathematics to try to insure that their ideas will mean the same to all persons, all the time in all places. It is a communication device—a language. It is also a model just as language is a model. A word is not the same as the thing or concept that the word stands for. Also just as a word has implications and the grammar of a language may force an attitude, in a similar manner the notation of mathematics may carry an implication or the system of mathematics may force an attitude about something in the "real" world that is not a part of the "real" world. In other words the mathematical model is no longer a model.

Science deals mostly with concepts while the inventor and engineer deal with facts. The scientist does have to make use of facts just as an architect may make use of bricks but facts do not make science any more than bricks make a building. The ideal of physical scientists is to develop a law and a single equation from which all physical phenomena may be deduced and described. Science then describes and relates physical phenomena. It deals with the how of mechanisms and not the why and must also exclude cause and effect as well as value and purpose.

Who is the man? The scientist? The word "scientist" was originated by William Whewell of Cambridge in 1840. It replaced the word "natural philosopher" and in a sense became its epitaph rather than a title. For towards the end of the nineteenth century and the beginning of the twentieth there came the intensification of specialization and the death of the cultured man of the Renaissance. Who was to wear the mantle of science and the title scientist? Each scientist was a specialist—geologist, geophysicist, biologist, plant physiologist—in a narrow domain precluding others. He usually prefers to be called by the field of his specialty—zoologist, biochemist—rather than by the generic title of scientist. If his work involves chemicals or blood—anything that might damage his suit—he will wear a long, white coat; but so does the foreman at a garage or a physician. There is nothing about his attire that sets him off from the crowd. Attend a meeting of scientists and you will find nothing to distinguish them from their fellowman—they look like businessmen, merchants, policemen in plain clothes, members of the syndicate, a gathering of society, or a union meeting. Some are women, while some are short or tall, fat or skinny, bald or hairy—rather like any collection of people you might care to imagine.

In order to become a lawyer one attends a law school, passes the bar examination and is admitted to the bar; to become a physician, one attends medical school, passes the state examinations and becomes licensed to practice medicine; to become a veterinarian or dentist there is a similar procedure. On the other hand, the procedure for becoming a scientist, in theory at least, is the same as that of a garage mechanic—he works at it. There is no set procedure in either case. All a person has to do to be a mechanic is set himself up in business and buy a set of tools or be hired by a garage, and a person becoming a scientist either sets himself up in business or is hired by a company doing science.

No one person has a knowledge of all of the sciences so that in a strict sense no person can be a scientist. Some have rejected the term.[3] The most fantastic portion of "Voyage to the Bottom of the Sea" and "Dr. Quest" is the claim that Admiral Nelson and Dr. Quest have the wide range of specialized knowledge that they are supposed to have encompassing all fields. It is entirely possible for a person to have specialized knowledge in several fields or even as Montagu claims some knowledge in many fields[4] but the idea of any one person possessing total knowledge must be rejected.

It might be pointed out that a person with expertise in one field by no means gives any individual any special insights into another field nor into public policy. The fact that Dr. Spock is a specialist in the care and feeding of babies does not give him special qualifications beyond that of an ordinary citizen for judging public action nor did the fact that Bertrand Russell was a great mathematician and philosopher give him particular insights that are not available to you or me. While judgments cannot be made, or should not be made, without facts, facts alone do not give enough of a basis. Facts are, or may be, the enemy of truth. Just as the statement is made that figures do not lie, but liars may figure, so also may this be said of facts.

Unfortunately the great majority of man has always been willing to let someone else do his thinking for him. Even today in our representative form of government the masses are led by the active 10%. The leaders do not have the expertise to obtain the facts so that they must rely on those who do. In the past man was oriented towards earning his daily bread and waiting for the life in the hereafter. The man who instructed him in both was the priest. In any calamity or catastrophe of a personal or parish nature it was to the priest that man turned for solution, solace or succor. The prophets spoke their interpretation of the Bible or what they had heard from God. Misfortunes were the punishments or tests of God and only the priest could intercede for man.

Science and the scientist have replaced religion and the priest in informing the leaders of man what needs to be done in solving a problem. This has come about because scientists have been reasonably successful in describing and predicting physical phenomena. Man has come to accept only the existence of the world he can touch, smell, taste, see, and hear—the sensory or material world. He has also come to accept as proof the evidence of his senses or the evidence of what the scientist tells him he is seeing. This success and acceptance of science has led other areas to want to be "scientific" and to be classed as sciences. Areas such as sociology, anthropology, psychology have eagerly sought to jump on the band wagon and to become classed as social "sciences" instead of "Humanities." We are perhaps more aware of the mental tools of the scientist than of the humanist.

If the leaders are to be led by the experts then there are dangers to beware of. Persons working in science are fond of stating that science does not deal with values—that it is amoral (without morals). Yet frequently the problems that are crucial to our lives and well being are dependent on what the expert or experts are saying. For example, should an antiballistic missile system be deployed? Or should a dam for power be built at a specific point? Or to what extent can we use atomic power plants in the future?[5] Each of these questions pose many value questions that go into the decision making process.

One of the dangers created by the science expert is the notion that the scientist is a superman—that he has been endowed with something that no one else has—that he is capable of viewing a problem with an air of detachment and uninvolvement. He has just as many hangups as anyone else and more than some because he may think he is cool and devoid of prejudices. Actually the scientist is afflicted with the spirit of a "true believer" especially in areas where he has been researching or where he has a personal interest.

Some of the factors affecting the scientist, as well as other persons, include: age—a young person is more apt to have a radical solution than a person of forty and want it faster; employment—those employed in industries seek to satisfy the managers and to do what is good for the company as opposed to those who are academically oriented seeking self satisfaction and their own pride of expertise and idealism; education—one may equate formal education and intellect and thus feel that because they have a lot of formal education that they are smarter than those who haven't (a sort of intellectual snobbery); affluence—those who have always been a part of the affluent society view differently than those who have recently acquired affluence.

How can the danger from the experts be overcome? One approach is to balance one expert against another. Find one who is for the idea under consideration and one opposed, then listen to both. Listen with a skeptical view and look for emotional bias in the presentation. Where do the facts end and the value judgments begin? How much is "fact" and how much is interpretation of facts. If an expert cannot be found who is in opposition to the idea then get one who is not concerned with the project and have him prepare a case against the idea. Frequently Congress and the public have been led down the primrose path by a strong advocate of an expert.

Early in our country's history a group of natural philosophers wanted to make an observation of a transit of Venus (the passage of Venus across the face of the sun) but they had no telescope with enough power. They convinced Congress that this observation would enable them to better establish the latitudes and hence aid navigation. They got their telescope. How much of the man in space program is a similar bill of sale?

This technique, of balancing one expert against the other, has come more into use in business and industry. Before sinking large sums of money into any idea, boards of directors and bosses have to be satisfied that something is going to come of the idea. Government is beginning to use this approach but still too often vast sums are wasted by the failure to examine ahead of time the total consequences of the idea and to explore more in the starting phases of a device, such as a new type of aircraft.

There is a fear in top management that scientific experts may take over their job. What scientists do in technological aspects and systems analysis is to reduce the petty problems that top management has heretofore had to deal with. These problems can then be relegated to computers and less skilled personnel. In so doing, top management is freed to deal with the big picture and long range planning where before they have been so busy with the trees that they couldn't see the forest.

The basic reason why scientists will not take over management is that most complex management problems cannot be analyzed quantitatively by science. Decisions that have to be made are based on long years of experience that have created the judgment of management. Computers have been employed to give a management trainee many years of business experience in making judgments in a very short time. This constitutes another reason why a take-over will not occur. The manager has a much better grasp of technical matters than did his forerunners. This is the purpose behind the technical part of a community college—to provide technical know-how and the vocabulary of technology to persons who can go into mid-management and management.

Vocabulary is an earmark of an educated man. Studies have shown that regardless of the level of education that a person in top management has had, he has a very high vocabulary level. An educated man also knows how to look and where for something that he does not know—the use of reference materials. He must be able to take ideas apart and restate, rephrase or synthesize a new idea. Then he must be able to communicate these ideas in writing or speech to his subordinates or superiors.

Complex management problems are on a small scale an example of social problems in general. Just as management problems do not always lend themselves to quantitative analyses the same may be said of many social problems. Social problems are much harder than those of physical sciences because there are so many variables and many of them interact. Also, man himself is involved in a social problem to a larger extent than in one of physical sciences. Physical scientists lack the specific training to deal with social problems and there seems to be no magic formula to handle a social problem. As has been remarked before the walls between disciplines tend to make the specialist unaware in other fields. Yet techniques, devices, and methods which have been developed in one area can find application in others. " . . . most of the tools of modern statistical analysis were developed in the biological and behavioral sciences. The theory of sampling was developed in the context of social surveys. Factor, latent attribute, and sealogram analyses, developed in psychology, are finding increasing use in the physical sciences."[6]

Perhaps, and this is a question which we seriously wish to consider, the methods of science is not the path to a solution of social problems at all. Is it the panacea for all the ills of mankind? Will it achieve the ideals of man?

To the ancients there were three ideals of man—truth, goodness, and beauty—to which we of the modern age have added plenty. Parts of these were proposed as the chief objectives of American and United Nations' policy as the four freedoms.[7] These were (1) Freedom of speech and expression, (2) Freedom of every person to worship God in his own way, (3) Freedom from want, and (4) Freedom from fear (worldwide reduction of armaments). It is questionable whether man can be persuaded to leave the racetrack of the Four Horsemen of the Apocalypse—Conquest, Slaughter, Famine, and Death.

The emphasis in the pursuit of ideals is on the word pursuit. In the perfect attainment the ideal itself would be lost. How can truth, beauty, goodness, and plenty be known in the absence of its opposite? Can the truth be known without the false, beauty without the ugly, the good without evil, plenty without want? Secondly, it is the pursuit of an ideal rather than attainment that provides man with his greatest motivations. This search for ideals is part of the identity of man which separates him from the rest of the animal kingdom and complete attainment would lead to the destruction of self. These are the reasons that man creates and without them man would not be man.

Truth is closely related to the first two freedoms—freedom of expression and freedom of worship—and it is also related to two basic drives of man; those of curiosity and order. Man needs order and explanation. These, in part, can be satisfied in science and are part of the value of science to the individual. As a matter of fact if a scientist is asked why he is working as a scientist, he will frequently respond that it is because he enjoys finding things out. The scientist is not working for the good of mankind, he is working because it provides self satisfaction. Another value of science is the freedom to doubt which was born in the early struggles against authority in the Renaissance and Reformation and from which modern science itself arose. Inquiry cannot begin until there is doubt. Today, however, science and the scientist seem to be assuming the role of authority, which chains the mind of man.

Science and its approaches are based on certain presuppositions lacking any proof and perhaps resting only on the desire of man. These are that the behavior in the universe is orderly, that this order has a cause, and that man can know the cause of this behavior. These foundations are increasingly coming under criticism, not only from scientists but from others as well, for reasons which will be explored later.

Since values increasingly pop up in social problems and since scientists deny that science deals with values perhaps science is not the approach to social problems or to studies involving man.

Humanities and sciences are both human enterprises and are predominantly intellectual. The fundamental difference seems to lie in the fact that humanities deals with values while sciences do not. However, values can be described, analyzed, appraised, and modified—these are all activities of the intellect. It is to the intellect that we must look for the greatest conquest of man, that is the conquest of the thought process.

Bridgman has shown that in twentieth century physics and mathematics there are implications to the development of new tracks. He states: ". . . the most important consequence of the conceptual revolution brought about in physics by relativity and quantum theory lies . . . in the insight that we had not been using our mind properly and it is important to find out how to do so."[8] From relativity can be seen the importance of semantics (the science of word meanings) and the properties of a word. In the example which he used the word "simultaneity" has a number of implications which may not apply in the case in which it is being used. He further indicates that linguistics may be a method for study to see if the words we have been using do not carry certain forced attitudes which we are not aware of and which influence our thinking and our way of thinking.

"It (the Quantum theory) forces us to realize that we cannot have information without acquiring that information by some method, and that the story is not complete until we have told both what we know and how we know it."[9] In other words we must always have an observer and in that case what we learn is going to be influenced by that observer.

Thirdly we have from Gödel's theorem as a consequence that a system of mathematics can never prove that the system is free from internal self-contradictions and hence, ". . . the realization that the human mind can never have certainty, by either logical, or metaphysical, or mystical methods."[10] We can see from this that certainly certainty cannot come from science.

On the other hand it could be argued as Conant does that science should be used and also leads to the third value of science.

Literally every step we take in life is determined by a series of interlocking concepts and conceptual schemes. Every goal we formulate for our actions, every decision we make be it trivial or momentous, involves assumptions about the universe and about human beings, to my mind any attempt to draw a sharp line between common-sense ideas and scientific concepts is not only impossible but unwise. Belief in the whole apparatus of a three-dimensional world and in the existence of other people is a policy essential for an individual's survival; for a physicist or chemist in his laboratory, a new working hypothesis is a policy guiding his conduct as an experimenter. Where is one to draw the line? The common-sense ideas of our ancestors before the dawn of modern science were the foundation of all their value judgments. If scientific concepts are now part of our common-sense assumptions, and who can doubt that they are, then to this degree, at least, the consequences of the actions of previous scientists now affect our value judgments. This much connection between science and human conduct seems to me quite certain.[11]

The third value of science, and first to the minds of most people, is that scientific knowledge enables us to do things. It is this applied part that in technology provides the device to drive out the pale horseman and is related to the third freedom, want. Only maldistribution allows famine to stalk the countries of the globe and hunger to walk our streets.

However, it is the third value of science that leads to the ethical and moral questions about science, the ideal of good for man. The scientist is apt to say that science is amoral and has nothing to do with ethics. He may say that every man is given the key to paradise but that the same key unlocks the gates of Hades. It may be added that the knife used for cutting a radish may be used for murder. Certainly the scientist's argument about the amoral nature of science is true *but* the same does not apply to the scientist. The scientist cannot be excused on the same grounds for the scientist should be a responsible citizen to the world since it is also argued that science is international. There are some projects that can only have evil ends and the scientist cannot argue for his own amoral position. He argues that if he doesn't do it that someone else will. This is the same weak and insipid argument that is used by businessmen to explain away shady dealings. It does not make right. The youth of this country, becoming ever aware of the three P's of the twentieth century —population, pollution, and poisons (cigarettes among others)—objected strongly at a recent scientific meeting.[12] They questioned the relevance of various work that scientists were conducting in the face of the tremendous questions facing man.

Many chose to hold that science does little towards the ideal of beauty. Beauty is a perception of man based on a value system which he already has and its arrangement is generally based on the creativity of man. Beauty is in the eye of the beholder. Yet pure science, as pure art, is creative and frequently the master in science is acknowledged by his peers as having formulated a "beautiful" equation or theory. It is a part of creativity. Creativity has its place in all fields and science is no exception. Scientific ideas cannot be made to appear by will or on order and no amount of money spent without a creative mind is going to produce a new idea. Governments and foundations frequently make the mistake that money can solve any problem and if a problem lags all that is needed is to supply a little more money.

Creativity does need money or at least the leisure that money can buy. It is impossible to create if one is living a hand-to-mouth existence and every hour is needed to gain subsistence. Certainly too no one creates in a vacuum. Creation depends on many factors and pieces of information being in the right position at the right time. New ideas are usually only one step away from an old idea. Another condition for creativity in addition to the leisure is to value a particular area of endeavor. If a man has wealth and leisure he may just lie about, join the jet set, or play polo all day. Leisure only provides time for worthwhile investigations if those investigations are valued by the investigator and this value is in some measure an indicator of what society values.

Creativity is any new idea or rearrangement of an old idea that comes from the mind of man or anything that is produced by the hand of man. These may be classed as inventions—in the case of an idea it is an invention of the mind. A man may discover a river—it was already there—but an idea is an invention of the mind. Some persons consider a discovery anything found accidentally but the creativity and hence invention is the realization of what one has. Laws thus are also inventions since they do not exist unless perceived by man—a river is discovered because it is seen but the law of universal gravitation was an invention of Newton because he perceived the law.

In exploring the invention of an idea or device it will be seen that the step is not a giant one but a culmination of the efforts of a number of persons over a period of time. It seems also that the time must be right for the invention to take hold. Sometimes an invention gathers dust waiting for the time and in other instances several persons independently invent the same thing at nearly the same time.

Perhaps too it is not totally a matter of creativity that gives the person the credit as an inventor but rather the wealth behind him and how good a public

relations man he has had. This will be rather obvious in some of the cases which will be covered.

The starting problem is one which has long perplexed man. What is truth and how can we know it?

FOOTNOTES

1. Russell L. Ackoff *Scientific Method,* John Wiley and Sons, N.Y. 1962, p. 439.
2. See pp. 71 and 76 in O. G. Sutton's *Mathematics in Action,* Harper & Brothers, N.Y. 1960 for several specific examples.
3. Faraday preferred to be called a philosopher and had an intense dislike for the term physicist. John Tyndall *Faraday As A Discoverer.* Thomas V. Crowell 1961, p. 5. Many historians refuse to be classified as social scientists.
4. Ashley Montagu *The Cultured Man,* World Publishing, Cleveland 1958.
5. These last two questions were discussed and debated at the 1969 meeting of the American Association for the Advancement of Science (AAAS) which met in Boston.
6. Ackoff, *Scientific Method,* p. 441.
7. Message to Congress by Franklin D. Roosevelt January 6, 1941.
8. Arons and Bork, *Science and Ideas,* p. 271.
9. *Ibid,* p. 273.
10. *Ibid,* p. 275.
11. *Ibid,* pp. 229-230.
12. AAAS, 1969 in Boston.

Further Suggested Readings:

A. R. Patton, *Science For The Non-Scientist,* Burgess Publishing Co., 1962. pp. 1-23.

Richard P. Feynman, "The Value Of Science" in A. B. Arons and A. M. Bork (eds), *Science and Ideas,* Prentice-Hall, 1964. pp. 3-9. In the same book, see James B. Conant, "Science and Human Conduct," pp. 217-231. And Percy W. Bridgman "Quo Vadis" pp. 270-278.

"The Ideals Of Science And Society" in Russell L. Ackoff 's, *Scientific Method.* John Wiley, 1962. pp. 429-445.

John Brademas, "Technology and Social Change: A Congressman's View" in A. Warner, et al (eds), *The Impact Of Science On Technology.* Columbia, 1965 pp. 142-171.

Norman Carlisle and Frank Latham, *Miracles Ahead,* Macmillan Company. 1944.

James B. Conant, *Science and Common Sense,* Yale University Press, 1951. pp. 1-42.

Erwin C. Schrodinger, "Science, Art and Play" in *Science Theory and Man,* Dover, 1935. pp. 27-38. In the same book, "Is Science A Fashion Of The Times?", and "Physical Science and Temper Of Age," pp. 81-132. Anthony Standen, *Science Is A Sacred Cow,* E. P. Dutton and Company, 1950.

Stephen Toulmin, "Introductory" and "Discovery" in *The Philosophy of Science,* Hutchinson University Library, 1953. pp. 9-56.

J. Bronowski, *The Common Sense Of Science,* Harvard University Press, 1966.

James B. Conant, *Modern Science and Modern Man,* Columbia University Press, 1952.

Harry Woolf (ed.), *Science As A Cultural Force,* Johns Hopkins Press, 1964.

STUDY AND DISCUSSION QUESTIONS

1. What is the "image" of science?
2. What slows down progress?
3. Why wouldn't a physical scientist consider anthropology and sociology as sciences?
4. Does science deal with "facts"?
5. Why do scientists use mathematics?
6. How does a biological science differ from physical science?
7. What is the difference between words and symbols?
8. Why can't a layman understand science, theology, and law?
9. Is mathematics always a model?
10. What are the first three presuppositions of science?
11. Why might an area wish to be considered a science?
12. What is the basis for the assumption that the universe is orderly?
13. What is the difference between science and technology?
14. What is science?
15. What areas of human experience are not covered by science?
16. How does a "scientist" differ from a "non-scientist"?
17. How is "why" used in science?
18. Can scientists disagree on "fact"?
19. Are scientists competent in all fields of science? Cite examples.
20. Is it true that given enough money, science can do anything? Give examples of this approach.
21. Define "natural science."
22. What is subjective about science?
23. What is "scientism"?
24. What is the scientist's attitude?
25. Can science create a Utopia?
26. How does travel create loneliness?
27. For what three reasons does the scientist not deal with social problems?
28. What is the first way in which science is of value? Is it always "good"?
29. What is the second value?
30. How can we tell that this is a scientific age?
31. What is the third value of science?
32. What should the applied sciences do?
33. What chains the minds of men?
34. How have we been conditioned by scientific discoveries?
35. How do "experts" make a danger for a free society?
36. How can the danger from experts be overcome?
37. What assumptions and presuppositions might affect science or one's attitude?
38. What are the differences and similarities between sciences and humanities?
39. What is the most important mark of an adequately educated man?
40. Can science be understood by itself?
41. Name and define the four ideals of man. How may they be achieved?
42. To which ideal has science least contributed?
43. What two factors have tended to limit the individual scientist?
44. What is the scientist's responsibility for social consequences?
45. Which science has the greatest prestige and why?
46. Why can't scientists replace management?
47. Who has the responsibility for morality in science?
48. Was George Washington Carver a scientist?

II

The Basis of Truth, Perception, and Reality

The purpose of this section is to examine the underlying assumptions of the scientist, as well as others, in order to see what is acceptable as proof and also to note that these assumptions are not true to all persons at all times. Most of the terms used here are abstract concepts and an abstract concept is usually difficult to define. Semantics, which deals with the meaning of words, uses up a large part of our life. Yet we must communicate.

Most will agree that truth is the degree to which an idea or "thing" conforms to reality, or truth is that which is (i.e. existence). If all can agree what truth is, then truth is an absolute, i.e. true and the same to all persons at all times and in all places. The big hang-up, however, is not with the nature of truth but with the nature of reality on which there is no common agreement.

The trouble with reality, and for that matter with nearly all human relations, is that each of us feels that there is only one reality—our own particular one. We feel that the way we think and look at something is not only true but the only way. These feelings are reinforced by dictums picked up as part of our cultural baggage from the logic of ancient Greeks, the beastiary and scholasticism of the Middle Ages, the manners and morals of the Victorian era, various folklore from many peoples. E.g. the principle of the excluded mean—something cannot be at the same time both black and white in its entirety; there is an order in the universe; for each action there is a cause which precedes it; the sum of the parts must equal the whole, etc. There are a number of "valid" realities and some of these we will examine later.

Let us examine some of the sources of truth to see where what we believe to be truth comes from. Some of the sources we will find are methods and some we will find are systems of truth itself. In our minds there is usually no clear distinction and in fact they are apt to be a mixed jumble. We are not going to try to make an orderly array but we do want to see their origins.

One of the sources for truth is authority. But then the question may naturally arise as to what or whom may be used as an authority. Authority is very useful to save time since if every person had to prove everything to himself, it would defeat the purposes of education and indeed would cause the downfall of civilization itself. To a child the source of authority may be his parents, some other adult, or his peers. On the other hand, an uneducated person, and indeed some educated ones as well, may regard the printed word as the authority. This printed word may be the newspaper, the Bible, or perhaps an almanac. Even the printed word is dependent on men—living and dead—and must be weighed carefully.

Another source for truth is social agreement. Social agreement does not necessarily make for truth, but it may make something accepted as truth. To use an example from language, if someone is called a "cool cat," a dictionary meaning (or authority) would give quite a different meaning than that which is socially agreed. The meaning of words, in general, is a matter of social agreement. But, although many persons agreed, at one time, that unicorns did exist; their agreement did not make unicorns exist, nor true.

This leads to two other criteria or truths which are founded in emotions. The first of these is belief which may be derived from one or the other sources and founded in faith. Examples of truths of this kind are Christianity, Communism, God, angels, flying saucers, and atoms. To cite a couple of cases of twentieth century beliefs let me choose Voodoo and Pow-Wow. Voodoo you may wish to dismiss as beliefs of uncivilized or uneducated Caribbeans, Africans or South Americans—this particular brand of witchcraft is far enough away that it seems unbelievable that any would believe. Pow-Wow, on the other hand, is closer than you may think. In the 1930's a man was murdered in Pennsylvania because he had placed a hex on another man. You will find barns in Ohio that have "hex" symbols on them to protect from the power of Pow-Wow. The second criteria is that of intuition. This seems to be possessed primarily by women and mathematicians. The quality of knowing truth without visible means of support.

Somewhat similar to the above is the truth statement that something is "self-evident." This is frequently used in mathematical systems and is supposed to be so obvious an assertion as to require no further explanation or it is asserted as a first principle or axiom. Mathematical systems and logic fall into the category of á priori truths because they follow rules which are set up ahead of time. Neither logic nor a mathematical system needs to conform to what we call the "real" world in order to be true. They only need to be consistent with the rules as set up on the system. As a matter of fact, as we shall see, sometimes when we apply a mathematical system to a problem that exists in our "real" world, it doesn't give us the solution that we want because there is not a correspondence between the mathematical system and the "real" world.

The source of divine truth is the diety. Knowledge about divine truth is obtained from divinely-inspired writings, such as the *Koran* or *Bible,* interpretation of these documents by those possessing Grace, or direct revelation from the diety (which may be sensory or non-sensory).

Most of the above sources do not stand by themselves and can only be valid if used with another or several others. The philosopher often judges what is the truth by how well it fits with a whole group of ideas. That is that it is consistent with a total system. He may use logic as a tool to seek inconsistencies to determine what is truth.

Truth, however, as we have already pointed out, is dependent on a particular concept of reality.

Plato, in his book, *The Republic,* is faced essentially with the problem of finding perfection in an imperfect changeable world. Not only the imperfection of things but also the problem of imperfect ideas and concepts. He argues that the concepts and things in our environmental world are only imperfect copies of perfect concepts and things in the "real" or ideal world. He uses the allegory of the cave in which he describes man chained to the floor of a cave in such a way that he can see the wall of the cave but not the outside so that what he sees is only the shadows of things that are. This view is adapted to Christianity by St. Augustine who makes heaven the ideal or "real" world. Plato's ideals are very important to the development of science in the 17th century; and to Christians during the Middle Ages for this was not the real world in which happiness existed. Plato also believed that man's mind already contained all knowledge so that to learn all one had to do was ask the right questions.

Sextus Empiricus, an early Greek, argued that knowledge of reality is impossible for our only knowledge is sensory and the senses can deceive. This is really the grounds for the various forms of skepticism which in effect denies the existence of any reality or

at least that if there is a reality that it is impossible to know what it is. Strands from the early skeptics are also important in later scientific development.

There are others who deny the existence of an environmental world except as it is sensed and interpreted in the mind. It is also possible to take another step and to insist that the real world is not environmental at all but only exists in the mind of man. Qualities like color, and heat do not exist in the object (if indeed there is an object—weight and extension).

It is only a short step to argue that the "real" world isn't in the mind of man but in the mind of God. We and the universe are only flickering thoughts in the mind of the diety.

Pythagoras, another early Greek, claimed that the "real" world was one of numbers and form—of geometry. This view, modified to include a diety, was also held by Newton and Kepler as we shall see. It is a view maintained today by Wheeler at Princeton— i.e. that there is no matter only a geometric arrangement of space. The geometry of Wheeler differs from that of Pythagoras and Euclid and is a different system of mathematics.

The "real" world or reality is not a problem to most of us in our everyday living. It is the world subject to our senses of which we are aware (sensory). Is the real world only in the mind or is it outside? All of our evidence of the outside comes from the senses and is perceived by the mind. Is the real world merely an image in the mind of God?

If reality is that which is based on perception then we need to examine perception. One of the first distinctions that we need to make is between "seeing" and "perceiving." Perceiving not only implies sight but also some mental process based on experience. A South Sea Islander on seeing ice for the first time would not perceive what he was seeing because he has no experience with ice. A child may reach for the moon as well as for his rattle because he doesn't have enough experience with distances. As a matter of fact, the image of what we see is focused on the retina upside down so that the first stage is to perceive rightside up. What we see is blotches of reflected light. What we perceive is patterns which we through experience identify with certain objects or relations.

We do not all really "see" the same things. Our eyes are not in the same position as another's nor is the distance between our eyes the same as another's so that from purely physical differences we do not see the same. The angle from which we are viewing something may be far different from the angle of another and this gives rise to having a different slant

Figure II-1. Birds or Rabbits?

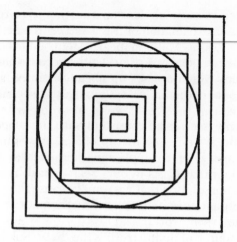

Figure II-2. Is the Circle Round?

on things. Something that looks square from my angle may look more rectangular from another angle.

All of us are familiar with deception in perception. We have spoken to perfect strangers in the belief that they were someone we knew. We have been suckered into mistaking the size of two objects because of what we have "seen." We have braked the car to avoid a large puddle that wasn't there. These are all experiences that most of us have had.

What is the basis for these "natural illusions"? In many cases particularly those of optical illusion it is the background relating to the object which has led to our confusion. In the case of the mirage what we see is actually someplace else and it is the refraction of the light which makes it appear to be where it is not. If we see a part of an identifiable pattern our mind will supply the rest of the object—this is known in psychology as closure—and we may be misled.

What is the stuff that dreams are made of? Or more importantly here how do we know when we are dreaming and when we are awake. Sometimes a dream will seem so "real" that it is some time after we are awake that we are aware that we were dreaming. A short time ago my dream was so real that I was still mad at breakfast time although by then I was aware that I had been dreaming. Usually a dream is detected as a dream because it does not match the

"real" world in some respect and so we may realize we are dreaming while we are dreaming but sometimes the dream is very "real." When that happens, how do we know it is a dream? When we wake up we are sure but how do we know when we are waking and when we are dreaming?

In addition to what we have referred to as natural illusions there are those which we might class as abnormal because they do not normally occur.

The taking of various drugs might fall under this general category. The effects of a number of drugs, including alcohol, have been known for some time while others have come into recent prominence, e.g. L.S.D. There are some rather mild drugs that are taken regularly whose effect on perception have not been investigated fully. A number of cola beverages contains caffeine as does coffee and tea. This is a mild stimulant but does it have an effect on perception? Natives in the mountainous parts of South America chew cocoa leaves which contain a stimulant and give a good feeling. The habitual taking of dexedrine for weight reduction or for the taking of examinations is prevalent and does affect the perception after a very few days.

The effects of the stronger drugs on perception have been investigated. L.S.D. and mescaline not only distort time but also all of the sensors—sight, hearing, touch, smell. They create in short hallucinations so real that to act on some of them may cause death. (*Scientific American,* April, 1964, pp. 29-37)

Other factors besides drugs may cause hallucinations. Various diseases may distort the senses, either temporarily or permanently. Most of us have experienced childhood fevers in which hallucinations which seemed very real occurred. Even when we were aware that they were hallucinations we were unable not to react to them.

Old age may bring with it hardening of the arteries and blood clots which may cause double vision and other hallucinations. The person may hear voices when no one is around. They may see persons or creatures that are not there. They may be able to read printing on plain walls or furniture and may have other delusions.

Fatigue is another factor which at any age may cause hallucinations. Driving long distances at night or in fog after a long period may cause the eyes to begin imagining shapes or objects on the road that are not there.

Deprivation of food or water for a long period of time may cause hallucinations—voices or sights seen only by the person. How many of the prophets who starved their bodies for purification or to separate the body from the spirit saw or heard things that were hallucinations rather than divine revelations?

With these various factors which can affect the senses, is it any wonder that some have denied completely their validity? Even those who would keep a real world as one known by the senses must be aware of their fallibility and limited usefulness. To go overboard on this line is to lead to complete skepticism and doubt of all sources of knowledge.

STUDY AND DISCUSSION QUESTIONS

1. Why do philosophers claim that our knowledge of the physical world is necessarily limited?
2. Where are qualities, such as heat, cold, and colors located?
3. Does everyone see the same?
4. Does an object exist if it is not being sensed?
5. What kind of conditions produce hallucinations?
6. What are hallucinations? How do we "know" when we are having one?
7. Are there differences between a "dream" and a hallucination? How do we "know" when we are awake?
8. What is the role of "experience" in "seeing"?
9. What is the common-sense view of perception? Is it always clear?
10. Can there be a noise if there is no one to hear it?
11. What is skepticism?
12. What is truth?
13. What is truth to a pragmatist? To the "ordinary man"? To a scientist?
14. What is "divine" truth?
15. How do mathematical truths differ from other truths?
16. How is language a model of reality?

Suggested Reading:

"Truth," *The Great Books of the Western World,* Vol. 3, pp. 915-922.

John V. Canfield and Franklin H. Donnell, Jr., (eds.) "Perception" in *Readings in the Theory of Knowledge.* Appleton-Century-Crofts, 1964, pp. 371-518.

Bertrand Russell, *The Problems of Philosophy,* Oxford University Press, 1912, pp. 119-140.

III

How Do We Look at the Stars?

STARS ARE THE WINDOWS OF HEAVEN

To the naked eye between 2,000 to 5,000 stars are visible depending on eyesight and location of the observer. We live in an age that has put a man on the moon but we haven't time to look at the stars. In earlier ages when man did have time to look he found, as we still do, that he could aid his memory by grouping objects and relating them. Most peoples grouped the stars and related them to myths. Possibly the storytelling of these groups of stars, called constellations, took the place of nightly T.V. watching. The position of the constellations change slightly each night or the time at which they are visible changes—depending on how you look at it. They were a means of telling time, of finding one's way over the trackless desert or ocean, a friendly guide against the terrors of the night, a way of passing time for shepherds or other watchers of the night, a means of knowing the seasons, a way of tracking the wandering stars (planets), to some a means of foretelling the future, and just nice to look at.

In the table below are listed the constellations visible in northern latitudes. Altogether there are about 88 recognized constellations.[1] The date given is that at which the constellation is highest above the horizon (the meridian) at 9 p.m. Standard Time. Each night the constellation is about 4 minutes earlier or about 2 hours each month.

Constellation	Date	Constellation	Date
Andromeda	Nov. 10	Leo (Lion)	Apr. 10
Antlia (Pump)	Apr. 5	Leo Minor (Little Lion)	Apr. 10
Aquarius (Waterbearer)	Oct. 10	Lepus (Hare)	Jan. 25
Aquila (Eagle)	Aug. 30	Libra (Scales)	June 20
Aries (Ram)	Dec. 10	Lupus (Wolf)	June 20
Auriga (Charioteer)	Jan. 30	Lynx	Mar. 5
Bootes (Herdsman)	June 15	Lyra (Lyre)	Aug. 15
Caelum (Bruin)	Jan. 15	Microscopium (Microscope)	Sept. 20
Camelopardalis (Giraffe)	Feb. 1	Monoceros (Unicorn)	Feb. 20
Cancer (Crab)	Mar. 15	Ophiuchus (Serpent Bearer)	July 25
Canis Venatici (Hunting Dogs)	May 20	Orion	Jan. 25
Canis Major (Great Dog)	Feb. 15	Pegasus (Flying Horse)	Oct. 20
Canis Minor (Little Dog)	Mar. 1	Perseus	Dec. 25
Capricornus (Goat)	Sept. 20	Phoenix	Nov. 20
Cassiopeia (Queen)	Nov. 20	Pisces (Fishes)	Nov. 10
Cepheus (King)	Oct. 15	Pisces Austrinus (Southern Fish)	Oct. 10
Cetus (Whale)	Nov. 30	Puppis (Stern of the ship Argo)	Feb. 25
Columba (Dove)	Jan. 30	Pyxis (Compass)	Mar. 15
Coma Berenices (Berenice's Hair)	May 15	Sagitta (Arrow)	Aug. 30
Corona Austrina (Southern Crown)	Aug. 15	Sagittarius (Archer)	Aug. 20
Corona Borealis (Northern Crown)	June 30	Scorpius (Scorpion)	July 20
Corvus (Crow)	May 10	Sculptor	Nov. 10
Crater (Cup)	Apr. 25	Scutum (Shield)	Aug. 15
Cygnus (Swan)	Sept. 10	Serpens (Serpent)	
Delphinus (Dolphin)	Sept. 15	Caput (Head)	June 30
Draco (Dragon)	July 20	Cauda (Tail)	Aug. 5
Equuleus (Colt)	Sept. 20	Sextans (Sextant)	Apr. 5
Eridanus (River)	Jan. 5	Taurus (Bull)	Jan. 15
Fornax (Furnace)	Dec. 15	Triangulum (Triangle)	Dec. 5
Gemini (Twins)	Feb. 20	Ursa Major (Big Bear)	Apr. 20
Hercules	July 25	Ursa Minor (Little Bear)	June 25
Hydra (Sea Serpent)	Apr. 20	Virgo (Virgin)	May 25
Lacerta (Lizard)	Oct. 10	Velpecula (Fox)	Sept. 10

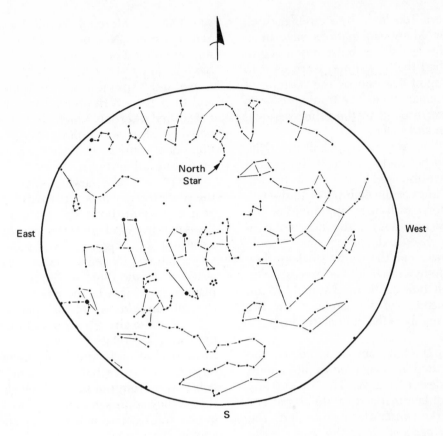

Figure III-1. The Constellations.

The path on which the sun seems to travel is called the ecliptic. Most of the ancients who used the sun to tell time, such as the Egyptians, Romans, Greeks, and Chinese, divided this path into twelve parts. Each part is 18° wide and 30° long (12 X 30° = 360°). Together these parts are called the Zodiac and are named for constellations—Aries (the Ram), Taurus (the Bull), Gemini (the Twins), Cancer (the Crab), Leo (the Lion), Virgo (the Virgin), Libra (the Balance), Scorpio (the Scorpion), Sagittarius (the Archer), Capricorn (the Goat), Aquarius (the Water Bearer), and Pisces (the Fishes). It is a little strange that our idea of dividing the circle into 360 parts came from the Babylonians, who used a lunar rather than a solar calendar, and the names of the parts of the Zodiac are Roman while the names of many of the stars are Arabic.

The stars seem to be imprisoned on a sphere, which is called the celestial sphere. This sphere is divided by a celestial equator, which is a projection of the earth's equator, and a meridian, which is a projection of a line around the earth from pole to pole cutting the observer. The point directly above the observer is called the zenith.

When the celestial equator is crossed by the ecliptic day and night are of equal length. This condition is called the equinox and occurs twice a year, on or about March 21 and September the 23. On the other hand, the points of greatest separation between the ecliptic and celestial equator are called the solstice and occur on or about June 22 and December 22. At these times the longest and shortest days of the year occur. Since the sun seems to move through the constellations of the Zodiac each year, the appearance of the first point of the sign of Cancer (the Crab) just after sunset signifies the summer solstice, June 22, and the appearance of the first point of the sign of Capricorn indicates the winter solstice, December 22. Thus the stars can tell us the seasons. One of the most fascinating of modern stories which blends together the tale of time-telling, the field of archeology, and the application of computor science in solving an ancient mystery is portrayed in *Stonehenge Decoded*.[2] Stonehenge is a group of large standing stones, weighing many tons, on Salisbury Plain, Wiltshire, England. There are other smaller stone groupings in other parts of northern Europe but Stonehenge is the largest found. The stones are arranged in four series enclosed by a circular ditch 300 ft. in diameter. Many explanations have been given in the past for the stones but today it seems fairly well established that the stones were used to verify the beginning of a season by the position of sunrise over a particular stone. The framed-in parts of the rocks also correspond to the other seasonal positions of the sun and moon.

To the priests of the various river cultures (Nile etc.) the stars could be used to forecast the annual floods. The stars were also used by the priests to forecast various happenings to the country or to the rulers. The power to forecast future happenings of any kind gives the predictor a hold on the masses of people if they believe. Even today since many people believe in the powers of the scientist then they tacitly submit problems and accept his forecasts. The position of a constellation can be used at night to tell the time. The movement of the handle of the Big Dipper will point roughly to the hour all through the night.

Aside from telling the time (annual and daily), the stars were and are used in navigation to plot a path across the sea, sky, desert, or arctic. The traveler any place who is without landmarks to guide his path or where the landmarks constantly change finds the stars to be friendly beacons along the way. As the science of navigation developed, it became possible not only to know one's way but also his exact position by taking exact positions of the stars and the sun at a known time of day and then making calculations.

Perhaps because most of the constellations in our culture have Roman names, the stories or myths that come with them are also Roman which they took from the Greeks. Other cultures have different myths to go with their star groups. Just to give an example we will use the first constellation on our list Andromeda.

Andromeda was the daughter of Cepheus and Cassiopeia, king and queen of Ethiopia. Cassiopeia boasted that her beauty exceeded that of the Nereids (sea nymphs). The Nereids prevailed on Neptune (God of the sea) to dispatch a sea monster to obtain vengence and an oracle declared that Andromeda must be sacrificed. She was accordingly chained to a rock in the sea. Perseus who was flying back after killing Medusa chanced to see her, killed the monster, and married her.[3] They and the memory of the story are preserved in the sky—Andromeda, Cepheus, Cassiopeia, Perseus, and Cetus.

In order to give a base for comparison when we study the past views, let us pause and examine our current beliefs about the nature of our solar system. We believe that there are nine planets circling the sun. They are Mercury, Venus, Earth, Mars, Jupiter, Saturn, Uranus, Neptune, and Pluto. Circling the planets are a total of 31 moons or satellites. The Earth has one, Mars—two, Jupiter—12, Saturn—9, Uranus —five, and Neptune—two. In addition to the planets and their moons, there are thousands of minor planets called asteroids which circle the sun in an orbit between Mars and Jupiter. There are also a large number of comets, all of which have not been discovered, and millions of meteors.

The planets take varying lengths of time to circle the sun. Mercury makes the trip in about 1/4 of our year and Venus takes about 1/2 a year while Mars takes nearly two of our years. The other planets take a considerably longer period of time. Jupiter takes 12 years, Saturn—29 years, Uranus—84 years, Neptune —165 years, and Pluto—248 years. Only the first five planets are visible to the naked eye and hence are the only ones known to the ancients.

Comets also circle the sun but they are not visible until they are relatively close to the sun. They are made of dust and ice crystals and may or may not have one or more tails or streamers which always point away from the sun and which stream out further as they approach the sun. They have caused fear in man during most of his history and indeed still do among those who are not aware of their nature. Comets which are visible to the naked eye are rare but most people can see at least one in their lifetime. With a telescope between five and ten may be viewed during a year.

Meteors or meteorites, the so-called falling stars, can be seen every night especially after midnight. Meteors are those that are large enough so that they do not burn up on their way down and are thus more rare than meteorites. In some seasons, fall being one, meteorites may be observed in large numbers and on occasion enough fall to be considered a shower. While comets were usually portents of evil a falling star could, and can, be wished upon.

We of today still see the same things in the sky as did the ancients but because they needed the stars more and were closer to nature than we, they observed more and better than we do. They did not have the glow of artificial lights to obscure their view nor did they have the smog and other air pollutants to cut their vision. They did not drive at high speeds in glass and steel coffins separated from nature and from their fellow man. Who has the time today to pause out of doors in the evening with their son or daughter and point out the constellations with their stories? Or if they did take the time would their son or daughter pry themselves from the television long enough to watch. If they live in the city, and most today belong to an urban culture, where would they go to look? The sky is almost the same but urbanized man has changed.

The nature of our planet gives us a different slant on things. Its poles are tilted from the vertical by 23-1/2°. It is this tilt that gives those away from the equator unequal days and nights, with the two exceptions already noted, and makes the seasons. The sun is actually closer to the earth in winter than in summer but the earth receives more light and heat in the summer because it is so inclined.

If you have ever looked at a spinning top, you can imagine the turning earth and, just as when the top is tilted, the top of the top turns more slowly than does the spinning top. This phenomena is called precession and takes about 25,000 years from the top of the earth. This means that the pole star is not always the same. Polaris is about 1° from true north and by 2095 it will be less than 1/2°. Five thousand years ago, when the pyramids were being built, the north star was Thuban in the constellation Draco. That this was not known in the near past, as astronomical time goes, is evident from Shakespeare's lines:

But I am constant as the northern star,
of whose true-fix'd and resting quality
There is no fellow in the firmament.[4]

Precession also causes a gradual change of the constellations. This change amounts to about 1° every 72 years and while not very noticeable it had been observed by Hipparchus about 125 B.C. when he compared his observations with those of earlier astronomers. When the sun crosses the celestial equator at the spring equinox, the position is called the 'first point of Aries' which it was in the time of Hipparchus although today the phenomena occurs in the constellation of Pisces so that if you were not aware of precession you might think there was something fishy about the terminology.

Even though the stars are frequently referred to as the fixed stars, as opposed to the movable stars or planets, they are in motion at high speeds. We are not aware of their motion because they are so far from us just as we may not be aware of the speed of a jet that is high in the sky. This motion of the stars means that the very make-up of the constellations and the relative positions of the stars within the constellations are changing. If we wonder today how the ancients perceived the figures for which the constellations were named think how much more wonder will be created in future generations of man as these position changes go on. Due to precession some of the constellations named by the ancient Greeks and Romans of the Mediterranean are no longer even visible from that region but can only be seen from the southern hemisphere. Some are believed to have been named three thousand years ago and are referred to in Homer's *Iliad* (Book 18, line 483) where Vulcan is making Achilles shield. On it he wrought

. . . the earth, the heavens, and the sea; the moon also
at her full and the untiring sun, with all the signs
that glorify the face of heaven—the Pleiads, the Hyads,
huge Orion, and the Bear, which men also call the Wain
and which turns round ever in one place, facing Orion,
and alone never dips into the stream of Oceanus.

From this quotation it would appear that Ursa Major, or what we call the big dipper, was the only circumpolar constellation, i.e. a constellation that appears to rotate about the polar star and never sets, at that time. Today, the circumpolar constellations are —Ursa Major, Ursa Minor, Cassiopeia, Cepheus, and Draco.

Having seen our current position relative to the Solar System let us now examine the route by which we got here. For although the sky has not much changed in appearance our perceptions and concepts of what we see has radically changed. We will examine the evolution and revolution of our concept of our physical universe.

FOOTNOTES

1. For additional elementary details of astronomy see:
   Herbert S. Zim and Robert H. Baker, *Stars,* Golden Press, New York, 1956.
   Hubert J. Bernhard, et al *New Handbook of the Heavens,* Mentor Books, New York, 1948 and later editions.
   A. E. Fanning, *Planets, Stars and Galaxies,* Dover, New York, 1966.
   Robert H. Baker, *When the Stars Come Out,* Viking Press, New York, 1954.
2. Gerald S. Hawkins, *Stonehenge Decoded.* Doubleday, New York, 1965.
3. If you are interested in these or other myths consult:
   J. Loughran Scott (ed) Thomas Bulfinch, *The Age of Fable,* Philadelphia, 1898.
   E. M. Berens, *The Myths and Legends of Ancient Greece and Rome,* New York, N.D.
   H. A. Guerber, *Myths of Greece and Rome,* New York, 1893.
4. William Shakespeare, *Julius Caesar,* Act III, Scene 1, Line 60.

STUDY QUESTIONS

1. Why are the constellations in the Zodiac not seen all the year?

2. Why should all the planets be found along the ecliptic?

3. How do the equinoxes and solstices correspond to the seasons?

FROM ARISTOTLE TO AQUINAS

Perhaps the title of this section will seem a misnomer since it will start about three centuries before Aristotle's time. But a little background may be useful.

Nearly all peoples in all times have tried to formulate some ideas about the world and the universe in which they lived. The Greeks certainly were no exception. Most of the Greeks, to be sure, just as most Americans, were engaged in the day to day businesses of raising food, buying and selling; but a few —teachers, and men of wealth—had noticed that everything seemed to be in a state of change. These men were seeking the unchanging essence of all things and most of them also had ideas of what the earth and universe was like.

These men, called by us the pre-Socratics because they lived before Socrates, each had their particular essence; for Thales the essence was water and the earth was drum-shaped; Anaximander, on the other hand, proposed a substance called apeiron and the earth was flat; for Anaximenes—air; for Heraclitus—fire; for Pythagoras—numbers; for Parminides—everything is an illusion and only logic leads to knowing; Empedocles proposes a four element theory—earth, air, fire, and water—in which all of the elements are in conflict to achieve their natural positions; for Leucipos and Democritus there is nothing but atoms and the void.[1]

As you can see, there is a wide variety of ideas and almost any concept about the earth or its elements have a forerunner in some Greek philosopher.

The Peloponnesian War (431–404), between Sparta and Athens and the plague which broke out in Athens in 430 undoubtedly were among the factors which diverted the course of philosophy from physics to ethics. This is reflected in the words of Socrates, (c. 469 B.C.–399 B.C.) (he, like Christ, wrote nothing himself) and the writings of Plato, (c. 428 B.C.–348 B.C.). Even though the main portion is on ethics, there are facts which are of interest here.

Plato is convinced that perception leads to illusion so that for him what we see are only imperfect copies of the ideal objects which are not to be found on this earth. Of perhaps more importance, at least for this section, is what he said on the nature of the universe.

> . . . and there in the midst of the light, they saw the ends of the chains of heaven let down from above: for this light is the belt of heaven, and holds together the circle

Figure III-2. Plato.

of the universe, like the under-girders of a trireme. From these ends is extended the spindle of Necessity, on which the revolutions turn. The shaft and hook of this spindle are made of steel, and the whorl is made partly of steel and also partly of other materials. Now the whorl is in form like the whorl used on earth; and the description of it implied that there is one large hollow whorl which is quite scooped out, and into this is fitted another lesser one, and another, and another, and four others making eight in all, like vessels which fit into one another; the whorls show their edges on the upper side, and on their lower side all together form one continuous whorl. This is pierced by the spindle, which is driven home through the centre of the eighth. The first and outermost whorl has the rim broadest, and the seven inner whorls are narrower, in the following proportions —the sixth is next to the first in size, the fourth next to the sixth; then comes the eighth; the seventh is fifth, the fifth is sixth, the third is seventh, last and eighth comes the second. The largest (or fixed stars) is spangled, and the seventh (or sun) is brightest; the eighth (or moon) coloured by the reflected light of the seventh; the second and fifth (Saturn and Mercury) are in colour like one another, and yellower than the preceding; the third (Venus) has the whitest light; the fourth (Mars) is reddish; the sixth (Jupiter) is in whiteness second. Now the whole spindle has the same motion; but, as the whole revolved in one direction, the seven inner circles move slowly in the other, and of these the swiftest is the eighth; next in swiftness are the seventh, sixth, and fifth, which move together; third in swiftness appeared to move according to the law of this reversed motion the fourth; the third appeared fourth and the second fifth. The spindle turns on the knees of Necessity; and on the upper surface of each circle is a siren, who goes round them, hymning a single tone or note. The eight together form one harmony; and round about, at equal intervals, there is another band, three in number, each sitting upon her throne; these are the Fates, daughters of Necessity, who are clothed in white robes and have chaplets upon their heads, Lachesis and Clotho and Atropos, who accompany with their voices the harmony of the sirens— Lachesis singing of the past, Clotho of the present, Atropos of the future; Clotho from time to time assisting with a touch of her right hand the revolution of the outer circle of the whorl or spindle, and Atropos with her left hand touching and guiding the inner ones, and Lachesis laying hold of either in turn, first with one hand and then the other.[2]

As you can see, if you read carefully, Plato's description leaves a number of problems. If there are a total of eight whorls (a whorl is a drum-shaped device in the spinning operations of textile works) and the stars are on the largest, how can the moon be eighth? Does the spindle go through the earth causing it to turn or is it stationary and the outer whorl of stars turning? There are also apparent problems from Plato's positioning of the planets.

Although Plato was interested in mathematics as an approach to real knowledge and, as a matter of fact, advocated the study of geometry in his academy (a school which he founded), he did little himself. Some of his students, especially Eudoxus and Euclid (Euclid's geometry), did more in the way of mathematics. Eudoxus also tries to eliminate some of the problems created by Plato's idea of the universe.

Eudoxus postulates a series of concentric spheres all turning on a major axis but linked by sub-axis at

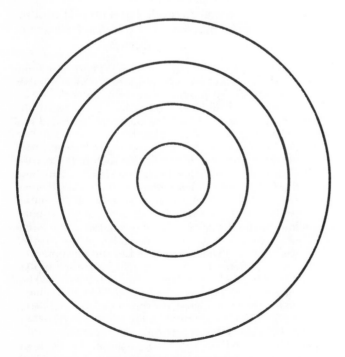

Figure III-3. Concentric Circles.

differing angles to the major axis in an attempt to explain the apparent backward movement of some of the planets at certain times of the year. This is the so-called retrograde motion of the planets. To explain the total motion of the heavenly bodies he advocated 27 spheres. One for the fixed stars, three for the sun, three for the moon, and four for each of the five other planets.

Figure III-4. Aristotle.

This explanation did not satisfy another of Plato's students, Aristotle (384 B.C.–322 B.C.) and he set out to build a comprehensive system of knowledge in which all motion would be explained. Indeed motion was only a small part of this total system which combines and relates the sum total of man's knowledge. This is one reason that Aristotle's works are and were so highly regarded. Here we shall be looking at only a small part and that part is only the physical portion.

Aristotle proposes that there are four elements (note that he is not the first to make this proposal) which would have a "natural" position in concentric spheres arranged: earth, water, air, and fire. He contends that the fact that they are not so arranged gives rise to natural motion on earth in a straight line toward the center or away from it.[3] The only other natural movement is circular and this is in the heavens. All other motion is unnatural.

For the perfect is naturally prior to the imperfect, and the circle is a perfect thing. This cannot be said of any straight line:—not of an infinite line; for, if it were perfect, it would have a limit and an end: nor of any finite line; for in every case there is something beyond it, since

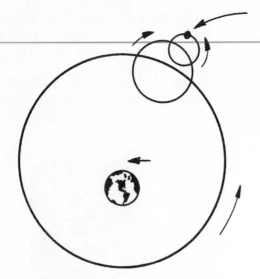

Figure III-5. Equants.

any finite line can be extended. And so, since the prior movement belongs to the body which is naturally prior and circular movement is prior to straight, and movement in a straight line belongs to simple bodies—fire moving straight upward and earthy bodies straight downward toward the centre—since this is so, it follows that circular movement also must be the movement of some simple body.[4]

Heavenly bodies are neither light nor heavy, are perfect, unchanging, spheres.

The reasons why the primary body is eternal and not subject to increase or diminution, but unaging and unalterable and unmodified, will be clear from what has been said to any one who believes in our assumptions. Our theory seems to confirm experience and to be confirmed by it. . . . For in the whole range of time past, so far as our inherited records reach, no change appears to have taken place either in the whole scheme of the outermost heaven or in any of its proper parts.[5]

That the heaven as a whole neither came into being nor admits of destruction, as some assert, but is one and eternal, with no end or beginning of its total duration, containing and embracing in itself the infinity of time, we may convince ourselves not only by the arguments already set forth but also by a consideration of the views of those who differ from us in providing for its generation.[6]

Aristotle's logical argument for the foregoing and following statement is too difficult to sum up here and not necessary for our purpose.

We have next to show that the movement of the heaven is regular and not irregular. This applies only to the first heaven and the first movement; for the lower spheres exhibit a composition of several movements into one.[7]

Aristotle's next argument is that the stars (remember that to the Greeks and others of this time the only

difference between stars and the planets was the planets were wandering stars) cannot move of their own accord but only move with the circles to which they are affixed.

. . . the remaining alternative is that the circles should move, while the stars are at rest and move with the circles to which they are attached . . . since the stars are spherical, . . . and since the spherical body had two movements proper to itself, namely, rolling and spinning, it follows that if the stars have movement of their own, it will be one of these. But neither is observed. (1) Suppose them to spin. They would then stay where they were, and not change their place, as, by observation and general consent, they do. . . . (2) On the other hand, it is also clear that the stars do not roll. For rolling involves rotation: but the 'face,' as it is called, of the moon is always seen. Therefore, since any movement of their own which the stars possessed would presumably be one proper to themselves, and no such movement is observed in them, clearly they have no movement of their own.[8]

Aristotle holds that the earth is stationary. If the earth were moving then things on the earth would have this motion, but every part moves in a straight line to its center. Also anything that moves has a circular motion and is observed to be passed by other bodies and yet the same stars always rise and set in the same parts of the earth. The center of the earth and the center of the whole universe are the same for they would both be the goal of heavy objects and this is observed. Again, heavy objects thrown straight upward return to the point they started from which could not be if the earth moved.[9]

Another topic that Aristotle takes up that is of interest to us is the movement of spheres. He reviews the work of Eudoxus and Callippus.

Callippus made the position of the spheres the same as Eudoxus did, but while he assigned the same number as Eudoxus did to Jupiter and Saturn, he thought two more spheres should be added to the sun and two to the moon, if one is to explain the observed facts; and one more to each of the other planets.

But it is necessary, if all the spheres combined are to explain the observed facts, that for each of the planets there should be other spheres (one fewer than those hitherto assigned) which counteract those already mentioned and bring back to the same position the outermost sphere of the star which in each case is situated below the star in question; for only thus can all the forces at work produce the observed motion of the planets. Since, then, the spheres involved in the movement of the planets themselves are—eight for Saturn and Jupiter and twenty-five for the others, and of these only those involved in the movement of the lowest situated planet need not be counteracted, the spheres which counteract those of the outermost two planets will be six in number, and the spheres which counteract the next four planets will be sixteen; therefore, the number of all the spheres—both those which move the planets and those which counteract these—will be fifty-five.[10]

At about the time of the death of Alexander (c. 323 B.C.) most of the Greek men of letters had moved to Alexandria. Some of these men had ideas about the universe but lacked the total system that Aristotle had.

Heraclides (c. 390–310 B.C.) stated that Mercury and Venus revolved around the sun, but that it and other planets revolved about the earth. And Aristarchus of Samos (who lived during the 3rd century B.C.) had argued that the universe was centered about the sun and that the earth moved. One of the great observational astronomers in Alexandria, Hipparchus (c. 160–125 B.C.) cataloged the position of 850 stars with an accuracy of 20 minutes. (360° in circle, 60 minutes to a degree.)

The works of Hipparchus and other observational astronomers are used by Claudius Ptolemaeus (c. 100–180 A.D.), better known as Ptolemy, to which he adds his own observations, made between 127–151 A.D., in compiling his work, *The Mathematical Composition.* Later the Arabs called it "The Greatest" or the *Almagest,* and it has been known by this title ever since.

Ptolemy is familiar with the works of his predecessors, including Aristotle. As a matter of fact, he accepts a great deal of Aristotle's statements: the heavens move spherically, the earth is spherical, the earth is in the middle of the heavens, the earth does not move—here he uses Aristotle's arguments and adds, "And if it had some one common movement, the same as that of other great weights, it would clearly leave them (smaller objects) all behind because of its much greater magnitude. And the animals and other weights would be left hanging in the air, and the earth would very quickly fall out of the heavens. Merely to conceive such things makes them appear ridiculous."[11] Remember that, according to Aristotle, heavy objects fall faster than lighter ones, and that there are two different prime movements in the heavens.

It follows that if Ptolemy does accept these ideas from Aristotle, he must reconcile these ideas with the observed facts and phenomena. Which is referred to as "saving the phenomena," and, of course, is merely a method of saving the theory. Aristotle, although he postulates the mechanism of the motion of the heavenly bodies, does not have any astronomical observations to go with his system. Ptolemy has the system, observations of others and his own observations to support the system, as well as detailed mathematical calculations to show that the observations fit the theory.

Since the next thing is to explain the apparent irregularity of the sun, it is first necessary to assume in general that the motions of the planets in the direction contrary to the movement of the heavens are all regular and circular by nature, like the movement of the universe in the other direction. That is, the straight lines, conceived as revolving the stars or their circles, cut off in equal times on absolutely all circumferences equal angles at the centers of each; and their apparent irregularities result from the positions and arrangements of the circles on their spheres through which they produce these movements, but no departure from their unchangeableness had really occurred in their nature in regard to the supposed disorder of their appearance.

But the cause of this irregular appearance can be accounted for by as many as two primary simple hypotheses. For if their movement is considered with respect to a circle in the plane of the ecliptic concentric with the cosmos so that our eye is the center, then it will be necessary to suppose that they make their regular movements either along circles not concentric with the cosmos, or along concentric circles; not with these simply, but with other circles borne upon them called epicycles. For according to either hypothesis it will appear possible for the planets seemingly to pass, in equal periods of time, through unequal arcs of the ecliptic circle which is concentric with the cosmos.[12]

Persons before Ptolemy had used the idea of off-centered circles to explain heavenly motions, but Ptolemy is the first to make a complete system, and to explain the motion of each planet in detail, as well as the motions of the moon and sun.

With the rise of Christianity, the fall of Rome, and the closing of the Academy of Justinian; the learning of the Greeks is largely unavailable to the Western world until the twelfth and thirteenth centuries. The Romans had only been interested in the popular aspects of learning so that this is all that was left to the west. It would be somewhat similar if our culture were destroyed and all that remained was *Life* and *Look* magazines. In addition to this factor, the early Christians were made more interested in the world hereafter and the "drama of salvation."

East and West trade and the crusades brings scholars from the West into contact with the Greek learning which the Arabs had accumulated, kept, and annotated. The Greek works are translated from Arabic into Latin (the language of the learned in Europe). The knowledge diffuses slowly—from copyist to copyist.

By the end of the twelfth century, the works of Aristotle and of Ptolemy had been translated. St. Thomas Aquinas (1224–1274) reconciled the philosophy of Aristotle to the theology of Christianity. But the works of Aristotle are soon under attack. Three years after the death of Aquinas, parts of Aristotle's works are condemned as heresy by the school of theology at the University of Paris. Aristotle's theory of unnatural movement, e.g. that a thrown stone continues to move because the air rushes behind the stone and continues to push it, is attacked by a number of persons—William of Ockham (c. 1285–1349), in England, John Buridan (d. 1360), and Nicholas of Oresme (d. 1390).

Nor is Ptolemy free from attack.

> In his book, *Du ciel et du monde,* Nicholas of Oresme discussed the subject of the earth's movement in relation to the sun; and he proposed several reasons to show that the Ptolemaic hypothesis was by no means certain. For example, it is not possible to show by observation that the heaven rotates daily, while the earth remains stationary; for the sun would appear to move even if it was really the earth which moved and not the sun. As for objections drawn from Scriptures against the hypothesis of the earth's movement, the Scriptures speak according to the common mode of speech and should not be regarded as scientific treatises. From the statement that the sun was stayed in its course at the prayers of Joshua, one is no more entitled to draw the scientific conclusion that it is the sun, and not the earth, which moves than one is entitled to draw from phrases like 'God repented' the conclusion that God can change His mind like a human being.[13]

Nicholas of Oresme was also Bishop of Lisieux as indeed were virtually all men of learning during the Middle Ages since the Church conducted all education and was the chief patron of learning until the late Middle Ages. Someone must foot the bill for thinkers or else they must be independently wealthy.

Even in St. Thomas Aquinas, there is room for doubt:

> Reason is employed in another way, not as furnishing a sufficient proof of a principle, but as confirming an already established principle, by showing the congruity of its results, as in astrology the theory of eccentrics and epicycles is considered as established because thereby the sensible appearance of the heavenly movements can be explained; not, however, as if this reason were sufficient, since some other theory might explain them.[14]

Thomistic thought welds the Greek view of the physical universe to the theological base of Christianity, and although St. Aquinas as well as St. Thomas doubted that holding the earth stationary was the only method of explaining heavenly motions this view came to dominate the thoughts of Western man.

Following the invention of printing ideas were more readily passed to those who could read. This would include churchmen, members of the courts of nobility in the late Middle Ages, and wealthy merchants and craftsmen of the Renaissance. Some selected parts of what they would be aware of are represented in the next section.

FOOTNOTES

1. Those interested in this area will find G. S. Kirk and J. E. Raven, The *Presocratic Philosophers,* Cambridge, 1962, pp. 445, very useful as we have merely skimmed over the topic.
2. Plato, "The Republic" in *Great Books of the Western World,* Vol. 7, pp. 438-439.
3. Aristotle, "On the Heavens" in *Great Books of the Western World,* Vol. 8, p. 359.
4. *Ibid.,* p. 360.
5. *Ibid.,* p. 361.
6. *Ibid.,* p. 375.
7. *Ibid.,* p. 379.
8. *Ibid.,* pp. 380-381.
9. *Ibid.,* pp. 387-388.
10. Aristotle, "Metaphysics" in *Great Books of the Western World,* Vol. 8, p. 604.
11. Ptolemy, "The Almagest" in *Great Books of the Western World,* Vol. 16, p. 11.
12. *Ibid.,* pp. 86-87.
13. Frederick C. Copleston, *Medieval Philosophy.* Harper Torchbook, 1961.
14. Thomas Aquinas "Summa Theologica," in *Great Books of the Western World,* Vol. 19, p. 177.

STUDY AND DISCUSSION QUESTIONS

1. What is the present belief of the positions of the planets in our solar system? How does this belief compare to Plato's?
2. How do each of the following explain the cause of motion of the planets? Plato? Aristotle? Heraclides? Ptolemy?
3. What is an equant? an epicycle? a deferent?
4. What kind of argument do Nicholas of Oresme and St. Thomas Aquinas use against Ptolemy?
5. Why does Aristotle last so long as the authority?
6. How were phenomena explained prior to the Greeks?
7. Why is a circle perfect?
8. What problems arise from Plato's explanation of planetary motion?
9. Why do objects fall according to Aristotle?
10. What circulates ideas during the Middle Ages?
11. What are Aristotle's arguments for the earth being stationary?
12. Can you prove that the earth moves?

THE VIEW FROM THE FIFTEENTH CENTURY

This view of the fifteenth century ideas of the physical world is from a work, *Image du Monde,* which had been read for two hundred years and was itself translated from a Latin work, according to the current editor, Oliver H. Prior, of this 1966 reprint of the English version. The English translation was probably printed in 1480 under the title *Caxton's Mirror of the World*.

English was not standardized as to spelling until the eighteenth century so that the spelling that follows may not be consistent. It is left in the original to provide some practice for future teachers in reading student papers. It might be noted that Johnny can spell, but in the wrong century. Although some parts have been "translated" from Middle English most remains in the original and so the following clues are provided. For "y" try "i" and for "u" the letter "v" is sometimes a better fit. The letter "u" sometimes will need to be supplied and the final "e" will sometimes need to be dropped. The "th" of "the" is sometimes added to the word which follows and sometimes it is necessary to put two words together. Do try to use your own imagination.

The Auncyent philosophres mesured the world on alle parties [ancient philosophers measured the world in all its parts] by their science, Arte and wytte [wit], vnto [unto] the sterres [stars] all on hye [high], of whiche they wolde [would] knowe the mesure ffor [for] to knowe the better their nature. But first they wolde mesure therthe [the earth] and preue [prove] his gretenes. And thenne [then], whan [when] they had mesured therthe al [all] aboute by a crafte that they knewe, and proued [proved] by right [straight] reson [reason], they mesured it rounde aboute lyke [like] as they sholde haue [should have] compassed it al aboute wyth a gyrdle [with a girdle], and thenne they stratched [stretched] out the gyrdle al alonge. And thenne that whiche wente out of lengthe of the gyrdle, they fonde [found] it in length xx·M·cccc· and · xxvii · [20,327 miles] myles; of whyche euery [every] myle conteyneth [contains] a thousand paas, and euery paas fyue [five] foot, and euery fote [foot] xiiii ynches. Somoche [so much] hath the erthe [earth] in lengthe round a boute [circumference].

By this fonde [found] they after how thycke [thick] therthe is in the myddle [its diameter]. And they fonde the thycknes therof, lyke as it shold ben clefte [as if it were cut] in the myddle fro [from] the hyest [highest] to the lowest or fro that one syde to that other, vi·M· and v·C· [6,500] myles. By this laste mesure, whyche is after nature right, they mesured iustely [justly] the heyght of the firmament; ffor they coude nowher fynde [could no where find] a gretter mesure tror textende [the extent of] the gretenesse of alle thynges whiche ben [are] enclosed wythin the heuene [heaven].

Therthe, as the auncyent philosophres saye, after they had mesured it they mesured the sterres [stars], the planetes [planets] and the firmament.

And first they mesured the mone [moon] & preuyd [proved] his gretnesse. And they fonde the body of therthe [the earth], without and withinne, that, after their comune [common] mesure, it was more grete than the body of the mone was by xxix tymes and a lytil [little] more. And they fonde that it was in heyght aboue the erthe xxiiii tymes and an half as moche as therthe hath of thycknes.

Also in lyke wyse [in a similar manner] preuyd they touchyng the sonne [sun] by very demonstraunce [demonstration] and by reson [reason], that the Sonne is gretter than alle therthe is by an hondred syxty and six sythes [sizes]. But they that knowe nothynge herof [here of], vnnethe [with difficulty] and wyth grete payne [pain] wyl byleve [believe] it. And yet it is suffysauntly [sufficiently] preuyd as wel by maystryse [mastery] of scyence as by verray connyng [very cunning] of Geometrye. Of whyche haue ben [there are] many, syth [say] the phylosophres that fonde [found] this first, that haue studyed and trauaylled [travelled] for to knowe the trouthe [truth], yf it were soo [so] as is sayd [said] or not; somoche that by quyck [quick] reson they haue preuyd that thauncyent phylosophres had sayd trouthe as wel of the quantyte of the Sonne as of the heyght. And as to the regard of hym that compyled this werke [work], he sette [set] all his entente [intent] & tyme, by cause [because] he hadde so grete meruaylle [marvelled] therof, tyl he had perceyuyd [perceived] playnly that of whiche he was in doubte; ffor he sawe appertly [apparently] that the Sonne was gretter than al therthe wythout ony [any] defaulte by an C·lxvi·tymes, and thre [three] parte of the xx parte of therthe, with al this that thauncyent philosophres sayde. And thenne byleuid [believed] he that whiche was gyue [given] hym to vnderstonde [understand]. And he had neuer [never] put this in wrytyng [writing], yf he had not certaynly knowen the trouthe & that he playnly had proued it. And it may wel be knowen that it is of grete quantyte, whan it is so moche ferre [very far] fro vs [from us] & semeth [seems] to vs so lytil [little]. Ne [nor] he shall neuer be so ferre aboue vs but in lyke wyse he shal be as ferre whan he is vnder or on that other side of vs. And for trouthe it is fro therthe vnto the Sonne, lyke as the kynge Tholomeus [King Ptolemy] hath prouyd it, ffyue [five] hondred lxxx [80] and v tymes as moche as therthe may haue of gretenes and thyckens thurgh [earth diameters].

Now wyll I recounte to you briefly of the sterres [stars] of the firmament, of whiche ther is a right grete nombre [number]; and they ben [are] alle of one lyke heyghte [height], but they ben not all of one gretenes. And it behoueth [behoves] ouer longe narracion [narration] that of alle them wolde [would] descryue [decry] the gretenes. And therfore we passe lyghtly ouer [over] and shortly; how wel I aduertyse [advertise] you and certefye [certify], that ther is none so lytil of them that ye may see on the firmament but that it is gretter than all therthe is. But ther is none of them so grete ne [nor] so shynyng [shining] as is the Sonne; ffor he enlumyneth [enlightens] alle the other by his beaulte [beauty] whiche is so moche noble.

Ffro therthe vnto the heuen [heaven] wherin the sterres ben [are] sette, is a moche grete espace [is a very great distance]; ffor it is ten thousand and · lv · sythes [sizes] as moche, and more, as is alle therthe of thycknes. And who that coude accompte [account] after the nombre and fourme [form], he myght knowe how many ynches [inches] it is of the honde [hand] of a man, and how many feet, how many myles, and how many Journeyes it is from hens [hence] to the firmament or heuen [heaven]. Ffor it is as moche way vnto the heuen as yf a may [man] myght goo the right way without lettyng [stopping], and that he myght goo euery day xxv myles of Fraunce [France], which is · l · [50] englissh myle, and that he taried not on the waye, yet shold he goo the tyme of seuen · M · i · C · and · lvii · yere and an half er [before] he had goon [gone] somoche waye as fro hens [hence] vnto the heuen where the sterres be inne.

Yf the firste man that God fourmed euer [ever formed], whiche was Adam, had goon [gone], fro the first day that he was made and created, xxv myles euery day, yet shold he not haue comen theder; but shold haue yet the space of · vii · C · xiii · [seven hundred thirteen] yere to goo, at the tyme whan this volume was perfourmed by the very auctour [author]: And this was atte Epyphanye in the yere of grace · i · M · ii · C · and · xlvi · [1242]. That tyme shold he haue had so moche to goo, er he shold comen theder.

Or yf ther were there a grete stone whiche shold falle fro thens vnto therthe, it shold be an hondred yere er it cam to the grounde. And in the fallyng it shold descende in euery hour, of whiche ther be xxiiii in a day complete, xliii myle and a half. Yet shold it be so longe er it cam to therthe. This thing hath be proued by hym that compiled this present volume, er he cam thus ferre in this werke. This is wel · xl · tymes more than an hors [horse] may goo, whiche alle way shold goo without restynge [which should go all the way without resting].

To the regard of the Sterres we shal saye to yow the nombre lyke as the noble kynge Tholomeus nombred them in his Almageste [Almagest]; to whome he gaf [gave] the propre names and sayd that ther were a thousand and xxii, all clere and that myght be all seen, without the vii planetes; and may be wel acompted [accounted] without ony paryll [parallel]. In alle ther be · i · M · and · xxix · [one thousand twenty-nine] whiche may wel be seen, withoute many other whiche may not wel be seen ne espyed. Ther may not wel moo [more] be espyed but so many as sayd is, ne apertly [apparently] be knownen. Now late hym beholde that wil see it; ffor noman [no man], trauaylle [travel] he neuer somoche ne studye, maye fynde nomore. Neuertheles ther is no man lyuyng [living] that may or can compte [count] so moche, or can so hye mounte in ony place, though he be garnysshid of a moche gentil instrument & right subtyl [except with the aid of a very good instrument], that shold fynde moo that the kynge Tholomeus fonde, by whiche he knewe & myght nombre them, and where eueryche [every one] sitteth, & how ferre it is from one to an other, be it of one or other or nygh [nigh] or ferre, and the knowlege of the ymages of them, the whiche by their semblaunce [semblance] fourmed them. Ffor the sterres whyche be named ben [are] all fygures on the heuene, and compassed by ymages and that all haue dyuerse beynges [have diverse beings]. [The constellations]. And euerych hath his fourme and his name. Of whiche ben knowen pryncypally xlvii within the firmament. And of them ben taken xii of the most worthy whiche ben called the xii Sygnes [the zodiac]. And they make a cercle round aboute the vii planettes, where as they make their torne [appearance].

We ben moche ferre from heuen merueyllously. [We are marvelously far from heaven.] Aud [and] late euery man knowe that he that deyeth [dies] in dedly synne [sin] shal neuer come theder. And the blessyd sowle [soul] whyche is departed fro the body in good estate, not withstondyng the longe way, is sone [soon] come thether, ye truly in lasse than half an hour, & vnto the most hye place to fore the souerayn iuge [sovereign judge] which sitteth on the right syde of God the fader [father] in his blessyd heuen; the whiche is so ful of delytes [delights] of alle glorye and of all consolacion [consolation] that ther is noman lyuyng that may ne [nor] can esteme [estimate] ne thinke the Joye & the glorye where this blessyd sowle entreth [enters].

And ther is no man that can esteme ne thinke the capacite & retnes [greatness] of heuene, ne may compare it ne valewe [value] it to the capacyte and gretnes of all therthe, or so moche as may compryse fro therthe to the firmament, as to the regard of the inestymable gretenes aboue the firmament; ffor that greteness is inestymable without ende and without mesure. Certes [Certainly] the firmamente on hye is so spacyous, so noble and so large, that of alle his wytte [wit] may not a man vnnethe [with difficulty] thinke or esteme the nombre of lyke masses as all therthe is that shold fylle [fill] it, yf they were alle in one masse. Who is he that coude or myght comprehende or compryse the gretenes of them, whan they alle be assembled, and euerich [each] as grete as all therthe? Neuertheles we shal saye to you therof as moche as we may wel ymagyne.[1]

This clerenesse [clearness] of whiche we haue spoken, whiche is callyd ayer spyrituel [called air spiritual], and where the angels take their araye and atourement [air and appearance], enuyronneth [surrounds in a circle] al aboute the worlde the foure elementis [elements] whiche God created and sette that one with in that other [made them concentric]. Of whiche that one is the ffyre [fire], the second is thayer [the air], the therde [third] is the water, and the fourthe is therthe; of whiche that one is fastned in that other, and that one susteyned [holds up] that other in suche mannere as therthe holdeth hym in the myddle. The ffyre, whiche is the firste, encloseth this ayer in whiche we bee. And this ayer encloseth the water after, the which hol-deth hym al aboute the erthe: Alle in liche [like] wise as is seen of an egge, and as the whyte encloseth the yolke, and in the myddle of the yolke is also as it were a drope of grece [grease], whiche holdeth on no parte; and the drop of grece, whiche is in the myddle, holdeth on neyther parte.

By such and semblable regard [In the same way] is the erthe sette in the myddle of heuen [heaven] so iuste and so egally [just and so equally] that as fer [far] is the erth fro heuen fro aboue as fro bynethe [beneath]; ffor, whersomeuer thou be vpon therth, thou art liche [like] fer fro heuen, lyke as ye may see the poynt of a compas [drawing type compass] whiche is sette in the myddle of the cercle; that is to saye that it is sette in the lowest place. Ffor, of alle fourmes that be made in the compaas, alle way [always] the poynt is lowest in the myddle. And thus ben [are] the foure elementes sette that one within that other, so that the erthe is alway in the myddle; ffor as moche space is alway the heuen from vnder therthe as it appiereth [appears] from a boue [above]. . . .

For as moche as therthe is heuy more [more heavy] than ony other of thelementis [the elements], therfore she holdeth her more in the myddle; and that whiche is most heuy abydeth [stays] aboute her; ffor the thynge which most weyeth [weighs] draweth most lowest, and alle that is heuy draweth therto. And therfore behoueth [behooves] vs to joyne [join]

to the erthe, and alle that is extrait [extracted] of therthe.

Yf so were and myght so happene that ther were nothing vpon therthe, watre [water] ne other thinge that letted [stopped] & troubled the waye what someuer parte that a man wold, [if there were nothing physical to stop a man] he might goo round aboute therthe, were it man or beste [beast], aboue and vnder, whiche parte that he wolde, lyke as a flye goth [goes] round aboute a round apple. In like wyse myght a man goo rounde aboute therthe as ferre [far] as therthe dureth [lasts or extends] by nature, alle aboute, so that he shold come vnder vs. and it shold seme to hym that we were vnder hym, lyke as to vs he shold seme vnder vs, ffor he shold holde his feet ayenst [against] oures and the heed to ward heuen [head toward heaven], no more ne lasse [nor less] as we doo here, and the feet toward therthe. And yf he wente alway forth his way to fore hym [if he always went in the same direction], he shold goo so ferre that he shold come agayn to the place fro whens [whence] he first departed.

Figure III-6. Figure 13.

And yf it were so that by aduenture two men departed that one fro that other, and that one went alle way to ward the eest and that other to ward the weste, so that bothe two wente egally [equally], it behoued [behooved] that they shold mete [meet] agayn in the opposite place fro where as they departed, & bothe two shold come agayn to the place fro whens they meuyd [moved] first; for thenne had that one and that other goon [gone] rounde aboute the erthe aboue and vnder, lyke as rounde aboute a whele [wheel] that were stylle [still] on therthe.

In lyke wise shold they goo aboute therthe, as they that contynuelly [continually] drewe them right to ward the myddle of therthe; for she fastneth alle heuy thyng to ward her. And that most weyeth [weighs], moste draweth and most ner [near] holdeth to ward the myddle; ffor who moche depper one delueth [delves] in therthe, somoche heuyer shal he fynde it. . . .

But for to vnderstonde the bettre, and more clerly [clearly] conceyue [conceive], ye may vnderstande by another ensample [example]: Yf the erthe were departed right in the myddle, in suche wyse that the heuen myght be seen thurgh, and yf one threwe a stone or an heuy plomette [ring] of leed [lead] that wel weyed [weighed], whan it shold come in to the myddle and half waye thurgh of therthe, there ryght shold it abyde and hold hym; for it myght nether go lower ne arise hyer, but yf it were that by the force of the grete heyght it myght, by the myght of the weight in fallyng, falle more depper than the myddle. But anon [soon] it shold arise agayn in suche wise that it shold abyde in the myddle of therthe, ne neuer after shold meue [move thence] thens; ffor thenne shold it be egally ouerall [overall] vnder the firmament whiche torneth [turns] nyght and daye. And by the vertue [virtue] and myght of his toryng nothyng may approche to it [the firmament] that is poysant [weighty] and heuy, but withdraweth alway vnder it; . . .

And yf the erthe were perced [pierced] thurgh in two places, of whiche that on [one] hole were cutte in to that other lyke a crosse, and foure men stoden [stood] right at the foure heedes [heads] of thise ii hooles, on [one] aboue and another bynethe, and in lyke wyse on bothe sides, and that eche [each] of them threwe a stone in to the hoole, whether it were grete or lytyl [little] eche stone shold come in to myddle of therthe wythout euer to be remeuid [removed] fro thens, but yf it were drawen away by force. And they shold holden them one aboute another for to take place eueriche [everyone] in the myddle of therthe.

And yf the stones were of like weight, they shold come therto alle at one tyme, as sone that one as that other; ffor nature wold suffre [permit] it none other wise. . . .

And yf their weyght and powers were not egall fro the place fro whens they shold falle, that whiche were most heuy [heavy], that sholde sonnest come to

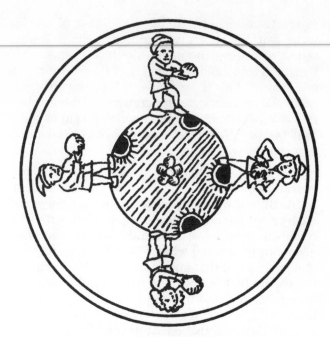

Figure III-7. Figure 15.

the myddle of therthe, and the other shold be al aboute her, . . .

Now thenne plese [please] it you to here for to deuyse [display] playnly to you how the erthe is rounde.

Who that myghte mounte on hye in thayr [the air] and who that myght beholde by valeyes & by playnes the hyenes [highness] of the grete montaynes and the grete and depe [deep] valeyes, the grete wawes [waves] of the See [sea] and the grete flodes [floods], they shold seme lasse tappere [seem less to appear] vnto the gretnes of the erthe than sholde an heer [hair] of a man doo vpon an apple or vnder his fyngre. Ffor neyther montayne ne valeys, how someuer [howsoever] hye ne depe it be, taketh not away fro therthe his roundenesse, no more than the galle [chestnut] leueth [lacks] to be round for his prickis [points]; ffor it behoueth the erthe to be rounde ffor to amasse the more peple [people]; and we shal saye to you here after how the world muste nedes [needs] be round.

God fourmed the world al round; ffor of alle the fourmes that be, of what dyuerse [diverse] manners they be, may none be so plenere [full] ne resseyue [receive] somoche by nature as may the figure round. Ffor that is the most ample of alle figures that ye may take example by. Ffor ther is none so wise ne so subtyl [subtile] in alle thinges, ne somoche can vnderstande, that may for ony thynge make a vessel, be it of woode or of stone or of metall, that may be

Figure III-8. Figure 18.

so ample, ne that may holde within it so moche in right quantite as shal do the rounde.

Ne fygure that ony may make may so sone meue [soon move] ne so lyghtly make his torne to goo aboute, that ony man can vnderstande, but that it muste take other place than this to fore, sauf [save] only the Rounde whiche may meue round without takyng other place; ffor she may haue non other than the firste, ne passe one only ligne [line] or Ray fro the place where she holdeth her in. Wherof ye may see the nature by a fygure squared sette within a round or another whiche is not round, and make them bothe to torne; the corners of them that ben [are] not rounde shal take dyuerce [diverse] places that the rounde secheth [seeks] not. . . .

Yet is ther another thynge: that ther is nothyng vnder heuen enclosed, of what dyuerse facion [fashion] it be, that may so lightly meue [move] by nature as may the rounde. And therfore God made the world round to this ende that it myght best be filled on alle partyes [in all parts]; ffor he wil leue [leave] nothyng voyde [void], and wille that it torne day and nyght; ffor it behoueth [behooves] to haue meuyng on [motion in] the heuen whiche maketh all to meue, ffor alle meuynges [motions] come fro heuen; therfore it behoueth lightly and swyftly to meue; and without it ther is nothyng may meue.

Owr Lord God gaf meuyng [gave motion] vnto the heuen whiche goth so swyftly & so appertly that no-

man can comprise in his thought; but it semeth not to vs for his gretenes, nomore than it sholde seme to a man, yf he saw fro ferre [far] an horse renne [run] vpon a grete mountayne, it shold not seme to hym that he wente an only paas [pace]; and somoche as he sholde be most ferre fro hym, somoche the lasse sholde he seme to goo.

And the heuen is somoche hye and ferre aboue vs that, yf a stone were in thayer [the air] as hye as the sterres be, and were the most heuyest of alle the world, of leed [lead] or of metall, and began to falle fro an hye aboue, this thyng is proued [proved] and knowen that it shold not come to therthe tyl thende [the end] of an hondred yere, so moche and ferre is the heuen fro vs, the whiche is so grete that alle the erthe round aboute hath nothyng of gretenes ayenst [against] the heuen, nomore than hath the poynt or pricke in the myddle of the most grete compaas [compass] that may be, ne to the grettest cercle that may be made on therthe. And yf a man were aboue in heuen, and behelde and loked [looked] here doun [down] in the erthe, & that alle the erthe were bren-nynge [burning] alle in cooles flammyng [in flaming coals] & lighted, it shold seme to hym more lytil [little] than the lest sterre that is aboue semeth to vs here in therthe, thawh [though] we were on a mon-tyne or in a valeye.

& therfore it may wel be knowen that the heuen muste lyghtly meue [move], whan it maketh his torne and goth round aboute therthe in a day and a nyght, lyke as we may apperceyue [perceive] by the sonne [sun] that men see in the mornyng arise in thoryent [the orient] or in the eest, and goth doun in the west; and on the morn erly [early] we see hym come agayn in the eest. Ffor thenne he hath per-fourmed his cours [course] round aboute therthe, whiche we calle a day naturel, the whiche con-teyneth [contains] in hym day and nyght. Thus gooth and cometh the sonne, the whiche neuer shal haue reste ne neuer shal fynysshe to goo wyth the heuen, lyke as the nayle [nail] that is fixed in the whele [wheel], the whiche torneth whan she torneth.

But by cause [because] that it hath meuyng [mov-ing] ayenst [against] the cours or tornyng of the firmament, we shal saye to yow another reson: Yf a flye wente round aboute a whele [wheel] that wente rounde it self, and that the flye wente ayenst it, the whele shold brynge the flye with her; and so shold it falle [happen] that the whele shold haue made many tornes whilis that the flye shold make one torne, and er [before] she had gon round aboute the whele vnto the first poynt. So ye muste vnderstonde that in suche manere goon the mone [moon] and the sonne by a way that is comune [common] to the vii planetes that ben on the heuen, whiche alle goo by the same way, alleway to ward the eest. And the heuen torneth to ward the weste, lyke as nature ledeth [leads] hym. . . .

Syth [Since] that the erthe is so lytil [little] as ye haue herd [heard] here to fore deuised [displayed], lytil maye we preyse [praise] the goodes therof vnto the regard of heuen, lasse than men do donge [give] ayenst fyn [against fine] gold or ayenst precyous stones; how wel that in thende [the end] that one and that other shal be of no valewe [value]. But fo somoche as we, beyng in this world, vs semeth that the erthe is moche grete, we haue declared to yow as wel the roundenesse as the gretenes to our power, and that shortly.

Syth we haue vnderstande how the erthe is rounde on all partes as an apple, neuertheles it is not enhabited [inhabited] in alle partyes [parts], whiche is wel knowen, of no peple [people] of the world. And it is not enhabited but in one quarter only, lyke as the philosophres haue enserched [searched], whiche put for to knowe it grete trauayll [travel] and estudye [study]. And therfore we shal deuyse [divide] it al about in foure partyes. Of whiche ye may take ensample [example] by an Apple whiche shal be parted by the myddle in foure parties right of lengthe and of brede [breadth] by the core. And pare a quarter and stratche [scratch] the parell, for to see and vnderstonde the facion [method], in playn erthe [make marks in sand or dirt] or in your hande.[2]

Ye ought to knowe that aboue thayer [the air] is the fyre. This is an ayer whiche is of moche grete resplendour [very great splendor] and shynyng, & of moche grete noblesse; and by his right grete subtylte [subtlety] he hath no moisture in hym. And is moche more clere than the fyre that we vse, & of more subtyl nature, than tharyer is ayenst the water or also the water ayenst the erthe.

This ayer in whiche is no maner moisture, it stratcheth [reaches] vnto the mone. And ther is seen ofte vnder this ayer somme sparkles of fyre, & seme that they were sterres. Of which men saye they be sterres whiche goon rennyng [go running], & that they remeue [remove] fro their places. But they be none; but it is a maner of fyre that groweth in thayer of somme drye vapour which hath no moisture within it, whiche is of therthe; & therof groweth by the sonne [sun] whiche draweth it vpon hye; & whan it is ouer hye [over high], it falleth & is sette a fyre like as a candel [candle] brennyng [burning] as vs semeth; & after falleth in thayer moyste [moist], and there is quenchid [quenched] by the moistnes of thayer. And when it is grete & the ayer drye, it cometh al brennyng vnto therthe.

Wherof it happeth ofte that they that saylle [sail] by the see or they that goon by londe [land] haue many tymes founden & seen them al shynyng & brennyng [burning] falle vnto therthe; & whan they come where it is fallen, they finde none other thing but a litil asshes or like thing, or like som leef [leaf] of a tree roten, that were weet [wet]. Thenne apperceyue [perceive] they wel, and byleue [believe], that it is no sterre; ffor the sterres may not falle, but they muste alle in their cercle meue [move] ordynatly [as ordained] & contynuelly nyght & day egally [equally].

The pure ayer is aboue the fyre, whiche pourpriseth [purposes] and taketh his place vnto the heuen.

In this ayer is no obscurte ne derknes [darkness], ffor it was made of clene purete [clean purity]. It resplendissheth & shyneth so clerly that it may to nothing be compared.

In this ayer ben [air are] vii sterres whiche make their cours al aboute therthe, the whiche be moche clene & clere [those which are very clean and clear], & be named the vii planetes; of whome that one is sette aboue that other, and in suche wyse ordeyned that ther is more space fro that one to that other than ther is fro the erthe to the mone [moon] whiche is ferther fyften [farther fifteen] tymes than al the erthe is grete; & euerich [every one] renneth by myracle on the firmament and maketh his cercle, that one grete and that other lytil, after that it is and sitteth more lowe. Ffor of somoche that it maketh his cours more nyghe [nigh] therthe, so moche is it more short; and sonner [sooner] hath perfourmed his cours than that which is ferthest. That is to saye that who that made a poynt in a walle, & with a compaas made dyuerse [diverse] cercles aboute, alway that one more large than another, [concentric circles] that whiche shold be next the poynt shold be leste of the other, and lasse shold be his cours; ffor he shold sonner haue don his cours than the grettest, so that they wente both egally [equally]; . . .

Thus may ye vnderstande of the vii planetes of whiche I haue spoken that that one is vnder that other, in suche wise that she that is lowest of alle the other is leest of alle, & that is the mone [moon]. But by cause [because] that it next to therthe, it semeth grettest & most apparaunt of alle the other; & for thapprochement [the approachment] of therthe, & by cause it goth so nygh, it hath no pure clerenes that cometh of hym self proprely, by cause therthe is so obscure.

But the clernes & lyght that it rendreth [renders] to vs she taketh alway of the sonne, lyke as shold a myrrour whan the rayes of the sonne smyteth [hits] therin, & of the reflexion the myrrour smyteth on the walle & shyneth theron as longe as the rayes of the sonne endure in the glasse; in lyke maner sheweth & lyghteth to vs the lyght of the mone; & in the mone is a body polysshyd [polished] and fair lyke a pommell [the spherical part of a saddle or handle of a sword] right wel burnysshed, whiche reflaumbeth [reflects] and rendrith [renders] lyght & clerenes whan the rayes of the sonne smyteth therin.

The lytil clowdes [clouds] or derkenes that is seen therin, somme saye that it is therthe that appereth within; and that whiche is water appereth whyte, lyke as ayenst a myrrour whiche receyueth dyuerse [receives various] colours, whan she is torned therto. Other thinke otherwyse and saye that hit happed and byfelle [it happened and befell] whan Adam was deceyued [deceived] by thapple that he ete [ate], whiche greued [grieved] alle humayne lignage [lineage], and that thenne the mone [moon] was empesshed [impressed] and his clerenesse lassed [lessened] and mynuysshid [diminished].

Of thise vii sterres or planetes that ben [are] there and make their cours on the firmament, of whom we haue here to fore spoken, ffirst were no moo [more] knowen but the tweyne [twain or pair], that is to wete [wit] the sonne and the mone; the other were not knowen but by Astronomye. Neuertheles yet shal I name them for as we haue spoken of them to yow.

Of thyse [these] ther ben tweyne [two] aboue the mone and byneth [beneath] the sonne, and that one aboue that other, of whom eche hath on therth propre vertues [each has on the earth their proper virtues]. And they be named Mercurie and Venus.

Thenne aboue the mone & thise tweyne is the sonne whiche is so clere, fayr [fair] & pure that it rendreth [gives] lyght & clerenesse vnto alle the world; and the sonne is sette so hye aboue that his cercle is gretter & more spacyouse [spacious] than the cercle of the mone, which maketh his cours in xxx dayes, xii sithes [sizes] somoche; ffor the sonne, whiche gooth more ferther fro the erthe than the mone, maketh his cours hath ccclxv dayes: this is xii tymes somoche & more ouer [moreover], as the calender enseigneth [designates], & yet more the fourth part of a day, that be vi houres. But for this that the yere hath dyuersly [year has variously] his begynnyng, that one begynneth on the daye & another on the nyght, which is grete ennoye to moche peple [which greatly annoys many people], this fourth part of a day is sette, by cause alle way in four yere is a daye consumed whiche is aboue in that space; the yere is named bysexte or lepe [leap] yere, whiche in iiii yere falleth ones [once]; and so is sette fro four yere to four yere always more a daye. And

thenne is the sonne comen agayn in his first poynt: and that is in myd Marche, whan the newe tyme recommenceth [recommences] and that alle thynges drawe to loue [love] by the vertue of the retorne [return] of the Sonne. Ffor in this season had the world first his begynnyng; and therfore thenne alle thinge reneweth and cometh in verdure [greenness] by right nature of the tyme and none otherwyse.

Aboue the sonne ther be thre sterres clere and shynyng, and one aboue another. That is to wete [wit] Mars, Jupiter and Saturnus. Saturne is hyest of the seuen [seven], whiche hath in his cours xxx yere er he hath alle goon his cerkle [circle] & thyse iii sterres reteyne [retain] theyr vertues in thynges here bynethe [beneath]; . . .

Thise seuen planetes ben suche that they haue power on thynges that growe on therthe; and habounde [abound] their vertues more than alle the other that ben on the firmament, and more appertly [expertly] werke, lyke as thauncyent sage philosophres haue enserched [searched] by their wittes [wits].

Of thise seuen planetes taken the dayes of the weke [week] their names, as ye shall here [hear]. The mone hath Monday, and Mars the Tewsday, Mercurye the Wednesday, Jupiter the Thursday, Venus the Vryday, Saturnus the Saterday; and the holy Sonday hath his name of the Sonne whiche is the most fair. And therfor the Sonday is better than ony of the other dayes of the weke, ffor this day is sette and reseruyd [reserved] from alle payne [pain] & labour. And on this day shold men doo thyng that shold playse [please] Our Lord.

But syth [since] in this chapytre we haue touched of the firmament, we shal speke after of somme caas [case] that come on the heuen and therthe.

The Sonday is as moche to saye as the daye of pees [peace] and of praysynge [praising], ffor the creatour of alle thynges cessed [ceased] this day, the whiche made and created all.

Aboue Saturne, whiche is the last planest [planet] & hyest from vs of alle the vii planetes, is the heuen that men see so full of sterres as it were sowen, whan it is clere tyme and weder [weather]. This heuen that is so sterred is the firmament whiche meueth [moves] and goth [goes] round. Of whiche meuyng [motion] is so grete Joye, so grete melodye [melody] and so swete [sweet], that ther is noman that, yf he myght here [hear] it, the neuer [that never] after shold haue talente ne wylle [nor will] to do thynge that were contrarye vnto Our Lord in ony thynge that myght be, so moche shold he desyre to come theder where he myght alleway here [always hear] so swete melodyes & be alway wyth them. Wherof somme were somtyme that sayde that lytil yonge [young] chyldren herde [heard] this melodye whan they lawghed [laughed] in their slepe [sleep]; ffor it is sayde that thenne they here [hear] the Angels of Our Lord in heuen synge, wherof they haue suche Joye in their slepe.

But herof knoweth noman the trouthe sauf [truth save] God that knoweth all, whiche setted the sterres on the heuen and made them to haue suche power. Ffor ther is nothynge withyn the erthe ne withyn the see [sea], how dyuerse [diverse] it be, but it is on the heuen fygured and compassed by the sterres, of whiche none knoweth the nombre sauf God only whyche at hys playsir [pleasure] nombreth them & knoweth the name of eueriche [every one] of them, as he that alle knoweth & alle created by good reason.

At the regard of the sterres that may be seen, they may be wel nombred & enquyred [numbered and inquired about] by Astronomye; but it is a moche maistryse [great mystery]; ffor ther ne is sterre so lytil but that it hath in hym hole [the whole of] his vertue, in herbe, in flour [flower] or in fruyt, be it in facion [fashion], in colour or otherwyse. Ther is nothing in erthe that ought to be, ne therin hath growyng, but somme sterre hath strengthe and puissaunce [power] by nature, is it good or otherwyse, suche as God hath gyuen to it.

Bvt syth [But since] we haue descriued [described] and apoken of the firmament in this scond partye of this volume, we shal speke of somme caases that come and happen on hye and also lowe. And shal speke of the mesure of the firmament ffor to vnderstande the better the facion [fashion], and how it is made and proporcioned, and of that whiche is aboue, and also we shal speke of heuen. . . .

In this thirde and last partye [part] of this present booke we shall fynysshe [finish] it wyth spekynge of the faites [facts] of astronomye. And I wyl declare to you first thw [how] the daye cometh and the nyght, and for to make you vnderstande of the eclipses, and also for to vnderstande other thinges, the whiche may moche prouffyte [profit] to them that wylle doo payne [will take the pains] to knowe them, ffor to gouerne [govern] them the better after the disposicion of the tyme.

Trouthe [Truth] it is that the Sonne maketh this torne [turn] & cours aboute therthe in the daye and nyght, and goth egally euery [goes equally every] houre. And also longe as he abydeth [stays] aboue therthe, so longe haue we the deduyt [delight] of the day; & whan he is vnder therthe, thenne haue we the nyght; lyke as ye went tornyng a brennyng [turning a burning] candell aboute your heed [head], or as ye

shold bere [bear] it a lytil ferther of Round aboute an apple, and that the candel were alway brennyng; thenne the partye that were alway ayenst [against] the candel shold alleway be lyght, and that other partye that is ferthest fro it shold be obscure and derke. Thus in lyke wise doth the sonne, by his propre nature, for to be day and nyght aboute therthe. He maketh the day to growe byfore [before] hym, and on that other parte the erthe is vmbreuse [shadowy] & derke by hynde [behind] hym and where as he may not shyne. And this is the shadowe of the nyght whiche the deduyt [delight] of the day taketh away from vs.

But for as moche as the sonne is moche gretter than therthe, the shadowe goth lytil and lytil tyl at thende [little until at the end] it cometh to nought, lyke the sown [sound] of a clocke endureth after the stroke.

But yf the sonne and therthe were of one lyke gretenesse, this shadowe shold haue none ende, but shold be all egal [equal] without declynyng. And yf therthe were gretter than the sonne, thenne the shadowe of the sonne shold go enlargyng and be more; . . . Take some derke thing that may reteine [retain] lyght within it [opaque], as of tree or of stone or other thynge what it be that may [not] be seen thurgh; thenne sette that to fore your eyen [eyes], ayenst that thing that ye wold see, is it the heuen or erthe or ony other thynge. Yf that thyng that ye holde is more bredder [broad] and larger than your two eyen be a sondre [asunder], it shal take away the syght ayenst [sight against] that whiche is no bredder. And yf the thynge be alle egale [equal] in lengthe as moche as ye may stratche [stretch] your two eyen, as moche shal it be taken fro you as the thinge shal haue of gretnes. . . . And yf the thinge haue lasse [less] of gretnes than the lengthe is bytwene [between] your bothe eyen, it shal take fro you lasse for to see, as wel nyghe as ferre [near as far], that it is of largenes of that whiche ye wold see. And whan ye put the thynge fro your eyen, so moche the more may ye see of that other part ouer and aboue you, so that ye may se [see] all. In lyke wyse is it of the sonne withoute ony doubtaunce [doubt] or variacion [variance]; ffor it passeth therthe in gretnesse, so that it seeth the heuen al aboute, the sterres, and all that is on the firmament.[3]

STUDY AND DISCUSSION QUESTIONS

1. How many miles in circumference was the earth found to be? Compare to today's value.
2. What is the diameter of the earth? By whose method and how? Compare to today's value.

3. What was the distance to the moon? Compare with today's value.

4. How does the year of Saturn compare to what we know?

5. Where are Mercury and Venus located?

6. How does the order of the planets compare to our ideas? To Plato's?

7. Why are Roman numerals used throughout?

8. When did Adam complete his trip?

9. When was the earth created?

10. What is the velocity of a soul? How does this compare to a stone's velocity?

11. How many stars are there?

12. In what manner are the elements arranged and how many are there?

13. What is the shape of the earth? The relation of men to each other and to it?

14. What of Columbus and his problems?

15. What happens to a stone dropped through the earth?

16. How does this compare to today's belief? What causes the difference?

17. What is the relation between weight and rate of fall?

18. For what reasons did God make the earth round?

19. What reason or reasons would be valid today?

20. Where do all motions come from?

21. Why don't the heavens seem to move swiftly?

22. How long would it take a stone to fall from heaven?

23. What example is used and how is retrograde motion explained?

24. How are meteors and meteorites explained?

25. How can it be told when a planet is close to the earth?

26. How does the moon give light and what accounts for the "face" or shadows in the moon?

27. What are the times given for planets to complete their orbits? Compare to today's values.

28. What is the length of a year? Compare to today's value and to that of Copernicus.

29. How much in error were they (c. 1200) as to their calendar?

30. To what does the "planets retaining their virtue on earth" refer to?

31. How are day and night explained? Eclipses?

FOOTNOTES

1. Oliver H. Prior (ed) *Caxton's Mirrour of the World.* Oxford University Press, 1966 pp. 169-174.

2. *Ibid,* pp. 50-61.

3. *Ibid,* pp. 122-133.

IV

The Crisis and Copernicus

It frequently comes as a surprise to modern man that there is a close relationship between religion and most historical aspects of his activities.

In the area of mathematics, the Hindu number system (which we call Arabic numbers) was originated to enable the priests to determine the dates of religious festivals. Algebra and the Hindu-Arabic numbers were introduced into Western Europe, in spite of the fact that they were regarded as a heathen device, to assist in the determination of the date of Easter.

The question as to when Easter occurs gives rise to our earliest scientific revolution. It might simply be asked as to why not look at a calendar but this leads to the question of which calendar and how obtained?

Historically the ancient Babylonians and Egyptians used the cycles of the moon for their annual calendars. At a latter date, the Babylonians were composed of Semitic peoples. The Hebrews are a Semitic people and also used a lunar calendar. The Egyptians later turned their worship towards Re, the sun god, and their calendar to a solar one. The Egyptian priests were quite learned and had calculated a solar calendar which Julius Caesar established as the official state Roman calendar in 46 B.C.

Even today not all the earth's people use the same calendar and indeed some, like the Christians, use two. To give some examples using a Gregorian calendar as a reference and 1971 as the comparison year: the Japanese year begins January 1 but it is the year 2631, the Mohammedan year begins February 27 in the year 1391, the Indian year 1893 begins March 22, the Grecian year 2283 begins September 14, and the Jewish year 5732 begins September 20. January 14 is January 1 on the Julian calendar.[1]

In order to see the origin of some of this problem let us look at facts which we know today but were not known then. The moon does not go around the earth at an even rate. It rises about 50 minutes earlier each night and travels through the constellations at the rate of about one every two nights. The time that the moon takes to travel from one star back in the heavens to the same star is called the "sidereal period"

and *on the average* is 27 days 7 hours 43 minutes 11.5 seconds. Another way that lunar time can be constructed is the position relative to our sun from one full moon to the next which is called the "synodic period" and is $29^d12^h44^m2^s.8$. The synodic period is the one which is closest to our calendar month.

The motion of the earth about the sun is also not regular nor is its rotation an even value. Under apparent sun time we should rotate every 24 hours, however the figure is closer to 23 hours 56 minutes but if compared to the background of the stars or what is called a "sidereal day" then it is 23 hours 56 minutes 4 seconds.

A question of great importance to persons of the fifteenth century is raised over the calendar. You might ask that since they weren't going anywhere in a hurry, why did they need a calendar? The Spring planting usually rested on some folklore and not on the calendar anyway. The question was, and still is for that matter, when is Easter? The Christian calendar is a lunar one, that is each month follows the cycle of the moon, which comes from Judaism since the early Christians were Jews. On the other hand, the official state calendar is a solar calendar, the Julian calendar. The lunar and solar calendar do not match. The solar calendar was assumed to have 365-1/4 days, with the normal year of 365 days, while the lunar year is approximately 354 days. It had also been assumed that the lunar and solar calendars could be made to match every 19 years. Actually the solar year is 365.2422 days which makes the Julian year about 11 minutes too long. Hipparchus had made this calculation and his year was about six minutes too long while that of Ptolemy made it 7-1/2 minutes too long. Even a few minutes a year difference do add up during the centuries and between Julius Caesar to the sixteenth century amounted to over ten days.

In early Christianity, since many of the Christians had been Jews, there was little difficulty in finding Easter which was tied to the celebration of Passover. Passover occurs on the 14th day of the first month, i.e. the lunar month which falls on, or follows next to, the vernal equinox. By the second century there was argument and most Christians wanted Easter to be on a Sunday while some did not. It was finally decided, at the Council of Nicaea (325 A.D.), that Easter must be celebrated on the Sunday following the first full moon after the vernal equinox, usually March 21, except when this would coincide with Passover when it would be postponed one week. Further it was decided that since the intellectual capital of the world was at Alexandria and the best persons for making the calculations as to when Easter would fall were there the bishop of Alexandria should notify the Christian world when Easter would occur. Since Easter must be proclaimed prior to Lent, forty days beforehand, it meant that the date had to be calculated well in advance and circulated. One-third of the Christian year is dependent and determined by the date of Easter.

With the decline of the Roman Empire, roads fell into disrepair and became unsafe so that coupled with the rise of Islam, Alexandria became less accessible to the west and in isolated areas bishops and priests made their own calculations. When a king married, his bride generally brought her own priest and confessor with her. Different priests with different methods of calculations meant that sometimes the king was celebrating Easter while the queen was fasting in Lent.[2] Easter is a spring festival which can vary, with the differences between the lunar and solar calendar, between March 22 and April 25. By the sixteenth century, even if the calculations were correct, the error of 10 days added puts Easter well into May.

After the crusades, with the rise in trade and travel one could go from area to area always celebrating Easter during most of the spring and well beyond the possible date. Many complaints poured toward the Pope. These and other types of complaints added to the fact that Louis XII had called his own church council caused Julius II to convoke a rival council at the Lateran in 1512. One of the critical problems was that of the reform of the calendar and it is now time to introduce the hero of our unfolding drama.

Niklas Koppernigk, Polish, (1473–1543) was born in Torun, the youngest of four children. His father died when he was nine and the children were adopted by an uncle, Lucas Watzelrode, who later became Bishop of Ermland.

His education included three years at the University of Cracow (1491–1494), where he studied Euclid's geometry, philosophy, astronomy and astrology. He left without taking the examinations for the degree and returned home. When he found that

Figure IV-1. Copernicus.

his uncle could provide him with a sinecure, he went to Italy to study (1496–1506). He had gone first to the University of Bologna where he prepared himself in canon law and continued his studies in astronomy. In 1497 he was appointed canon of the cathedral of Frauenburg but had taken leave to continue his studies. He lectured informally on mathematics and astronomy in Rome during the jubilee year 1500. (Century years saw many pilgrims in Rome.) Between 1501 and 1505, Copernicus studied medicine at the University of Padua except for the year 1503, when he completed his doctorate in canon law at the University of Ferrara. By the time he returned to Poland, he was a humanist who had studied Greek, mathematics, and astronomy, and also a physician and jurist. As a matter of fact, during his lifetime, he was better known as a physician than as an astronomer.

Partly because of his lectures in Rome on astronomy and partly because of the praise that appeared about Copernicus in the foreword of a book that he translated written by a college friend, he was invited by the Lateran Council in 1513, to give his opinion on the proposed reform of the calendar. He declined on the grounds that the then current astronomical theory was not good enough to calculate a new calendar.

He had begun to develop his astronomical theory in 1506, when he returned to Poland. It was in 1530 that he presented an outline of the heliocentric theory in a book called *Commentariolus*. The new doctrine was lectured about in Rome by Johann Albrecht Widmanstadt and Pope Clement VII gave his approval so that Cardinal Schonberg begged Copernicus to publish his whole thought on the subject. The first general account was published by Joachim Rheticus, a professor who was also a student of Copernicus', under the title *Narrato Prima* in 1540. The book was very favorably received and Copernicus was encouraged to permit the publication of his entire work—Nicolai Copernici Torinensis, *De revolutionibus orbium coelestium.* Norimbergae, 1543. *(On the Revolutions of the Heavenly Spheres.)*[3]

> Copernicus owes his distinguished place among the founders of modern science almost entirely to the one great book of his which was published in the last year of his life. This book marks an epoch in the history of human thought. As a monument of scientific genius, it ranks with the *Almagest* of Ptolemy, with Newton's *Principia* and with Darwin's *Origin of Species.*[4]

The book itself is divided into six sections most of which contains calculations in plane and spherical trigonometry, tables of trigonometry, and the positions of stars and constellations from his observations or the observations of others. The first part of the book indicates how he came to his ideas and the last part to detailed accounts of the motions of the planets including the earth.

In this work Copernicus retains most of the Ptolemaic system except that the earth is placed in motion, like the other planets, around a fixed point *near the sun.* He constructs an interlocking mathematical system which has fewer assumptions than the Ptolemaic system. Since he retains circular uniform motion, he also has to retain epicycles, and since he places the center of motion near, rather than at the sun, he also has to retain an equant. The motion of the planets in the same direction makes the total motions easier to explain. His system also better explains the year. So well did he work that his estimate of the star year (365^d 6^h 9^m 40^s) was only thirty seconds greater than that value accepted today.

His acceptance of Ptolemy's tables and much of Aristotle shows that Copernicus, as well as any other revolutionary, steps but little away from tradition. This aspect can be seen in many of the religious revolutionists—Christ, Luther, Buddha, and Wesley for example. Copernicus is almost scholastic in his approach by first quoting authorities. He says that he first had his idea from a comment in Cicero that Nicetas thought that the earth moved and later from Plutarch that Philolaus the Pythagorean stated that the earth moved like the sun and moon. What seems to be the clincher for Copernicus, and has since become a scientific dogma, was given two systems that both explain the same phenomena, the simpler is chosen.

According to legend, Copernicus saw an advance copy of his book on his deathbed. At any rate he was dead before his book was published and he did not reform the calendar, but gave an accurate basis for the reform.

To tie up a loose end, the calendar was reformed with the mathematics of Aloysius Lilius, a learned astronomer and physician of Naples, and verified, compiled, explained, and published by Clavius in 1603. The old calendar was abolished by the decree of Pope Gregory XIII in March, 1582 in which October 5, 1582 was to become October 15. The calendar was rectified with the time of the council of Nicaea. The difference between the Gregorian Calendar or New Style (N.S.), in 1700 became eleven days, and the Julian or Old Style (O.S.) because it was a leap year in the Julian but not in the N.S. The calendar calculations devised by Lilius ironically do not depend on astronomical observations at all but are purely computational.

It should be noted that Copernicus, as others that we shall be examining, was an example of the man of the Renaissance. His education included languages, the humanities, science, medicine, and religion. This pattern remains among the educated until the nineteenth century and we shall see it repeatedly. The other factor worthy of note is that any person engaged in creative activity must have the income to assure leisure time. In Copernicus' case the Church is the source of income as is the case through all of the middle ages. This shifts to royalty, nobility and industry as patrons of learning and also to the universities which serve the "haven" function. In the twentieth century the great patron has come to be governments and foundations also fill the function in part.

STUDY AND DISCUSSION QUESTION

1. What is the origin of mathematics?
2. How was our calendar obtained?
3. What caused the calendar crisis?
4. How does solar time differ from lunar and stellar time?
5. When does Passover occur? Easter? What is the connection between them?
6. When did Easter and Passover occur in 1968? Significance?
7. What association does the use of eggs and rabbits at Easter have?
8. Why did the day for Easter have to be calculated?

9. What makes Copernicus a man of the Renaissance?
10. What shows that Copernicus was a traditionalist?
11. Why is George Washington's birthday given as February 11 and February 22, 1732.

FOOTNOTES

1. ————*The American Emphemesis and Nautical Almanac for the year 1971,* Government Printing Office, 1969, p. 1.

2. An interesting example of this is found in Stahl's *Roman Science,* pp. 220-232.

3. A translation is available in the *Great Books of the Western World,* volume 16.

4. Angus Armitage *The World of Copernicus,* Signet paperback, 1951. A brief and useful account of the life and work of Copernicus, p. 99ff.

V

As the World Turns

How and why is an idea accepted? What does the acceptance of an idea mean? These are questions that we will want to explore in this chapter.

In dealing with an accepted revolutionary idea from the past of a currently unacceptable idea, each of us has the hang-up of not being able to divorce ourselves from our fixed ideas. For example, it is very difficult to get a Christian to ask why was Christianity accepted because he cannot envision any other way. Most formally educated people of today accept the idea that the earth is in motion, but cannot prove it if asked to do so. Further, most of the greatest revolutionary ideas from science have no "practical" value at all. When celestial navigation is taught to pilots it is assumed that the earth is fixed and that the celestial sphere is in motion. As we have seen, the final solution of the calendar did not rest on Copernicus and I suspect that it could have been rectified without his contribution.

This particular revolution has more than usual relevance for us because today we are passing through a very similar one. I shall draw some parallels later in more detail than in this present chapter to support my contention.

Before proceeding it should be pointed out that ideas do not spread overnight, the world did not go to sleep one night as a stationary one and waken in motion, nor to the same degree in all strata of the same culture. As far as the growth is concerned, we are speaking only of the educated, Western, European. Ideas tend to trickle down through a culture from the apex of the social pyramid to the mass base —from the 10% or less who lead to the majority (in the seventies, designated as silent—perhaps dumb in the sense of mute would be better) who follow.

Having examined the difficulties and purposes of this exploration let us examine first the opposition to the Copernican ideas.

The major argument for us today and for many at that time was that the earth does not appear to move. There is no sensory evidence that we have to support the idea of a moving earth. If we walk, ride a horse, ride a cart—in all of these we are aware of motion.

In addition, if we examine a moving wheel with mud on it we can see the mud fly off the wheel as it moves faster and faster, and if the earth moved everything would fly off of it in the same manner. In running or riding a horse, a person would feel the air moving past him and he does not constantly feel a strong wind as he would if the earth were moving. If a person jumps into the air, or throws a rock straight up, both the person and the rock come down in the same place which could not happen if the earth moved.

It was also argued that if the earth moved around the sun, then at two times, six months apart, observations could be made of the stars, and if the Copernican idea was correct, the stars would appear to have moved. This is the principle of parallax. It can be illustrated easily. Using your two eyes as the position of the earth six months apart, close first one eye and then the other alternately. The object at which you are looking will appear to move back and forth. An object which is close will appear to move more than a distant object. Look at a pencil at arm's length as a close object and chose an object across the room for a distant object. Observations of the stars revealed no parallax as would be expected if the earth moved.

Light things such as fire rise whereas heavy objects move toward the earth. The earth is heavy and ponderous whereas celestial bodies are light and ethereal, so that it is easier to expect that they will move than that the huge earth could be turned. It was well known that celestial bodies differed from the earthly ones because the earth was full of imperfections and change whereas the celestial bodies were perfect and changeless, eternal. This could be called the idea of economy—one is easier to do than the other.

This had also been one of Aristotle's arguments. Aristotle's works had been attacked in parts before but the whole had withstood. One reason that Aristotle's works were great for such a long period of time was that it was a total system which was integrated in all its parts covering the whole field of man's knowledge. If the view of Copernicus were to prevail then the total structure of Aristotle would crumble. For a Christian world that had interwoven theology with Aristotle, this could have far-reaching effects.

The universe had a religious as well as a physical significance. Hell was located in the geometric center of the universe which was the center of the earth, while heaven and God's throne was placed beyond the eighth sphere of the fixed stars. The motion of the planetary spheres and the epicycle motion of planets was supplied by angels.

Only in Dante's *Divine Comedy* do we see the extent to which the Aristotelian-Ptolemaic system had its relation to theology.

Taken literally, Dante's epic is a description of the poet's journey through the universe as conceived by the fourteenth-century Christian. The journey begins on the surface of the spherical earth; descends gradually into the earth via the nine circles of Hell which symmetrically mirror the nine celestial spheres above; [the ninth sphere was added by Moslem astronomers to account for the procession of the equinoxes, and the motion of the heavenly pole. It rotates every 24 hours.] and arrives at the vilest and most corrupt of all regions, the center of the universe, the appropriate locus of the Devil, and his legions. Dante then returns to the surface of the earth at a point diametrically opposite the one where he had entered, and there he finds the mount of purgatory with its base on the earth and its top extending into the aerial regions above. Passing through purgatory, the poet travels through the terrestrial spheres of air and fire to the celestial region above. At the last he journeys through each of the celestial spheres in turn, conversing with the spirits that inhabit them, until finally he contemplates God's Throne in the last, the Empyrean sphere. The setting of the *Divine Comedy* is a literal Aristotelian universe adapted to the epicycles of Hipparchus and the God of the Holy Church.

For the Christian, however, the new universe had a symbolic as well as literal meaning, and it was this Christian symbolism that Dante wished most of all to display. Through allegory his *Divine Comedy* made it appear that the medieval universe could have had no other structure than the Aristotelian-Ptolemaic. As he portrays it, the universe of spheres mirrors both man's hope and his fate. Both physically and spiritually, man occupies a crucial intermediate position in this universe filled, as it is by a hierarchical chain of substances [see Lovejoy's *Great Chain of Being* for a discussion of the importance of this] that stretches from the inert clay of the center to the pure spirit of the Empyrean. Man is compounded of a material body and a spiritual soul: all other substances are either matter or spirit. Man's location, too, is intermediate: the earth's surface is close to its debased and corporeal center but within sight of the celestial periphery which surrounds it symmetrically. Man lives in squalor and uncertainty, and he is very close to Hell. But his central location is strategic, for he is everywhere under the eye of God. Both man's double nature and his intermediate position enforce the choice

Figure V-1. Dante's Universe.

from which the drama of Christianity is compounded. He may follow his corporeal, earthy nature down to its natural place at the corrupt center, or he may follow his soul upward through the successively more spiritual spheres until he reaches God. As one critic of Dante has put it, (Charles H. Grandgent, in his *Discourses on Dante,* p. 93) in the *Divine Comedy* the "vastest of all themes, the theme of human sin and salvation, is adjusted to the great plan of the universe." Once this adjustment had been achieved, and change in the plan of the universe would inevitably affect the drama of Christian life and Christian death. To move the earth was to break the continuous chain of created being.[1]

Finally, for the Christian world, there is the authority of scripture in which not only is there no mention of the earth moving, but also there are several passages which indicate that the earth does not move.

In essence, there may be considered five kinds of arguments against the heliocentric or Copernican view and these are the sensory, lack of parallax, economy, unity, and authority. Let us turn to those arguments that favor Copernicus.

A heliocentric world is not a new idea since others can be cited who have thought so—in other words, it is not mere novelty—as Copernicus himself stated.

The revival of ancient learnings placed a greater emphasis on the sun which could also be seen in the Franciscan's interests earlier (St. Francis' Hymn to the Sun e.g.) which has spread. This emphasis and interest in the sun would tend to center it.

Mathematically, the Copernican system was easier for calculations than were those of Ptolemy and it could in some sense be considered simpler. As we now know it gave a more accurate year and it also brought a closer correspondence between calculation and observation.

The positive arguments for the Copernican theory may be summed up as authority, ideas from antiquity, a renewed interest in nature, and ease of calculation.

The choice among theories, ideas, or intangibles is for an individual purely a matter of belief. We shall examine cases for individual beliefs and how gradually the Copernican view came to be supreme over the Ptolemaic idea. Any idea usually requires at least one generation to take hold. It is necessary for those who have spent a lifetime with the old idea, and hence have a vested interest in supporting the old ideas, to die and for the proper dissemination of the new idea.

When the lifework of Copernicus was published, there was little to indicate that the idea would spread. It was a highly mathematical work written in Latin and thus only available to a relatively few persons. It is quite possible that had it not been for the printing press, the Copernican revolution, like the Protestant reformation, might never have occurred. But once it was in print, it was almost impossible to suppress.

There were factors which favored the spread of the idea other than the technological one of printing. One of these, already mentioned, was the emphasis which humanists of the renaissance placed on the sun. Then, too, there was the Protestant reformation itself which created ferment in men's minds and caused doubts to appear. If there is no doubt then another idea cannot intrude. Further, more had been discovered about the earth itself.

To draw a few parallels, they had a means of disseminating information more rapidly than previously (the printing press c. 1454) and we, in this century also have expanded (radio c. 1920, television c. 1950). They had discovered a new world, Columbus, 1492; and we also have with the moon landing, 1969. Copernicus, Galileo, and Newton overturned old concepts; and in the twentieth century Einstein,

Planck, Heisenberg and Schrödinger have done likewise. From the past the thought revolutions gave rise to modern science, materialism, and the idea of progress. By an examination of the effects of the thought revolutions of the past we may get a forecast of the magnitude of the effects to come in the future from the thought revolutions of this century. Note, for example, the much more rapid succession of concept revolutions in this century as compared to those prior to the eighteenth century. We shall return to this theme in a later chapter.

The explorations of Columbus, (1492, 1493, 1498, 1502); John Cabot, (1497, 1498); Alonso de Ojeda with Amerigo Vespucci, (1499); Pedro Álvares Cabral, (1500); Gaspar Corte-Real, (1500); and a number of others culminating with the circumnavigation of the globe by Ferdinand Magellan, (1519–1522) showed great flaws in the geography of Ptolemy which had been the standard work up to that time.

Ptolemy had taken from Hipparchus the idea of dividing the equator into 360 parts and had erected lines of latitude and longitude around the earth, but from there his text becomes a comedy of errors. Eratosthenes (276–196 B.C.) had measured the earth using geometry at 250,000 stadia, or about 25,000 miles around at the equator and this figure was accepted by most other geographers. However, about 135–150 B.C., the geographer Posidonius reduced this to 180,000 stadia, or 18,000 miles and this was the figure that Ptolemy used. Most geographers had assumed that most of the earth's surface was water which surrounded a central land mass, but Ptolemy extended the land masses so that they took up most of the globe and further he shortened the distance between Europe and Asia by 50°. This last error was the one that encouraged Columbus in his belief. Ptolemy's *Geographia* was printed in Latin with maps about 1462 and by the 1470's, Columbus had been convinced of his idea of sailing west.

Another factor which encouraged the spread of the heliocentric idea was the rise of neoplatonism. There were two ideas from the writings of Plato which contributed to the Copernican revolution and to the rise of science. One of these was the emphasis on mathematics (geometry) and the other was concerning the nature of reality. We shall save the part on mathematics to a later chapter, although Copernicus certainly does use geometry in his calculations.

For Plato, and his followers, the real world did not consist of the world that we can perceive with our senses. What we perceive with our senses is merely an imperfect copy of that which exists in the real world. The real world is located someplace else and for Christendom, of course, the location is heaven.

We may summarize the factors contributing to the spread as printing, revolution against authority, explorations, and neoplatonism.

There was a strong reaction before Copernicus' book was printed (1543). Martin Luther (1483–1546) in one of his "Table Talks" of 1539 said:

> People gave ear to an upstart astrologer who strove to show that the earth revolves, not the heavens or the firmament, the sun and the moon . . . This fool wishes to reverse the entire science of astronomy; but sacred Scripture tells us (Joshua 10:13) that Joshua commanded the sun to stand still, and not the earth.[2]

Even Luther's humanist lieutenant, Melanchthon (1497–1560), wrote in 1549:

> The eyes are witnesses that the heavens revolve in the space of twenty-four hours. But certain men . . . have concluded that the earth moves; and they maintain that neither the eighth sphere nor the sun revolves . . . Now it is a want of honesty and decency to assert such notions publicly, and the example is pernicious. It is the part of a good mind to accept the truth as revealed by God and to acquiesce in it.[3]

Not only the Lutherans, but soon other Protestant leaders joined in their condemnation and rejection of Copernicus. They had a great stake in this argument for an important part of the reformation was to remove the Church between man and God and to revert to the authority of Scripture so that if the Scriptures were wrong in one part, then how could common men interpret the remainder?

John Calvin (1509–1564), in his *Commentary on Genesis,* cited the opening verse of the ninety-third psalm—"the earth also is stablished, that it cannot be moved"—and then demanded, "Who will venture to place the authority of Copernicus above that of the Holy Spirit?"[4]

Large numbers of clergymen searched the Scriptures for lines which would refute the Copernican doctrines by the seventeenth century and some still were describing the Ptolemaic system in the eighteenth. Ideas change very slowly.

It was also in the seventeenth century that the Roman Catholic Church became aroused (1616) and placed *De Revolutionibus* and all other writings that affirmed the motion of the earth on the Index. Catholics were forbidden to read or teach any versions of the Copernican doctrines except those which had been amended to leave out all mention to the moving earth and central sun.

Very likely the Church was more disturbed by the fuel that the theory added to the discord caused by the Reformation rather than an actual concern to disagreement with Scriptures. The Roman Catholic Church had always permitted considerable dissent and divergence of opinion within the Church.

It was not only clerics who attacked Copernicus, but intellectuals in many fields. Michel Eyquem Montaigne (1533–92) in his *Essais* (1580) discusses the Ptolemaic and Copernican systems and then writes: ". . . what use can we make of this, except that we need not much care which is the true opinion? And who knows but that a third, a thousand years hence, may overthrow the two former?"[5]

In a posthumous work Jean Bodin (1520–1596), famous as one of the most advanced and creative political philosophers of the sixteenth century, discards the Copernican theory on physical grounds:

> No one in his senses, or imbued with the slightest knowledge of physics, will ever think that the earth, heavy and unwieldy from its own weight and mass, staggers up and down around its own center and that of the sun; for at the slightest jar of the earth, we would see cities and fortresses, towns and mountains, thrown down. A certain courtier Aulicus, when some astrologer in court was upholding Copernicus' idea before Duke Albert of Prussia, turning to the servant who was pouring the Falernian, said: "Take care that the flagon is not spilled." For if the earth were to be moved, neither an arrow shot straight up, nor a stone dropped from the top of the tower would fall perpendicularly, but either ahead or behind . . . Lastly, all things on finding places suitable to their natures, remain there, as Aristotle writes. Since therefore the earth has been allotted a place fitting its nature, it cannot be whirled around by other motion than its own.[6]

A long poem, published in France in 1578, and immensely popular there and in England during the next 125 years, provides a typical description of the Copernicans as:

> Those clerks who think (think how absurd a jest)
> That neither heav'ns nor stars do turn at all,
> Nor dance about this great round earthly ball;
> But th' earth itself, this massy globe of ours,
> Turns round-about once every twice-twelve hours;
> And we resemble land-bred novices
> New brought aboard to venture on the seas;
> Who, at first launching from the shore, suppose
> The ship stands still, and that the ground it goes . . .
> So, never should an arrow, shot upright,
> In the same place upon the shooter light;
> But would do, rather, as, at sea, a stone
> Aboard a ship on ward uprightly thrown;
> Which not wighin-board falls, but in the flood
> Astern the ship, if so the wind be good.
> So should the fowls that take their nimble flight
> From western marches towards morning's light; . . .
> And bullets thundered from the cannon's throat
> (Whose roaring drowns the heav'nly thunder's note)
> Should seem recoil: since the quick career,
> That our round earth should daily gallop here,
> Must needs exceed a hundred-fold, for swift,
> Birds, bullets, winds, their wings, their force, their drift.
>
> Arm'd with these reasons, 'twere superfluous
> T'assail the reasons of Copernicus;

Who, to save better of the stars th'appearance,
Unto the earth a three-fold motion warrants.[7]

Although John Donne (1572–1631) (adventurer, secretary, poet, priest), is part of the renaissance tradition that places emphasis on the sun (e.g.),

Bussie old foole, unruly Sunne,
Why dost thou thus,
Through windows, and through curtaines call on us?
Must to thy motions lovers seasons run? . . .[8]

he was no advocate of Copernicus' system. This is evident from his statement:

Moving of th' earth brings harmes and feares,
Men rekon what it did and meant,
But trepidation of the speares[9]
Though greater farre, is innocent.[10]

Even after Donne himself conceded the probability of the earth's motion, he found it very disturbing.

[The] new philosophy calls all in doubt,
The Element of fire is quite put out;
The Sun is lost, and th' earth, and no man's wit
Can well direct his where to look for it.
And freely men confess that this world's spent,
When in the Planets, and the Firmament
They seek so many new; then see that this
Is crumbled out again in his Atomies.
'Tis all in Pieces, all coherence gone,
All just supply, and all Relation:
Prince, Subject, Father, Son, are things forgot,
For every man alone thinks he hath got
To be Phoenix, and that then can be
None of that kind, of which he is, but he.[11]

More than a century after the publication of *De Revolutionibus,* a well educated English poet, John Milton (1608–1674) who had visited Galileo before his death, could not decide between the two theories. He discusses both in *Paradise Lost* (1667) but concluded:

Whether the Sun predominant in Heav'n
Rise on the Earth, or Earth rise on the Sun,
Hee from the East his flaming rode begin,
Or Shee from West her silent course advance
With inoffensive pace that spinning sleeps
On her soft Axle, while she paces Eev'n,
And bears thee soft with the smooth Air along,
Sollicit not thy thoughts with matters hid,
Leave them to God above, him serve and feares; . . .[12]

Other poets and writers continued their attacks or disclaimers of the Copernican system. Henry Vaughan (1622–1695) shows his leanings towards Ptolemy in his lines from "The World":

And round beneath it [eternity], Time, in hours, days, years,
Driv'n by the spheres,

as does Sir Thomas Browne (1605–1682), a physician, who in spite of his contributions to the Royal Society

Figure V-2. Milton.

and his experimental works could write in his *Religio Medici* of 1642:

. . . but in divinity I love to keep the road; and, though not in an implicit, yet an humble faith, follow the great wheel of the church, by which I move; not reserving any proper poles, or motion from the epicycle of my own brain. By this means I leave no gap for heresy, . . .

Ben Jonson (1572–1637), English dramatist, lyric poet, actor; used satire against the Copernican view and like his friend Shakespeare rejected it. In a few passages of *Troilus and Cressida, King John* and *The Merry Wives of Windsor* the earth is described as motionless in the center of the universe. Henry More (1614–1687), English philosopher, who while Neoplatonic still said, "sense pleads for Ptolemee."

Nor were the attacks limited to the literary and religious writers. Mathematicians and astronomers/astrologers joined in the assault on the heliocentric view even though a number used the tables in Copernicus' book.

Erasmus Reinhold, of Wittenberg University used the tables in 1544 to cast a horoscope for Martin Luther. In Holland, Gemma Frisius admired the book but doubted in 1555 that the earth moved. Johannes Kepler's teacher, Michael Maestlin, used the Copernican tables without the theory. Another who used the observations was the Jesuit Christopher Clavius who reformed the calendar while calling Copernicus' hypothesis absurd. G. A. Magini (1555–1617), professor of mathematics at Bologna, did likewise. Fr. Maurolico, published his mathematical works in Venice in 1575 and wrote in the preface of one: "May also Copernicus be destroyed, who makes the sun rest and the earth turn like a top and who deserves a whip, rather than a reprimand."[13]

Francis Bacon (1561–1626), who introduced inductive methods to experimental science, attacks the Copernican theory in detail in his *Novum Organum* of 1620[14] and in general in his *Essayes:* "There is an abruse Astrologer that saith; If it were not for two things that are constant, (the one is, that the Fixed Starres ever stand at like distance, one from another, and never come nearer together no goe further asunder; the other, that the Diurnal Motion perpetually keepeth Time); no individuall would last one moment."[15] The father of statistics, Jerome Cardan (1501–1576) supported the geocentric scheme in his *Commentaries on Ptolemy* (1552) and attacked Copernicus on the basis of mathematics and astrology.

With all of the objections to the Copernican view and considering the lack of positive evidence it would seem unlikely that any person would give serious consideration to such a view. It is interesting to note that there was an attempt at compromise.

Tycho Brahe (1546–1601) was perhaps the greatest observational astronomer of all time, and certainly prior to the invention of the telescope. His very accurate measurements of the star angles and changing planet positions were used by Kepler and others.

Brahe made astronomical observations of the New Star of 1572, and of six comets between 1577 and 1596.[16] The variable brightness of the "New Star" was helpful in breaking-down the Aristotelian concept of the immutability of the heavens. Novas (or new stars) and comets had happened before and yet none had doubted before. Previously the explanation had been that these flares of light, along with meteors, meteorites and comets were exhalations from the earth which condensed in the region of fire, which was located between the earth and the moon, and put aflame.

> [But Tycho] . . . measured the position of the star relative to a known star in Cassiopeia, but could detect no change within 1' of arc in circumstances in which, were the star below the orbit of the moon, or even in the spheres of the planets, a change must have been detectable. He concluded therefore, despite Aristotle . . . that 'this star is not some kind of comet or of fiery meteor . . . but that it is a star shining in the firmament itself, one that has never been seen before our time, or in any age since the beginning of the world.'[17]

Brahe could find no evidence that the earth moved —no parallax for the stars—so he denied the Copernican view. His own figures did not agree with Ptolemy and there was the evidence of change in the heavens. He reverted to a compromise previously held by Herakleides of Pontos, in antiquity, that the planets revolve about the sun and that the system then revolved about the earth.

Because Tycho Brahe's system also "saved the phenomena" why was the Copernican system eventually accepted rather than the Tychonic system? "That the Tychonic and Copernican systems were geometrically equivalent, was shown by Kepler in a most ingenious argument involving the idea of relative motion."[18] The following not only answers this question, but also provides us with a clue as to the reason for the acceptance of the Copernican system by astronomers not concerned with the Tychonic system. ". . . the Tychonic system did not convert those few Neoplatonic astronomers, like Kepler, who had been attracted to Copernicus' system by its great symmetry."[19] This is our "Kuhnian hinge"—the importance of Neoplatonism—which will later prove fruitful in "opening the door" to an understanding of the Copernican Revolution.

There, of course, remained those who refused to be committed to any system. One of these, Blaise Pascal (1632–1662), is of particular interest because he places an emphasis on sensory evidence and as he states:

> . . . in the discourse of natural reason on the motion or stability of the earth all the phenomena of the motions and retrogradations of the planets follow perfectly from the hypotheses of Ptolemy, of Tycho, of Copernicus and from many others that could be framed, of all of which only one can be true. But who will venture to discern so far, and who can without risk of error support one to the prejudice of the others . . .[20]

How was the Copernican system accepted?[21] By examining various astronomers and mathematicians, their books reveal how the Copernican system was slowly assimilated into the science of the times, and thus supports the statement that ". . . the final victory of the *De Revolutionibus* was achieved by infiltration."[22] The question of "how" also ties in with the question of "why" and as has already been indicated, this varies with the individual.

In England, the first reference to the Copernican theory is found in *The Castle of Knowledge* (1556) by Robert Recorde (1510?–1588), physician in ordinary to Bloody Mary. It is not clear if he believed in the theory or not[23] but his leaning toward Neoplatonism is shown in his *Whetstone of Witte:*

> . . . for knowledge and certaintie . . . there is noe possibilitie without nomber. It is confessed emongeste all men, that knowe what learnyng meaneth, that besides the Mathematecalle artes, there is noe unfallible knoweledge, excepte it bee borowed of them.[24]

Another Englishman about whom there is doubt was John Dee (1527–1608), mathematician, astrologer (he fixed the date for Queen Elizabeth's coronation), alchemist, wizard.[25] "Record and Dee were universally recognized as the two founders of the school of able mathematical scientists which arose in England during the last half of the sixteenth century."[26] Although Dee adopted the Copernican system for astronomical calculations because of the mathematical superiority of the system, it is not known whether ". . . he ever completely accepted the physical reality of the Copernican system . . ."[27] Nevertheless, he helped to open the minds of the English scientists to the Copernican system. As Johnson states, "The credit for spreading a knowledge of the new Copernican astronomy among English scientists is due chiefly to Robert Recorde and John Dee."[28]

The acceptance of the Copernican system took place rapidly in England. ". . . an intelligent knowledge of the Copernican theory was spread among all the classes of practical scientific workers before 1600 . . ."[29]

In 1576, Thomas Digges (1545–1595) published his important treatise "A Perfit Description of the Caelestiall Orbes according to the most aunciente doctrine of the Pythagoreans, latelye reuiued by Copernicus and by Geometricall Demonstrations approued." The complete treatise is contained in a paper by Johnson,[30] and provides us with a valuable source from which we obtain a new insight into Digges' reasons for accepting the Copernican system —an insight which has been missed in all the previous work done on Digges.

As with the works of Recorde and Dee, Digges influenced later scientists. The influence of Digges' treatise on contemporary astronomical thought can hardly be overestimated. "It . . . was one of the most popular works of the period."[31] The book went through seven editions between the year 1576 and 1605.[32] Consequently, it caused later reactions because "From 1576 onward, references to Copernicus became much more frequent in English books."[33]

Digges' treatise is, in reality, a translation of certain parts of Book I of *De Revolutionibus*. However, Digges has added certain passages of his own to support the idea of an infinite universe, and it is this aspect of Digges work which has received the most attention. He was ". . . the first to advance the idea of an infinite universe as a corollary to the Copernican system."[34]

Digges also emphasized the importance of experiment. For example, in the treatise he adds the fact that a plummet dropped from the top of the mast of a ship in motion will hit at the bottom of the mast. It appears that Digges actually performed this experi-

ment because it is known from his other works that he did some experiments at sea. ". . . to verify his mathematical conclusions regarding the errors in navigation."[35] The importance of this experiment cannot be over-emphasized because it was the argument that the plummet would *not* hit at the bottom of the mast which was used to oppose the Copernican system.

Digges also made observations of the New Star of 1572 with the intention that these observations would provide an experimental test of the movement of the earth. He assumed that the change in brilliance of the star was due to the earth's motion with respect to the star; and that six months after it reached its brightest point, it would reach its darkest point.[36] Although the test failed, instead of denying the motion of the earth, Digges assumed—as Copernicus had previously assumed—that the distance to the stars was extremely far—too far to be detected by parallax.[37] Thus he was convinced of the validity of the Copernican system and was willing to modify the results of an experiment in order to "save the phenomena."

Why did Digges accept the Copernican theory? Upon reading his treatise, the first point of interest is the choice of material from the *De Revolutionibus* which Digges translated; and also the order in which he put it. It is of interest to note here that the previous writings on Digges completely overlooked this point, which is unfortunate because it is Digges' selection and ordering of material which is the focal point for our new interpretation of his work.

Before beginning our new interpretation of Digges, it will prove valuable to briefly outline Book I of *De Revolutionibus*.[38] Book I is divided into 14 chapters. Chapters 1-6 discuss the following: the spherical shape of the earth and the universe, the circular motion of the Heavenly Bodies, the diurnal motion of the earth, and the comparative sizes of the earth and the Heavens. Chapters 7 and 8 discuss the ancient theories supporting the view that the earth is at rest and arguments against these views, respectively. The motion of the earth around the sun is presented in Chapter 9. Chapter 10—"Of the Order of the Heavenly Bodies"—is considered by Kuhn as "crucially important" because of its emphasis upon the "admirable symmetry" and the "clear bond of harmony in the motion and magnitude of the spheres.[39] In Chapters 11-14 Copernicus begins his mathematical treatment of his system—the previous chapters contain no mathematics.

Turning to Digges' translation, the first point which strikes us is that Digges only translates Chapters 7-10 and he orders them in the following manner: 10, 7, 8, and 9. Also, he gives no reference to the fact that he is translating Copernicus and consequently the reader is unaware of Digges' particular selection. It therefore seems obvious that Digges was trying to convey the idea of his own to the English reader, the majority of which were unfamiliar with the original work of Copernicus.

In presenting a translation of an original work, the usual approach of a translator is to present the translation in the original order. Because Digges did not keep the chapters in their original order, it seems reasonable to assume that *he must have had some fundamental reason* for presenting his translation in this manner. Our conjecture is as follows: He placed Chapter 10 at the beginning of his treatise either because he intended to emphasize it or because he felt that this chapter presented the most appealing argument for the Copernican system. Apparently it was the Neoplatonism of Copernicus which is particularly exemplified in Chapter 10, that appealed to Digges. To prove this point, consider the following statement of Digges from the Preface to the treatise —a particularly vivid account of his Neoplatonism: "Why shall we so much dote in the apparance of our sences, which many wayes may be abused, and not suffer our selves to be directed by the rule of Reason, which the great GOD hath given vs as a Lampe to lighten the darcknes of our vnderstandinge and the perfit guide to leade vs to the golden braunche of Vertiy amidde the forrest of errours."[40]

Digges concludes the translation by adding the following sentence expressing his belief in the Copernican system. "So if we bee Mathematically considered and wyth Geometrical Meusurations euery part of euery Theoricke examined; the discreet Student shall fynde that Copernicus not without greate reason did propone this grounde off the Earthes Mobility."[41]

Digges belief in the motion of the earth around the sun is further shown by the fact that after presenting Chapter 10, he then presents Chapters 7, 8, and 9. That he skipped the first six chapters of Book I of *De Revolutionibus,* an extremely significant point as shown above, has never been noted prior to this time. Also we must not neglect noticing that Digges likewise skipped Chapters 11-14—the mathematical chapters. Recalling that Digges was a mathematician, this may at first seem odd; however, we must keep in mind that he was writing this treatise for a wide spectrum of English readers, and he probably felt that the mathematics was too difficult for the majority to comprehend.

The reason for presenting the following is their emphasis on the following three points: the beauty and simplicity of the Copernican system, the problem of the two years and the precession of the equinoxes, and the use of the Copernican system for making calculations.

In 1560, John Stradius praised Copernicus for his work on the precession of the equinoxes.[42] In 1573 Giutini wrote *Mirror of Astrology* in which he discussed solar and lunar eclipses, comets, and the movements of the planets using the calculations of Copernicus.[43] In this work, he also praised Copernicus for explaining why the two years were unequal.[44] Father Zuniga in 1579 wrote a commentary on the cosmology in the Book of Job in which he commented on Copernicus' theory emphasizing the clarity and simplicity of the system,[45] as well as the fact that it explained the phenomenon of precession better than the Ptolemaic system.[46] In Padua, in 1589, Magini published a work entitled "New Theories of the Celestial Orbs Agreeing with the Observations of Copernicus."[47] While writing a treatise on Ptolemy, Jacques Peletier (1517–1582) said that he would someday like to write a work on the "beautiful arguments" of Copernicus.[48] In 1564, Sevastian Theodoricus of Winsheim used Copernicus' calculation for his work on the movement of the ecliptic.[49] Gemma Frisius (d. 1555) ". . . expected a new heaven and a new earth from it [The De Revolutionibus] . . ."[50] He attacked the Ptolemaic system because of its errors in predicting the motion of mercury and its mistaken calculations of the vernal equinox.[51] Benedetti (1530–1590), an anti-Aristotlean who influenced Galileo—"it is possible that his influence was more than negligible in Galileo's development toward Copernicanism,"[52] emphasized the theory that it was easier for the earth to rotate than the sun and the other Heavenly Bodies.[53] This argument was also expressed by the Frenchman Francesco Patrizio (1530–1597).[54] The German, Michael Maestlin (1550–1631) said that his observations of comets led him to believe the Copernican system.[55] He also made the point that the starry sphere in the Ptolemaic system would have to travel at an "incomprehensible speed."

In 1600 William Gilbert (1540–1603), a physician, proclaimed his belief in the Copernican system in the publication of his famous work *On the Magnet.*

So, then, we are borne round and round by the earth's daily rotation . . . as a boat glides over the water . . . for nature ever acts with fewer rather than with many means; and because it is more accordant to reason that the one small body, the earth, should make a daily revolution than that the whole universe should be whirled around it. . . . And, indeed, nature would seem to have given a motion quite *in harmony* with the shape of the earth, for the earth being a globe, it is far easier and far more fitting that it should revolve on its natural poles, than they of the whole universe . . .[57]

Also, Gilbert did not believe in the motion, or the existence, of the eighth sphere.

> . . . it is a fiction believed in by some philosophers and accepted by weaklings . . . Besides, what genius ever has found in one same sphere those stars which we call fixed, or even has given rational proof that there are any such adamentine spheres at all? . . . [The number of stars] are many and that they never can be taken in by the eye, we may well believe . . . Astronomers have observed 1022 stars; besides these, innumerable other stars appear minute to our senses . . .[58]

He obviously was familiar with the infinite universe of Digges!

In a concluding statement on the motion of the earth, Gilbert introduces his magnetic theory: "The sun (chief inciter of action in nature), as he causes the planets to advance in their courses, so, too, doth bring about this revolution of the globe by sending forth the energie of his spheres—his light being effused . . . So the earth seeks and seeks the sun again, turns from him follows him, by her wonderous magnetical energy."[59]

On the continent in Spain:

> . . . the new regulations for the University of Salamanca, issued in 1561, required that in the three-year course of astrology and mathematics, the students should study first astrology, then Euclid, and then the Ptolemaic or the Copernican system . . . according to the choice of students.[60]

and by 1594 the university led the European world in teaching the Copernican system as part of its syllabus. One of its staff members, Diego de Zuniga, was expounding and defending Copernicus ten years before this. More importantly: "No index of the Spanish Inquisition ever prohibited a work of Copernicus, and the works of Galileo, Kepler and Tycho Brahe had never been forbidden."[61]

The claim has been made repeatedly that Giordano Bruno (1548–1600) was burned at the stake for espousing the Copernican System and hence was the first martyr for science. It's true that he did defend the Copernican system all over Europe but he had many heresies. His greatest crime seemed to be that he upset the establishment by calling knowledge and "learned" men into question. He could be considered a martyr to free thinking or today we would say academic freedom.

Galileo (1564–1642) it is said, gave thirty years of thought to the Copernican system.[62] "Galileo's first private declaration in favor of Copernicanism occurred in two letters: one dated May 30, 1597, addressed to Jacopo Mazzoni, professor of philosophy at the University of Piza . . . and the other dated August 4, of the same year addressed to Johannes Kepler."[63] However, Koyre feels that Galileo believed in the system during his sojourn at Piza (1589–1592). Whoever is correct, we do know that Galileo declared himself in favor of the Copernican system by 1597 because in the letter to Kepler he states: "I came round to the opinion of Copernicus many years ago and from his theory I have found the causes of many natural phenomena which doubtless cannot be explained by the ordinary theory."[64]

Note that it was before Galileo had made or looked through a telescope that he had made up his mind.

His observations with the telescope provided some indirect support for the Copernican theory. In 1609, a friend wrote to Galileo describing the telescope. Galileo made one and began his observations of the heavens. In 1610, he published some of his observations in *Message from the Stars*. He examined the face of the moon and discovered mountains and using their shadows was able to measure the height of some of them. These mountains were similar to mountains on earth and would tend to show that perhaps the moon was similar to the earth—that the nature of heavenly bodies was the same as the earth. He found four moons, or satellites, circling Jupiter. It was sort of a model solar system and they were celestial bodies that were not circling the earth. He observed sun spots; splotches or imperfections on the perfect heavenly bodies. The phases of Venus, like the phases which can be seen of the moon, were observed by Galileo. Copernicus had predicted this, whereas the Ptolemaic system had predicted only the crescent phases would be visible. These "proofs" are damning only if the evidence of the senses as modified through a telescope are acceptable. In addition, a person who does not have training with the use of an optical instrument is not going to see what he is supposed to see. Also the early telescopes were full of imperfections due to the state of glassworking; every object seen had a color fringe. In other words, you had to believe in order to see. Those who didn't believe did not see even when they looked through the telescope; and there were those who would not look because they said it was a device of the devil.

Strangely enough Galileo makes little use of these sensory observations in his most popular and powerful discussion of the Ptolemaic and Copernican systems, *Dialogue Concerning the Two Chief World Systems* (c. 1629), in which the arguments always seem to lead to the advantage of the Copernican view. Although this book was published in Italian and hence the view he was leaning toward had a wide audience—so large that although it was placed on the

Index five months after publication there were few copies to be found at the bookstores—he never taught it to his students at the university. Instead his course always presented the Ptolemaic view.

The Roman Catholic Church was never really bound to observe only the Ptolemaic view as the question about motion had been raised earlier by a number and by St. Thomas Aquinas in particular.

> Reason is employed in another way, not as furnishing a sufficient proof of a principle, but as confirming an already established principle, by showing the congruity of its results, as in astrology the theory of eccentrics and epicycles is considered established because thereby the sensible appearances of the heavenly movements can be explained; not, however, as if this reason were sufficient, since some other theory might explain them.[65]

In spite of censure by both Protestant clergy and the Roman Catholic Church the ideas of Copernicus were taught and spread. In spite, too, that the heliocentric view did not agree with the senses nor with common sense. What did the spread of this idea mean?

> Medieval theology, buttressed by the Ptolemaic system, held that man was at the center of the universe and that he was the apple of God's eye for whom God had specially created the sun, moon, and stars. By putting the sun at the center of the universe, the heliocentric theory denied this comforting dogma. It made man appear to be one of a possible host of wanderers on many planets which in turn, were drifting through a cold sky. He was an insignificant speck of dust on a whirling globe instead of chief actor on a central stage. It was unlikely, therefore, that he was born to live gloriously and to attain paradise upon his death, or that he was the object of God's ministrations. The sacrifice of Christ for insignificant man appeared pointless. The sky as the seat of God, the destination of the saints and of a Deity ascended from the Earth, and the paradise to which good people could aspire, was shattered by the passage of a speeding Earth. In short, the undermining of the Ptolemaic order of the universe removed cornerstones of the Christian edifice and threatened to topple the whole structure.[66]

The theological implications are perhaps of less importance to us today but even for today's world there is considerable importance derived from Copernicus.

> First, Copernican theory has done more to determine the content of modern science than is generally recognized. The most powerful and most useful single law of science is Newton's law of gravitation ... [which] depends entirely on heliocentric theory.

> Second, this theory is responsible for a new trend in science and human thought, barely perceptible at the time but all-important today ... the new theory rejected the evidence of the senses. Things were not what they seemed to be. [Reason was more reliable than sense data and] ... thereby set [ting] the precedent that guides

modern science ... that reason and mathematics are more important in understanding and interpreting the universe than the evidence of the senses. Vast portions of electrical and atomic theory and the whole theory of relativity would never have been conceived if scientists had not come to accept the reliance upon reason first exemplified by Copernican theory ...

> By deflating the stock of Homo sapiens, Copernican theory reopened questions that the guardians of Western civilization had been answering dogmatically upon the basis of Christian theology. Once there had been only one answer; now there are ten or twenty to such basic questions as: Why does man desire to live and for what purpose? Why should he be moral and principled? Why seek to preserve the race? It is one thing for man to answer such questions in the belief that he is the child and ward of a generous, powerful, and provident God. It is another to answer them knowing that he is a speck of dust in a cyclone.[67]

René Descartes (1596–1650) was the first of a long string of philosophers who attempted to answer the questions in a way that would place man again at the center of his world. He began by doubting all except that he existed which he knew, he said, because I think therefore I am. In his system he tries to show that the action of the mind proves reality and that God is the link between the mechanical senses and

Figure V-3. Descartes.

the rational mind. Parenthetically it might be added that the philosopher Sartre is today still trying to place man as the center of his world after the twentieth century revolution.

By far the greatest value of the heliocentric theory to modern times is the contribution it made to the battle for freedom of thought and expression. The treatment that the heliocentric hypothesis received at the outset illustrates one fairly safe generalization: the reaction to change is reaction. Because man is conservative, a creature of habit, and convinced of his own importance. Moreover, the vested interests of well-entrenched scholars and religious leaders caused them to oppose it. The most momentous battle in history, the battle for the freedom of the human mind, was joined on the issue of the right to advocate heliocentrism.[68]

STUDY AND DISCUSSION QUESTIONS

1. What is the Copernican system?
2. What is Neoplatonism? Why is it important?
3. How did Brahe know the position of the comets and Novae? What difference did it make?
4. Why was the Copernican system accepted by those who followed Copernicus? Is the reason the same in all cases?
5. If the Copernican system is accepted today why do we not use the phrase—"a beautiful earth turn"—instead of sunrise and sunset?
6. What arguments can be used against the idea of turning earth? For? What proofs can you offer that the earth moves? How do you know that it does?
7. Why was it not possible to suppress the ideas of Copernicus?
8. How could Renaissance astronomers use the works of Copernicus without accepting the idea of the earth's motion?
9. What is the effect of Copernicanism on the Christian world? Men on other worlds? Man's position with regard to angels and devils? An infinite universe?
10. Compare Donne's and Milton's reaction to Copernicanism and what it implied.
11. Why are Protestants more strongly opposed to Copernicanism than are the Catholics?
12. How were the works of Copernicus indirectly used by the Catholic Church?
13. What did Brahe use to show that the nova of 1572 was beyond the moon?
14. What observations of Galileo provided *direct* evidence for the Copernican theory? Indirect?
15. What were Francis Bacon's objections to the Copernican theory? Henry More's Comment?
16. Did the Copernican theory fit the facts?
17. How did the Copernican theory help shape modern times?
18. What is the scientific declaration of independence? Why?
19. In what ways were the observations of Galileo supporting evidence for the Copernican system?
20. Why wouldn't some of Galileo's contemporaries accept the evidence of their senses by looking through the telescope?
21. Why did some members of the Roman Catholic Church feel that the doctrines of Galileo should be condemned?
22. What was the Tychonic system? Why wasn't it as acceptable as the Copernican?
23. What role did Descartes play in the aftermath of the Copernican revolution?

FOOTNOTES

1. Thomas S. Kuhn, *The Copernican Revolution,* (Harvard University Press, 1966), pp. 111-112.
2. Translated and quoted by Andrew D. White in *A History of the Warfare of Science with Theology in Christendom* (Appleton, 1896), I, p. 126.
3. *Ibid.,* pp. 126-127, from Melanchthon's *Initia Doctrinae Physical.*
4. *Ibid.,* p. 127.
5. Essay II. 12, p. 276. *Great Books of the Western World,* Vol. 25.
6. Translated and quoted by Dorothy Stimson in *The Gradual Acceptance of the Copernican Theory of the Universe* (New York, 1917), pp. 46-47, from *Universae Naturae Theatrum* (Frankfort, 1597).
7. Quoted by Francis R. Johnson in *Astronomical Thought in Renaissance* England. (Johns Hopkins Press, 1937), pp. 188-189 from a translation (1605) by Joshua Sylvester.
8. *John Donne: A Selection of His Poetry,* John Hayward, ed. (Penquin, 1950), pp. 26-27.
9. This phrase refers to the precession of the equinoxes (changes in time sun arrives in Aries) which was explained in the Ptolemaic system by a swaying or oscillation of the ninth crystalline sphere.
10. Hayward, *op. cit.,* p. 54
11. "The Anatomy of the World" in *Complete Poetry and Selected Prose of John Donne,* John Hayward, ed. (Nonesuch Press, 1929), p. 202.
12. *Great Books of the Western World,* Vol. 32, p. 235.
13. Other similar references will be found in Hermann Kesten *Copernicus and His World* (Roy Publishers, 1945), pp. 315-322.
14. Available in *Great Books of the Western World,* Vol. 30, esp. pp. 139, 165-167, 178.
15. *The Essays of Francis Bacon* (Heritage Press, 1944), p. 183.

16. Kuhn, *op. cit.,* pp. 207-208.
17. W. P. D. Wightman, *The Growth of Scientific Ideas* (Yale University Press, 1953), p. 47. Also see Kesten, *op. cit.,* pp. 333-345.
18. *Ibid.,* p. 47.
19. Kuhn, *op. cit.,* pp. 204-205.
20. In *Great Books of the Western World,* Vol. 33, p. 369.
21. Kuhn is the best source.
22. *Ibid.,* p. 184.
23. J. L. E. Dreyer, *A History of Astronomy,* (New York, 1953), p. 347 and Kesten, *op. cit.,* p. 321 and take your choice.
24. Francis R. Johnson, *Astronomical Thought in Renaissance England* (Johns Hopkins, 1937), p. 134.
25. For the mystical side see Arkon Daraul *Witches and Sorcerers,* (Citadel Press, 1966), pp. 89-101.
26. Johnson, *Astronomical Thought,* p. 76.
27. *Ibid.,* p. 135.
28. *Ibid.,* p. 120.
29. Francis R. Johnson "The Influence of Thomas Digges on the Progress of Modern Astronomy in Sixteenth Century England." *Osiris,* Vol. 1, Jan., 1936, pp. 390-410.
30. Johnson and Lockey, pp. 69-117.
31. Johnson, *op. cit.,* p. 167.
32. *Ibid.*
33. *Ibid.,* p. 180.
34. Johnson and Lockey, p. 69.
35. *Ibid.,* p. 99.
36. *Ibid.,* p. 110.
37. *Ibid.,* p. 112.
38. *Great Books of the Western World,* Vol. 16.
39. Kuhn, *op. cit.,* p. 180.
40. Johnson and Lockey, p. 80.
41. *Ibid.,* pp. 94-95.
42. Lynn Thorndike, *A History of Magic and Experimental Science,* (New York, 1959), Vol. 6.
43. *Ibid.,* p. 130.
44. *Ibid.,* p. 43.
45. Stemsen, *op. cit.,* p. 44.
46. Dreyer, *op. cit.,* p. 353.
47. Thorndike, *op. cit.,* p. 56.
48. *Ibid.,* p. 33.
49. *Ibid.,* p. 34.
50. *Ibid.,* Vol. 5, p. 410.
51. Grant McColly "An Early Friend of the Copernican Theory": Gemma Frisius, *"Isis,"* Vol. 26, 1937, p. 323.
52. Ludovico Greymonet, *Galileo Calilei: A Biography and Inquiry into his Philosophy of Science,* (New York, 1965), p. 13.
53. Thorndike, *op. cit.,* Vol. 6, p. 51.
54. Dreyer, *op. cit.,* p. 349.
55. Thorndike, *op. cit.,* Vol. 6, p. 46.
56. Dreyer, *op. cit.,* p. 349.
57. William Gilbert, "On The Lodestone and Magnetic Bodies and On the Great Magnet the Earth," *Great Books of the Western World,* Vol. 28, pp. 109-110.
58. *Ibid.,* pp. 107-108.
59. *Ibid.,* p. 112.
60. Kesten, *op. cit.* p. 322.
61. Henry Kamen, *The Spanish Inquisition,* p. 299.
62. Galileo Galilei, *Diologue on the Great World Systems,* (Chicago, 1957), p. xi.
63. Geymonat, *op. cit.,* p. 23.
64. *Ibid.,* p. 230.
65. From his *Summa Theologica* as printed in *Great Books of the Western World,* Vol. 19, p. 177. For a more elaborate argument also see *The Provincial Letters of Blaise Pascal* in *Great Books of the Western World,* Vol. 33, pp. 163-165.
66. Morris Kline, *Mathematics in Western Culture,* (Oxford University Press, 1953), p. 117.
67. *Ibid.,* p. 121.
68. *Ibid.,* p. 123.

Acknowledgement

For the argument of Digges contribution the author is indebted to David Topper. See Theodore W. Jeffries *Guide to Physical Science,* Section Seven.

VI

My Mother Was a Witch

Any creative person must be considered within the time of his life—all formative and directive forces must be considered. In the writings of modern science there is a tendency to close the closet door on the non-logical skeletons of the thought processes of creative persons of the past. We purpose, in this chapter, to open the door of the closet for one man and one of his ideas and to trace some of the factors which are not frequently given when his laws are stated.

Johannes Kepler, Keppler, Khepler, Kheppler, or Keplerus . . . was conceived on May 16, A.D. 1571, at 4:37 A.M. , and was born on December 27 at 2:30 P.M., after a pregnancy lasting 224 days, 9 hours, and 53 minutes. The five different ways of spelling his name are all his own, and so are the figures relating to conception, pregnancy, and birth, [neither father nor mother used a stopwatch] recorded in a horoscope which he cast for himself. The contrast between his carelessness about his name and his extreme precision about dates reflects, from the outset, a mind to which all ultimate reality, the essence of religion, of truth and beauty, was contained in the language of numbers.[1]

Figure VI-1. Kepler.

The unimportance of spelling prior to the eighteenth century has already been mentioned as well as the extreme importance of astrology. In order to cast a good horoscope the dates under which an individual came into the world were necessary. Astrology came to the forefront during the renaissance to a greater degree than at any time during history save for the twentieth century that is seeing a tremendous revival and also gives another comparative historical link to those already mentioned.

One of the duties of several persons we have already mentioned, (Tycho Brahe and Cardan) to their patrons was to prepare horoscopes and to tell what the future had in store. Current writers of science tend to say that Kepler, for example, really did not believe in astrology but the above quotation from Kepler's work shows otherwise. The fixed and moveable stars both had a direct influence, as shown by horoscopes and an indirect influence as the Copernican revolution progressed.

Kepler, as did other figures of the renaissance, showed that he retained some of the influences from the middle ages. He was trained by the church—only at this time and in Germany the church was Lutheran—to be a pastor but he never reached the pulpit for he instead was called to teach mathematics. Like others of his time he saw his country pulled apart by the religious strife of the reformation and counter-reformation as well as the start of the Thirty Years War.

In addition to being deeply religious, he came under the other influences of his time. He wrote poetry which was expected of any educated man and was strongly under Neoplatonic influences. Also like some he was strongly a Pythagorean—that is he believed that number was reality. How this last affected him is one area we wish to explore.

The third law of celestial motion that Kepler evolved is usually stated in most texts as follows: the square of the period of revolution in orbit about the sun of a planet is proportional to the cube of the distance. This means that there is always a ratio between any planet's distance to the sun and the time

Table for Kepler's Third Law

If we let P stand for the period of revolution and R stand for the distance of the planet from the sun then K (a constant) should equal $\dfrac{P^2}{R^3}$

Planet	P	P^2	R($\times 10^6$)	R^2	K
Mercury	0.24 yr.	0.058	36	46656	1.25
Venus	0.61	0.38	67	300763	1.27
Mars	1.9	3.6	142	2863288	1.25

it takes to go around the sun. This law holds fairly well even with today's figures, as shown above. Even today this law helps astronomers find a planet in its orbit and the law can be said to have introduced the basic tool of celestial mechanics.

We want to see however just how Kepler came to this law that sounds so modern and seems to fit so well. Both the periods of the planets and their distances had been known for considerable time. The distances had been obtained through triangulation. Since number was reality, a Pythagorean idea, and the most important portion of mathematics was geometry, a Neoplatonic idea, there should be a relation between mathematical figures and the paths of the planets. He seeks a harmony—a harmony which will expose and relate the totality of creation and a reflection of the creator's mind.

Obviously then the starting place is with the most perfect of the figures—the circle—and it will be related to other perfect figures. He draws strongly from Proclus' commentaries on Euclid.

> According to it the mathematical ideas are the essence of the soul and inversely the soul is the essence for them. Thus the corresponding members of pure harmony are not supplied by the circle vouchsafed by the senses, but rather by the purely intellectually thought circle of pure harmony. . . . Geometry, being part of the divine mind from time immemorial, from before the origins of things, being God Himself . . . has supplied God with the models for the creation of the world and has been transferred to man together with the image of God. Geometry was not received through the eyes.[2]

There is also within the mind of God the other perfect figures. The square of the side of a square is half the square of the diagonal of such a square. Then next is the equilateral triangle with three equal sides; following which there were a large stated number of polygons that could be constructed with ruler and compass and hence were perfect figures.

These ratios that Kepler was deriving for his ideas of a harmonious universe he explored in geometry,

astronomy and music. So that in addition to relating number and figure to astronomy, he also sought to relate the basic ratios to music. As you may recall from the view of the 15th century, it was believed that the crystalline spheres carrying the planets made celestial music which the ear of man could not hear.

He cannot seem to make the relationship between the figures of geometry he has chosen and the five planets when it suddenly occurs to him that the solar system is three dimensional and not two so that what he is looking for is not five figures but five solids for the five planetary spaces between the six planets. There are only five perfect solids, i.e. solids whose faces are identical, and can be inscribed in a sphere with all its corners on the surface of the sphere or can be circumscribed around a sphere so that the sphere touches every face at its center. These are: the tetrahedron, a pyramid whose sides are four equilateral triangles (representing the element fire); the cube (representing the element earth); the octahedron, whose eight sides are equilateral triangles (representing air); the dodecahedron having twelve pentagon sides (representing the ethereal firmament since there are only four elements); and the icosahedron which has twenty equilateral triangles (represents element water).

Into the orbit of Saturn he inscribed a cube, while Jupiter's orbit lay inside this cube, into which was inscribed the tetrahedron with the spherical orbit of Mars lying within. Between the orbits of Mars and the earth was the dodecahedron while between the earth and Venus was the icosahedron and finally between Venus and Mercury was the octahedron.[3] These relations proved was God had made only six planets just as He had made only five perfect solids. In addition, these ratios provided clues to the celestial music.

In his *Harmonice Mundi* he began to compare some ratios of scale tones and musical chords to the

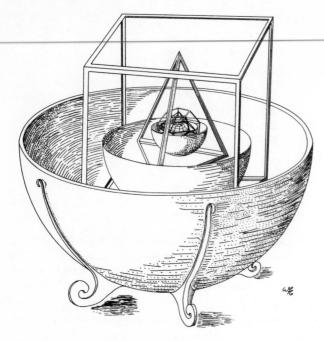

Figure VI-2. Five Solids.

planets' angular velocities (speed and direction in orbit). He noticed that when the planet was farthest from the sun (at aphelion) its angular velocity was least while at the point nearest the sun (a perihelion) it was the greatest. By gradual reduction of these ratios he was able to obtain relationships (even if he had to force the observations a little) which he made to agree with the lengths of chords which produced harmonic tones. A planet's perihelion would be compared to a major musical key while its aphelion was related to a minor key.

In music there was already a long tradition going back to the Greeks of relating weight ratios attached to strings which produced harmonic notes and also with the weight of metal bars which gave certain notes when struck. Since the music of the Renaissance involved an attempt to return to the music of the Greeks, Kepler's ideas fit the times well and made his mathematical arguments all the more valid. Then too an educated person of the Renaissance would be expected to know music.

The relationships that Kepler worked out regarding orbital velocities would later be used by Newton in formulating the Universal Law of Gravitation. Yet, as we have seen, this formulation begins from religion and mystical relationships that are all too much a part of the times.

Katharina

Kepler's mother had been raised by an aunt who had been burned at the stake for witchcraft and in August, 1615 she herself was accused of being a witch.

Ursula had been a crony of Katharina's for many years, but the friendship had recently soured. The reason was that Ursula . . . had become pregnant with a man other than her husband and had submitted to abortion to escape scandal. She had confided in Katharina, and Katharina had told Kepler's younger brother Christoph. . . . Christoph also sounded off in public about Ursula's indiscretion and his mother foolishly gave confirmation. To cover up the truth, Ursula attributed her debility to an evil spell and blamed Katharina Kepler for casting the spell.[4]

The Magistrate had a grudge against the Kepler family for he had loved and lost Johannes' sister in a humiliating manner. Then there were the stories.

Katharina had asked the sexton of the churchyard for the skull from her father's grave to silver and give to her son as a drinking cup. Her epileptic son told that his mother rode a calf to death and fed him a veal chop from it. She had a mysterious tin cup about which tales were told.

The schoolmaster remembered that he had become lame following a drink from the tin cup ten years before and another neighbor recalled that his wife had withered and died after drinking from the same cup. Katharina didn't help matters when she continued her rounds giving free advice and offering her homemade remedies from the notorious tin cup.

A charge of witchcraft was no joke. In Kepler's native region, which had only a few hundred families, thirty-eight women were condemned to death for the exercise of extraordinary powers during the years between 1615 to 1629.[5] To aid his mother, Kepler had left his home in Upper Austria and proceeded to Württemberg, arriving September 28, 1620 where she was awaiting trial. He spent some months preparing her defense and engaged a lawyer to defend her in court and to appeal to the Duke of Württemberg.[6]

It was perhaps only Kepler's position at court and his influence that kept his mother from the stake and even then she was questioned under threat of torture and in the torture chamber. Even after her release her neighbors would have put her to death themselves had not Kepler finally moved her.

Although many denied witches most lawyers, preachers, physicians, and philosophers believed and accepted their existence. Witches were part of the facts of the times.

Kepler had remained a dutiful son. As a child he had been sickly and had poor vision which prevented him from making observations of the type that Tycho had made. His family was poor so that his education was delayed and probably it was only his poor health that directed him into scholarly pursuits instead of a life of manual labour.

After the university, Kepler had been summoned to teach instead of preach and his job was also called district mathematician of Graz (1594–1600). His duties included teaching, answering mathematical questions for the nobility and the making of calendars.

Strangely, perhaps to us, his first successes were with the calendars.

In those days of belief in the influence of stars, calendars played a different role from nowadays. High and low were permeated with the belief that future fate could be predicted from the path of the heavenly bodies. So the calendar men ... were expected to furnish information about weather and harvest prospects, about war and the danger of pestilence, about religious and political events. People wanted to know on what day to sow and harvest, when to bleed, when to expect hail and storm, cold and heat, illness and hunger.[7]

In his first calendar he had hit the mark with his prophecies.

He had ... predicted bitter cold and invasions of the Turks. Both happened. Many herdsmen in the mountains were rumored to have died from the cold; upon arriving home, many lost their noses when they blew them; [the latter statement may be like the one about brass monkeys] ... This success turned attention to the young district mathematician and soon procured him prestige in the land, so that the lords gladly sought his astrological opinions and nativities.[8]

The beginning of the counter-reformation forced Kepler to look elsewhere for a position and for a time he aided Tycho Brahe until the latter's death when he assumed the mantle of Imperial Mathematician to the Emperor Rudolph at court in Prague (1600–1612). One might well ask, "What sort of thing does one do for an emperor?" Several of Kepler's essays have been preserved that deal with the types of questions he was asked.

[In these essays] ... information is given about the nativity of the Emperor Augustus, about the nativity of Mohammed, and the fate, which, according to the stars, is to be expected for the Turkish kingdom, supplementing and criticizing astrological arguments which were presented to the emperor from other sides, about the judgment of the then pending Hungarian question in accordance with the arrangement of the stars, about an astrological calendar which the emperor had read, about a prognosis in the fight in which the Republic of Venice was involved with Pope Paul V, about a pump without valves, which Kepler had devised. ... Two further propositions concern the so-called 'fiery trigon.' This deals with an astronomical phenomenon, to which astrologers had always imputed special significance and which just at that time gave rise to the rashest and most absurd prophesying. The astrologers divided the twelve

signs of the zodiac into four groups of three each, by putting together in one group the first, fifth, ninth, then the second, sixth, tenth and so forth. With these groups, in order, were associated the qualities fieriness, earthiness, airiness and wateriness. The first group, the 'fiery trigon,' was considered especially distinguished because it contains the beginning of the zodiac. The division is connected with the shifting of the so-called great conjunction, that of Jupiter and Saturn, which since time immemorial played an important part for the believers in the influence of the stars. The two consecutive conjunctions of these two planets sometimes occur about 117° from each other, so that ten consecutive conjunctions always fall in one of these three trigons. Since there are approximately twenty years between two such conjunctions, the designation fiery, earthy, etc., trigon is also carried over to the corresponding period of two hundred years and so an eight hundred year period, in which the series of the conjunctions repeats, is arrived at. Now since such a period began at the end of the year of 1603, a great many works about this subject were printed. The powerful figure of Charlemagne had come to the scene eight hundred years before, and eight hundred years still earlier Christ was born. And so, what sort of epochal event would come now. That was the question which occupied the minds of those who believed in the influence of the stars and thus also most actively that of the emperor.[9]

Astrology and astronomy did not occupy all of Kepler's time for like other men of the Renaissance he had concerns with practical matters. Hopefully at some stage such may occur to overcome the academic gangrene of our own times. For the Elector of Cologne Kepler composed a detailed study concerning the systematization of weights and measures and argued strongly for decimal division. Some years later he was asked to help in the regulation of weights and measures for the city of Ulm.

Kepler had previously demonstrated, in his Archimedes' Art of Measuring, that he was quite at home in the confusion presented at that time by this field, so important for trade, since each country, each imperial city possessed its own units of measure. ... In a lengthy opinion he analyzed his principles, and accordingly had an attractive basin molded, from whose interior measurements of width, height, weight and capacity it should be possible to conclude, with an exactness adequate for trade, the units for the length, weight and capacity measures. In this manner, the size of foot, ell, hundredweight, pail, imi (an old corn measure) was established.[10]

The observations which Kepler had inherited from Tycho not only enabled him to formulate his laws of planetary motions but also to prepare a set of astronomical tables to be called the Tabulae Rudolphinae after the Emperor who had authorized their preparation. From these tables Kepler wanted to prepare ephemerides (year books which give the positions of the sun, moon, and planets for each day of a year) for eighty years starting with 1582 because that was the year that Brahe had begun his observations.

Such ephemerides were very much in demand by astronomers who used them in their scientific researches, by seafarers who needed them for place determination, and last but not least also by calendar makers and astrologers to whom they were indispensable as supports for their prophecies.[11]

Kepler uses such charts to ascertain a horoscope. Even though he believes strongly in astrology he does not feel that free will is destroyed for the stars do not compel they merely stamp the individual with a certain set of characteristics which the person carries throughout his life.

Duke Albrecht von Wallenstein was Kepler's last patron and was strongly addicted to astrology. In or about 1627, Wallenstein had asked Dr. Stromair to get Kepler to cast his horoscope. Although Kepler was not told who the horoscope was for he guessed and wrote one that so nearly approximated the actual ambitions, fears and even the faults of the subject that a great impression was made. Some sixteen years later, Wallenstein in the meantime had become famous as a general in the Thirty Years War, Kepler again was asked for a horoscope of Wallenstein and this time without a coverup. The second horoscope, dated 1624, stops in 1634—Wallenstein was murdered on February 25 of that year.

Kepler died Novermber 15, 1630 and left behind rich contributions to modern science as well as his own epitaph:

I measured the skies, now the shadows I measure
Sky-bound was the mind, earth-bound the body rests.

In spite of his contributions his works are frequently ignored because:

He is more evidently rooted in a time when animism, alchemy, astrology, numerology, and witchcraft presented problems to be seriously argued. His mode of presentation is equally uninviting to modern readers, so often does he seem to wander from the path leading to the important questions of physical science. . . . We (of the modern age) are trained on the ascetic standards of presentation originating in Euclid, as reestablished for example in Books I and II of Newton's *Principia*, and are taught to hide behind a rigorous structure the actual steps of discovery,—those guesses, errors, and occasional strokes of good luck without which creative scientific work does not usually occur.[12]

STUDY AND DISCUSSION QUESTIONS

1. Why couldn't man hear the Celestial Music?
2. What characteristics of a renaissance man did Kepler have?
3. What were the duties of a mathematician?
4. What practical things did Kepler do?
5. How does his third law of motion come about?
6. On what grounds was Kepler's mother accused of being a witch?

FOOTNOTES

1. Arthur Koestle, *The Watershed* (Doubleday Anchor, 1960), p. 17. For another excellent popular version of Kepler's life see David C. Knight's *Johannes Kepler and Planetary Motion* (Franklin Watts, 1962) and for a reasonably complete bibliography see Louis A. Kenney *Johann Kepler Bibliography* (San Diego State College Library, 1970).
2. Max Caspar, *Kepler* (Abelard-Schuman, 1959), pp. 270-271. This translation by C. Doris Hellman is the most authoritative book, in English, for Kepler.
3. See Kepler's work itself in *Great Books of the Western World,* Vol. 16.
4. John Lear, *Kepler's Dream* (University of California Press, 1965), p. 29. However, about this book see the criticisms of Edward Rosen "Kepler and Witchcraft Trials" *The Historian,* 28:447-450 (May, 1966).
5. Lear, *op. cit.,* p. 30. Also see Jules Michelet *Satanism and Witchcraft* (Citadel Press, 1939) for an excellent account and history of the whole problem or *The Painted Bird.*
6. Rosen, *op. cit.,* pp. 448-449.
7. Casper, *op. cit.,* p. 58.
8. *Ibid.,* p. 60.
9. *Ibid.,* p. 153.
10. *Ibid.,* p. 324.
11. *Ibid.,* pp. 178-179.
12. Gerald Holton, "Johannes Kepler's Universe: Its Physics and Metaphysics" *American Journal of Physics,* 24:340-351 (1956). Also see E. J. Aiton, "Kepler's Second Law of Planetary Motion" 60:75-90 (1969).

VII

The Scientific Method

Most of you, if you have had any kind of a science course during your previous schooling, have been exposed to the scientific method. Before continuing this part, see if you can write the steps to the scientific method. Now see if you can write the answers to these questions. What is a fact? What is a theory? What is a hypothesis? What is a law?

On the pages that follow the answers to the above questions are given by a number of chemists. Read *thoughtfully* through these looking for common agreement in their answers. What are their definitions to the above questions? Write them for each. When you have finished reading this section, rewrite your answers to the above questions and compare the first set of answers to the second. Did you find any assumptions made by any of the authors? If so, what were they? Would any of the definitions apply to *all* sciences? If so, which of them would apply?

The fifteen sources chosen for your examination are all first year chemistry books so that they are elementary and are all in the same field. Each is a direct quote or a reasonable paraphrase so that no words are added that the authors have not chosen. The only criterion for their selection was that they happened to be on my shelves readily available—a slanted view was not intended. The paragraphs following the fifteen sources are merely intended for illumination.

A speculation based on observation which tries to give a general explanation is called a hypothesis. It should suggest further observation and raise questions as to its own validity.

When enough observations have been correlated to make the hypothesis seem reasonable and general, it graduates to the status of theory. A theory is not necessarily a true explanation; it should, however, explain all or most of the observations which have been made. Even if it does not account for all of the known observations, it may be useful in correlating a great deal of information, in suggesting new experiments, or in leading scientists to suggest alternative theories.

Every theory involves assumptions, and if the theorist is willing to assume enough, he can explain almost anything.

As a theory becomes established it gradually becomes regarded as a law, but even what is regarded as a law may not be completely true, for someone may sometimes observe a scientific fact which is contrary to it.[1]

Like other scientists, the chemist seeks to understand a large number of observed facts in terms of a few broad principles. The uncovering of these principles is the ultimate aim of scientific research. Once the principles have been established, they not only account for experimental observations, but also make possible the systematic organization of knowledge and the prediction of behavior in new situations. However, these principles can only be as good as the experimental observations on which they are based.[2]

For ease in remembering and handling experimental facts, scientists find it useful to correlate them in the form of generalized statements, or natural laws. A natural law is simply a description of the uniform behavior of nature.

Not content with the mere statement of laws, however, the true scientific adventurer strikes out into the realm of the imagination; he seeks to coordinate and to explain a group of interrelated laws by setting up a mental picture which would account for the behavior of nature as described by the laws. Such a picture, or set of assumptions, concerning the nature and structure of some portion of the universe is called a hypothesis. The various laws describing the behavior of gases, for example, led to a set of assumptions as to the discontinuous nature of gases, called the kinetic-molecular hypothesis. Each hypothesis sets into motion a new series of investigations. Scientists, reasoning deductively, conclude that if a given hypothetical explanation is correct, then certain other experimentally observable facts must inevitably follow. Experiments devised to test these deductions lead to new data and often to the discovery of additional laws.

Once a hypothesis has been tested by a wide variety of experiments and has been found to be in harmony with all existing knowledge on the subject, it assumes the status of a theory. In the hands of the scientist, a theory not only explains and correlates; it also predicts, and therefore serves as a powerful tool in the further search for truth. It is by applying theoretical concepts to increasingly complex problems that scientists have brought about the ever-accelerating scientific progress of today.[3]

Chemistry is fundamentally an experimental science. The chemist experiments to establish facts. He may then combine these facts with certain assumptions to formulate a theory, in which a large body of information is coordinated into a concise and unified picture. A theory is valuable in that it points out the relationship between otherwise disconnected facts and suggests further experiments which may yield more information. New facts may show that some of the assumptions of the theory are unfounded. As a result, the theory may be changed to include the new facts, or an entirely new theory may be formulated.

A hypothesis is the philosophical conjecture without any definite factual basis. It looks as if certain things should be true. In some instances a hypothesis is useful in suggesting further experiments, even if the validity of the hypothesis cannot be established or disproved. In many instances hypotheses have been advanced and for a number of years have proved useful, but later, as more information accumulated, they were found to be unsound and as a result were abandoned. In other cases, subsequent discoveries show the validity of the hypothesis. Many times hypothesis grow out of theories. Suppose that three facts, which we shall designate a A, B, and C, have been demonstrated in the laboratory. The scientist makes a statement which explains and interrelates these facts. This is a theory. On the basis of the facts and the theory he reasons, 'If this explanation is valid, it should be possible to go into the laboratory and discover another fact, D which is at present unknown.' This is a hypothesis. The subsequent discovery of fact D does not prove the theory, but it does give support to the assumptions of the theory.

A law is a statement of the invariable behavior of nature under the conditions which affect the phenomenon. The more exactly we know these conditions, the more exact our statement of the law becomes. Many times, statements which have been called laws are found to be inaccurate because one or more conditions affecting the phenomenon were unknown at the time the statement was formulated. For all practical purposes, a law is a statement of the behavior of nature under specified conditions in our statement because its effect is negligible.[4]

As facts accumulated and as knowledge of material things increased through observation and experiment, classification of these data became necessary not only to make possible a systematic, description of objects and phenomena but also to point out the direction in which investigation might lead to important advancements and new discoveries. By the inductive method scientists reason from particulars to the general. This method, according to Bacon, consists in concluding that what has been confirmed with respect to a part may be affirmed of the whole to which it conforms. Scientific laws result from securing data and systematically arranging those which are like and separating those which are unlike, in order to reach a generalization from particulars. However, care must be taken to reach a generalization from an unbiased point of view. As the distinguished Russian scientist Mendeleoff has said: 'Willingly or not, in

science, we all must submit not to what seems attractive to us from one point of view or another, but to what represents an agreement between theory and experiment. . . .'

Often no general law can be formulated; then theories are advanced to explain the data, and further observations are made to substantiate the deduction from theory. The deductive method is the opposite of the inductive and consists in reasoning from a broad principle or generalization to specific applications. Hence, laws become guiding principles in answering specific questions. Few scientific investigations go very far before both methods of thought are used in the solution of problems involved.[5]

When it is first found that a picture or a mathematical equation correlates or explains a number of empirical facts, the picture or equation is accepted as a hypothesis, to be subjected to further tests and experimental checking of deductions that may be made from it. If it continues to agree with the results of experiments, the hypothesis is dignified by the name of Theory or Law.[6]

The process by which men learn what the materials of the world must be like in order to behave as they do is greatly aided by the scientific method of thought. This approach to any problem may be broken down into four distinct steps.

1. Observation of facts regarding the problem.
2. Formulation of an explanation based on the facts observed. (Inductive thinking—from the particular facts to a general hypothesis.)
3. Use of the explanation (hypothesis) to figure out consequences not yet observed. (Deductive thinking—consequences.)
4. Devising of experiments to establish or disprove the expected consequences.

. . . This method is an application of directed common sense, and it has provided a powerful attack on problems outside the field of science as well as within it.[7]

The process of systematizing experimental facts into theories or laws upon which predictions of future studies and events can be based can be considered the essence of science.

The vast body of knowledge arising from centuries of experimental studies led to broad divisions of science, each of them closely allied to the others. One classification of these divisions could be physical, biological, behavioral, and social sciences. They all attempt to apply the scientific method to the study of their fields; they differ only in that they focus attention of different aspects on science.[8]

A field of study which deals with the observation and interpretation of reproducible facts is properly termed a science. A key word in this definition is reproducible. The facts studied must be such that they can be demonstrated to any interested person. This limits the scope of science a great deal, because if an observation of a fact is to be accepted, any interested person must be able to repeat it again and again . . .

There are four words which are most useful in discussing scientific subjects: fact, hypothesis, law, and theory . . .

A fact is an event or occurrence. A hypothesis is a guess, based on a few facts, which can be used as a basis for reasoning or for planning experiments to gather more information. A law, as the scientist defines it, is a concise statement which summarizes a large number of related facts ... A theory is a statement, based on facts and reason, which is designed to explain the facts and laws of nature.[9]

The scientific method furnishes a simple and logical way of finding the answer to questions which can be subjected to inquiry and investigation. The first step in applying the scientific method to the solution of a problem involves the carrying out of experiments to gain facts which give information about all phases of the problem. The second step consists of an attempt to formulate a simple generalization which will correlate a number of these facts. If this attempt is successful, the simple generalization becomes a law. Usually, however, no general law can be formulated to correlate the facts, and then a provisional conjecture, known as a hypothesis, is advanced to explain the data. A hypothesis may be tested by further experiments and if it is capable of explaining a large body of facts in a given field, it is dignified by the name of theory. Theories serve as guides for further work by serving as the basis for predicting new information or the direction in which additional information must be sought. Finally, a theory must be established or modified in such a manner that it can be accepted as a general truth, which is often referred to as a law.[10]

When we "do" science, we do two things. First, we make observations and then we try to account for what we observe. These two activities rely heavily on each other. From experiment and observation we develop new concepts; in turn these new concepts suggest further experiments and observations. It is this unending cycle which constitutes most scientific activity.[11]

Application of the method:

1. The scientist describes an observed situation or phenomenon.
2. He looks up all the available information about this phenomenon.
3. He then proposes a hypothesis, or possible explanation, which is based on logic and the previously recorded information.
4. He next plans and conducts laboratory experiments in order to obtain as much additional data as possible on the subject.
5. These new data are compared with all previously recorded data and then carefully analyzed. If the evidence supports the original hypothesis, the scientist works out a theory.
6. When enough conclusive data becomes available to prove that the theory is true, it may be restated as a law.[12]

The Scientific Method has the following steps: First step—collection of facts; Second—organize facts; Third—stating the law; Fourth—explain the law by a theory (e.g., Boyle's law—molecular theory of gases); Fifth—prediction of new facts from the theory; Sixth, are the facts as predicted.[13]

In his search for knowledge, the scientist makes use of two fundamental precedures—observation and reasoning. The application of these two procedures is known as the scientific method. The first step in the scientific method of approach to the solution of a problem is to conduct well-planned experiments from which observations can be made and recorded. From his observations, the scientific law or a principle describes a particular behavior of nature, but a law differs from a principle in that laws may usually be expressed mathematically. The scientist sometimes develops a theory or hypothesis to account for the observations upon which the principle or law is based. Such theories and hypotheses are useful in anticipating further experimental work, which may verify the theory or hypothesis. Or if the scientist has not reasoned correctly, the theory or hypothesis may have to be modified or even abandoned.[14]

The collection of the requisite experimental data constitutes an important first step in the so-called scientific or experimental method.

The second stage is the organization of these facts into generalizations called laws, which describe the natural phenomena observed. Laws made by legislative action are designed to regulate; the laws of science are merely descriptions of natural phenomena. ...

The laws of nature often suggest a hypothesis or a theory by which the observations can be explained in terms of some basic picture or model of the structure of matter and of the changes that matter undergoes. It may develop that several of the natural laws uncovered by scientists bear on the same subject ... and the model proposed is the same for each law. When this occurs, scientists gain confidence in the model and are more prone to accept it. Sometimes these models suggest that other experiments can be conducted to verify (or disprove) the theory.[15]

One of the foremost contemporary American physicists, P. W. Bridgman, has said that the scientific method is nothing more than doing one's level best with his mind, no holds barred. Dr. J. B. Conant, noted chemist and past president of Harvard University, defined it as the "tactics and strategy of science." "What most distinguishes scientific thinking is the mental attitude of the adventuring investigator, who pursues his intellectual quest with intense curiosity, utmost thoroughness, and a healthy skepticism of his own results. He is driven by an impelling desire to unveil nature's secrets, to push back the frontiers of the unknown, and to enlarge the horizon of knowledge. The technological developments that distinguish the present era constitute convincing proof that practical progress follows closely on the heels of a broader understanding of the natural universe.

The inquiring scientist approaches each problem with an open mind and with the highest type of intellectual integrity, willing and eager to follow wherever the facts may lead. Whenever possible, he analyzes a complex problem into its simplest components and devises carefully controlled experiments which clearly reveal cause and effect relationships. He is critical of the validity of his own data and subjects his hypothesis to exhaustive experimental tests before arriving at even tentative conclusions. In all his works, he relies upon a faith that natural phenomenons are reproducible and that the universe is orderly."

FOOTNOTES

1. J. C. Bailer, T. Moeller and J. Kleinberg, *University Chemistry,* (D. C. Heath and Company, 1965), pp. 4-5.
2. M. J. Sienko and R. A. Plane, *Chemistry,* 2nd ed., (McGraw-Hill, 1961), p. 4.
3. Harry H. Sisler, Calvin A. VanderWerf, and Arthur W. Davidson, *College Chemistry,* (Macmillan Company, 1961), p. 4.
4. J. W. Barker and P. K. Glasoe, *First Year College Chemistry,* (McGraw-Hill, 1951), pp. 5-6.
5. M. Sneed and Lewis J. Maynard, *General College Chemistry,* (D. Van Nostrand and Company, Inc., 1944), pp. 15-16.
6. Linus Pauling, *General Chemistry,* (W. H. Freeman and Company, 1947), p. 20.
7. O. W. Nitz, *Introductory Chemistry,* 2nd ed., (D. Van Nostrand and Company, (1961), p. 9.
8. P. Ander and A. J. Sonnessa, *Principles of Chemistry,* (Macmillan Company, 1965), pp. 1-2.
9. J. H. Wood, *et al, Fundamentals of College Chemistry,* (Harper and Row, 1963), pp. 1-2.
10. W. H. Nebergall, F. C. Schmidt, and H. F. Holtzclaw, *College Chemistry,* (D. C. Heath Company, 1963), pp. 10-11.
11. D. I. Hamm, *Chemistry,* (Appleton-Century-Crofts, 1965), p. 1.
12. C. E. Dull, H. C. Metcalfe, and J. E. Williams, *Modern Chemistry.* (Holt, Rinehart and Winston, 1958), pp. 3-4.
13. E. C. Weaver and L. S. Foster, *Chemistry For Our Times,* (McGraw-Hill, 1960), pp. 4-6.
14. Paul R. Frey, *College Chemistry,* 3rd ed., (Prentice-Hall, 1965), p. 2.
15. George W. Watt, Lewis F. Hatch and J. J. Lagowski, *Chemistry,* (W. W. Norton and Company, 1964), pp. 137-138.

Suggested Reading:

James B. Conant, "Concerning The Alleged Scientific Method" in *Science and Common Sense.* Yale University Press, 1951, pp. 42-62.

Ralph M. Blake, et al, *Theories of Scientific Method: The Renaissance Through the Nineteenth Century.* University of Washington Press, 1960.

Stephen Toulmin, *The Philosophy of Science.* Hutchinson University Library, 1953.

Some specific cases of application of the scientific method may be found in Louis F. Fieser's, *The Scientific Method.* Reinhold Publishing Company, New York, 1964.

Two rather more difficult, but more elaborate, studies are Karl R. Popper, *The Logic of Scientific Discovery,* Basic Books, 1961, and Richard B. Braithwaite, *Scientific Explanation,* Cambridge, 1964.

VIII

Methods and Logic of Science

All too frequently the creative, inventive phase of science is left without discussion because it does not lend itself to analysis.[1] Material discoveries like Fleming's of the antibacterial power of penicillin or Hyatt's of celluloid or inventions such as Goodyear's vulcanization of rubber are more examples of serendipity[2] (or discovery by accident) than planned investigations. Here the distinction should be made between inventions which are products of mental creation and discoveries which is the finding of something already in existence. For example, Newton invented the Universal Law of Gravitation since it was the product of his creative genius while Columbus or Leif Ericson discovered America since it was already here.

Seldom are either inventions or discoveries, strictly speaking, the work of one man. They depend rather on a series of individuals, or as is the case today, on groups of men and their inventions or discoveries as steps to new inventions or discoveries. There are also predisposing factors in the environment and culture that are determinants in invention and discovery.

We have already examined two inventors—Copernicus and Kepler—and we have seen that neither conforms to any pattern in their approach to invention.[3] Some inventions are the product of dreams while others are based on empirical perspiration—as Edison expressed it "genius is 99% perspiration and 1% inspiration." In the creative process, one can find many parallels among art, science, music, literature. It is almost impossible to speak of the differences between art and science at the creative level.

The major break-throughs of science have involved conceptual differences in perception of a set of facts rather than the discovery of new facts. The emphasis that Bacon, and following his lead led science in Britain and the United States, placed on experimental science has not led to major changes. Rather it has been a new slant on the old facts. The manner that these new slants are obtained do not seem at present, subject to a method but are more of an inspirational nature.[4]

Generally speaking, the two methods or approaches to science revolve about either an inductive or deductive approach. In the inductive the process is from many specific cases to the generalization.

One of the early steps of inductive science is to collect and classify. This is not restricted to modern science alone since most civilizations had methods of arranging or classifying various categories of things. Such things might include plants, animals, stars, etc. The basis of the classification system might, and indeed, did vary. Some had a functional basis (i.e., the use to which the plant or animal was placed)—hence categories as pot herbs (herbs placed in cooking pots) would be used. Other systems had a structural relationship base or a heirarchy based on the extent of development.

The process of classification is one of the earliest stages of a science to produce an order in the types of things with which one is to deal. Many persons, both scientists and non-scientists, sometimes hold the mistaken idea that to name or classify is the purpose of science. It is, of course, a part of the total process and synthesis of science.

Boyle's Ideal Gas Law comes the closest as an example of the inductive approach. Measuring the volume of a gas as the pressure was increased he found that there was a direct relationship that as the pressure increased, the volume decreased. At very high pressures the gas became a liquid so that the law did not hold nor did it hold exactly for any specific gas but only for an "ideal" gas. Here again we find that the law in science may not conform to what is seen.

As in the case of Copernicus where there was no physical or sensory evidence to support his theory, a law does not necessarily conform to what we see. Later when we look at Galileo's work with motion we will again come up with an "ideal" situation for the laws of motion only work in a vacuum and we do not live in a vacuum so that for practical purposes they are not true.

I suppose that one of the important differences between an engineer and a scientist is that the engineer must work with the "real" sensory world whereas this is not necessarily the case with the scientist. Another difference in the make-up of the two is that the scientist usually wants to describe how something works whereas the engineer is merely interested in making something work—he tends to be more pragmatic.

Explanation has always been a part of the process although modern science has changed the basis of the questions from why to how. If you ask a scientist the question, "Why is the sky blue?", the answer he will give is "how" the sky has color. If you then say "yes, but what I wanted to know is why the sky is blue instead of green or pink?". This he cannot answer. The explanation based on how explains by describing; it does not give a reason in the sense of a why answer. In the past 50 years, science has begun to give up the idea of an explanation on the basis of cause and effect. Today an explanation is apt to associate two events or phenomena rather than stating that one is the cause of the other. In fact, just because one event follows the other does not even mean that the two events are related. It is also possible that both are related to a third unobserved event and hence are only indirectly related to each other.

A scientific law does not however necessarily explain a relationship:

Perhaps the best-known example of a pure empirical formula is Bode's Law in astronomy. The planets in the solar system move in nearly circular orbits whose mean (average) distances from the sun do not follow any very obvious rule. Yet there is a rule, although it is not quite exact. Take the numbers 0, 3, 6, 12, 24, 48, 96, 192, 384 that is, a sequence in which every number, except the first, is exactly half the number which follows it. Add to each the number 4, divide by 10 and call the mean distance of the earth's orbit 1. The resulting numbers agree fairly closely with most of the relative mean distances of the planets, thus:

The law breaks down for Neptune but, curiously, the predicted mean distance for Neptune is almost exactly that found for the planet Pluto . . . Bode's Law was discovered [invented]—or perhaps stumbled on—as long ago as 1778, and has remained an enigma ever since. Today no one can say with certainty whether it is a mere accident of figures or an indication of a deep physical law as yet undiscovered.[5]

On the other hand, the deductive method ignores sensory "reality" and goes from some general principle to specifics. The basis of the Copernican Revolution is the application of such a method. In this the sun-centered system was regarded as simpler and more appealing than any other to the Neoplatonists. Newton's Universal Law of Gravitation is based on the Copernican system and is also more deductive than inductive in form. Generally, the deductive approach tends to be used more on the Continent while the inductive-experimental approach is used in England and the Anglized parts of the world.

Both of the above forms of reasoning may be written in symbols rather than in words and such an approach is called symbolic logic. The use of symbolic logic makes the examination of the validity of arguments easier. However, the application of this form of analysis is comparatively recent in origin.[6]

Another system of logic used by both approaches is called Aristotelian which includes the syllogism and the idea of the excluded middle. The idea of the excluded middle is so familiar that we cannot imagine without it. Something cannot be both black and white at the same time nor can something belong to two categories at the same time. This doctrine we have been led to believe in very strongly and it is almost a part of our nature.

Many times in science models, symbols and analogies are used instead of words to describe phenomena. Mathematics is a language that is frequently used by the scientist. You should keep in mind that just as words describing the sunset are not the sunset, mathematics used to describe a phenome-

Planets		Mercury	Venus	Mars	Asteroids
Actual relative distance (ARD)		0.39	0.72	1.52	2.77
Bode's Law		0.40	0.70	1.60	2.80

Planets	Jupiter	Saturn	Uranus	Neptune	Pluto
(ARD)	5.20	9.54	19.19	30.07	39.60
Bode's Law	5.20	10.00	19.60	38.8	fails

non is not the phenomenon. Scientists have sometimes made the mistake, and still do, that their model or mathematics is the reality.

When the scientist makes a model or mathematical formulation which is supposed to describe phenomena, he is also supposed to test his model. This test, if it agrees with his model, then makes him place more reliance on his model. Sometimes, however, the scientist is so convinced that his model is right that he will ignore evidence which does not agree with his model. This was true of Copernicus, as we have seen, whose model did not agree with evidence of the crucial experiment (i.e., the test of his model). The true scientist is supposed to throw out the model if it does not agree with the outcome of the crucial experiment.

The public thinks that the scientist is a man without emotion, who holds only to what he can see and prove by experiment. Such is not the case. The scientist may believe in many things that he has not seen (e.g., the electron, the atom, the structure of the moon's surface, flying saucers). Some scientists have been known to lie about data in order to protect their theory; and some have stolen data or ideas from their fellow scientists. A Russian scientist prostituted ideas in biology to make them conform to Marxist political thought. German scientists warped the facts to make the German "race" seem superior. Most scientists are honest as are most people, but because a man is a scientist does not make him more honest or less subject to bias.

Part of the reason that the scientist believes in things not seen is the rather indirect method of a number of scientific proofs. This is also one of the reasons that the educated public does not try to understand science. There is often a lengthy abstract chain of logic between the facts and the proof of the idea. The scientist cannot see the electron, but he can explain certain things that happen if he uses the idea of an electron. He can sometimes see or photograph the trail or tracks left by the electron as it moves through something. Like the trail left by a jet plane as it travels out of sight through the sky.

Another reason that the scientist believes in things not seen and a reason that he proceeds as he does is the assumptions that he makes. Many of these assumptions were made long ago so that he may not even think of them or if he does, he does not consider them assumptions. He assumes that there is a knowable reality, although he will not say what that reality is. He assumes that there is an order to the universe and that man can discover the laws that give rise to this order. He assumes that there are no discontinui-

ties—i.e., that existence is constant, true miracles do not occur. He assumes that what is a true law in one system will hold in all, and that it will hold under similar conditions everywhere. He assumes that true laws have always and will always be the same, e.g., that the decay rate of radioactive materials is constant. He assumes that anything which can be quantitatively measured or reduced to a quantitative measurement can be investigated with validity. He assumes that various models can approximate reality. He assumes that nature proceeds in the simplest possible manner. He assumes that if the part is studied and understood, and each part similarly studied that from this the whole can be understood, i.e., that the sum of the parts is equal to the whole.

In these assumptions, you will find some of the limitations of science. If the weight of a live person and the same person immediately after death remain constant, does this mean that there is no immortal soul? Is the mind more than the brain in action? In no manner is it possible to obtain a value judgment from science. What is beautiful? What is good? What is the best government?

These questions as well as many others are excluded from science because they are not subject to quantitative measurement. Scientists sometimes talk about or attempt to answer these and other questions but they have no special expertise when they are so engaged.

The scientist, or the person claiming to be a scientist—there are no licenses for scientists, has gradually replaced the priest as the person to whom the public turns in times of troubles or catastrophe. Science is the religion of the twentieth century and the scientist is the high priest. A plague of insects or locusts no longer brings forth the priest in an attempt to exorcise the disaster, today the exterminator is called or the scientist to devise a chemical or means of relieving the difficulty. People no longer pray for rain, they ask to have the clouds seeded and to have reservoirs built.

Because of the successes of the sciences in managing the environment more and more the sciences are asked to take over in increasing measure other areas of human concern. Today there is the beginning of the realization that frequently one solution from science or technology causes another problem in another area. Solving the problem of wholesale deaths in children has caused famine and shortages of food in later years. Killing insects with DDT has resulted in deaths of birds and other animals. The basic concept that has to be brought to the forefront is that when one casts a stone into a pool, ripples go forth to every bank and the effects are not localized. The motto must be to look before casting. To examine each facet of effects in an interdisciplinary manner before action is taken should be the rule.

Science is based as much on faith and belief as is religion as we have seen from the number of unseen things that science believes. In both there is a doctrinal basis that must be accepted and there is just as much authority. Many cases could be cited of attempts to introduce new ideas and have them rejected because they were not a part of the belief structure. Just as religion was institutionalized by the church so science has become institutionalized through the universities, scientific organizations and journals, as well as government and private research facilities.

In summation, the individual scientist does not use "a" method or logic but rather uses a mixture of the inductive and deductive. He accepts many things on faith and from authority just as does a person in any field for there in not enough time to repeat all the work done in a field. He continues to use models and classification systems for which there are exceptions for the lack of anything better and he constantly strives to "save the phenomena."

STUDY AND DISCUSSION QUESTIONS

1. Why isn't the creative phase of science subject to analysis?
2. Choose several scientific inventions and show how they came about.
3. What are the differences between invention and discovery?
4. What are the methods of science?
5. What assumptions does the scientist make?
6. How does the scientist differ from others?
7. What limitations are there on science?
8. How important a role does belief play in science?
9. What is the "real" world?
10. In what ways are models used in science and what are some?
11. How are science and religion alike? Different?

FOOTNOTES

1. Phillip Hanson, *Patterns of Discovery,* Oxford University Press. Thomas Kuhn's, *Structure of Scientific Revolutions,* (University of Chicago Press, 1962).
2. Finding valuable or agreeable things not sought for.
3. See Koestle's thesis in *The Sleepwalkers,* (Macmillan Co., 1959).
4. See for example Robert P. Crawford, *The Techniques of Creative Thinking* (Fraser Publishing Company, 1954).

5. O. G. Sutton, *Mathematics in Action* (Harper and Brothers, 1960), p. 11.
6. Hartly Rogers (M.I.T.) states that symbolic logic is a language invented by Frege, Russell and Whitehead about 1900. See also Carnap's *Meaning and Necessity* (Phoenix Paperback, 1956) for contemporary work on symbolic logic.

Suggested Readings:

John A. Mourant, *Problems of Philosophy.* Macmillan Co., 1964, pp. 517-563.

Henri Poincare, "The Selection of Facts" in *Science and Method.* Dover Publishing Co., pp. 15-24.

Stephe Toulmin, *The Philosophy of Science.* Hutchinson University Library, 1953.

Norman Campbell, *What Is Science?* Dover Publishing Co., 1952.

A book somewhat advanced, but which has readable portions is Stanley W. Jerons, *The Principles of Science: A Treatise on Logic and Scientific Method.* Dover Publishing Company, 1958.

Robert K. Nerton "Singletons and Multiples in Scientific Discovery: A Chapter in the Sociology of Science" *Proceedings of the American Philosophical Society,* 105:470-486 (Oct., 1961).

Peter Caws "The Structure of Discovery" *Science* 1375-1380 (Dec. 12, 1969).

A nice explanation of scientific method may be found in Uno Kask's *Chemistry Structures and Changes in Matter,* (Barnes & Noble, 1969), pp. 3-7.

A Partial Bibliography On Scientific Method Its History and Philosophy

Braithwaite, R. B. *Scientific Explanation.* Harper and Row, 1960. (1953) Difficult without advanced mathematics.

Burtt, E. A. *The Metaphysical Foundations of Modern Science.* Anchor, 1954. (1924) Readable.

Butterfield, H. *The Origins of Modern Science.* G. Bell & Sons, 1957. (1949) Difficult to uncultured reader.

Campbell, Norman. *What Is Science.* Methuen & Co., 1921. Readable.

d'Abro, A. *The Evolution of Scientific Thought.* Dover, 1950. (1927). pp. 343-442. "The Methodology of Science." Medium readability.

Davies, J. T. *The Scientific Approach.* Academic Press, 1965. Readable.

Hall, A. R. *The Scientific Revolution (1500–1800): The Formation of the Modern Scientific Attitude.* Beacon, 1956. (1954). Readable.

Hanson, Phillip. *Patterns of Discovery.* Cambridge, 1965. (1958). Chapters 1-5. Difficult in places.

Hobson, E. W. *The Domain of Natural Science.* Dover, 1968. (1923). Readable.

Lewinsohn, Richard. *Science, Prophecy and Prediction.* Fawcett, 1961. Readable.

Standen, Anthony. *Science is a Sacred Cow.* E. P. Dutton, 1950. Readable.

Taylor, Sherwood F. *A Short History of Science and Scientific Thought.* Norton, 1963. (1949) especially pp. 337-359. Readable.

Walker, Marshall. *The Nature of Scientific Thought Spectrum.* 1963. Chapter I, "The Scientific Method" and Ch XIII "Models in Social Science" can be managed with a little thought but the whole requires some mathematics.

Wartofsky, M. W. *Conceptual Foundations of Scientific Thought.* Macmillan, 1968. Most is quite readable but there are a few difficult places. Very useful book.

Whitehead, Alfred N. *Science and the Modern World.* Mentor, 1948. (1925). Readable.

Wightman, W. P. D. *The Growth of Scientific Ideas.* Yale University Press, 1951. Chapter I especially. Readable.

IX

Mathematics Teaches Science to Count

We have already seen that mathematics is used as a model in science. Some have claimed that the rise of modern science can be dated from the application of quantitative techniques to natural phenomena. Yet the importance of mathematics had been noted long before as can be shown by citing Caxton's, *Mirrour of the World.*

The fourth scyence is called arsmetri-que [arithmetic]. This science cometh after rethoryque [rhetoric], ande is sette in the myddle of the vii sciences. And without her may none of the vii sciences parfyghtly ne weel [perfectly nor well] and entierly be knowen. Wherfore it is expedyent that it weel knowen and conned [to examine or study closely]; ffor alle the sciences take of it their substaunce in suche wise that without her they may not be. And for this reson was she sette in the myddle of the vii sciences, and there holdeth her nombre; ffor fro her procede alle maners of nombres, and in alle thynges renne, come and goo [for from her come all numbers; by means of numbers all things run, come and go], and no thyng is without nombre. But fewe perceyue [perceive] how this may be, but yf he haue [if he would] be maistre [master] of the vii artes so longe that he can truly save the trouthe [truth]. But we may not now recompte [recount] ne declare alle the causes wherfore; ffor who that wolde dispute [discuss or argue] vpon suche werkes, hym behoued despute and knowe many thynges and much of the glose [a continuous commentary accompanying a text].

Who that knewe wel the science of arsmetrique he myght see thordynance [the ordering] of alle thynges. By ordynance was the world made and created, and by ordynance of the Souerayn [sovereign] it shal be deffeted [destroyed]. . . .

The fyfthe is called geometrye, to whiche more auaylleth [that which has more meaning] to Astronomye than ony of the vii other; ffor by her is compasses and mesured Astronomye. Thus is by geometrye mesured alle thingis where ther is mesure. By geometrye may be knowen the cours of the sterres [stars] which alleway go and meue [move], and the gretenes of the firmament, of the sonne [sun], of the mone [moon] and of the erthe. By geometrye may be knowen alle thynges, and also the quantyte [quantity]; they may not be so ferre [far], yf they may be seen or espyed with eye, but it may be knowen [however far objects may be, if they can be seen they can be measured and their magnitude known].

Who wel vnderstode geometrie, he myght mesure in all maistryes [mastery]; ffor by mesure was the world made, and alle thinges hye [high], lowe and deep.[1]

Thus it is clear that even in the middle ages it was believed that mathematics gave the user control over many things and indeed for the Pythagoreans numbers were things. By the Renaissance, a number of persons may be found who felt that in mathematics could be found the explanation of God's glorious scheme and that God was the great mathematician. At the same time the true nature of mathematics can be seen, not by them at the time, but by their actions for us.

Mathematics is a lot of silly games invented and played by man. Or you may prefer to read what mathematicians themselves have said about their subject. ". . . mathematics is the science of skillful operations with concepts and rules invented just for this purpose. The principal emphasis is on the invention of concepts."[2] The late Bertrand Russell has said that ". . . mathematics may be defined as the subject in which we never know what we are talking about, nor whether what we are saying is true." This comment is in agreement with the statement by Henri Poincaré that ". . . mathematics is giving the same name to different things" or with Benjamin Peirce's remark that ". . . mathematics is the science which draws necessary conclusions."[3]

One of the facets of mathematics developed in the Renaissance was the posing of problems as a challenge or two mathematicians would engage in a public debate—it was similar to professional baseball or football today. Each town had its champion and was cheered on by the multitudes during the contest. Just as today the spectator did not have to know the rules in order to enjoy the sport—the important thing was that your champion won. It took the place of jousting and scholastic debate of the Middle Ages. There were also poetry contests. This idea of challenging problems remains a part of mathematics today. For example, "In a famous address to the international congress of mathematicians at Paris in 1900, David Hilbert proposed thirty mathematical problems which were easy to formulate, some of them in elementary and popular language, but none of which had been solved nor seemed immediately accessible to the mathematical technique then existing."[4]

A problem in understanding for pragmatists of to-day is that a challenging problem for a mathematician does not have to be a practical one—that is it does not have to be useful.

"... When Alan finds a situation he doesn't like, he never tries to use his reason or ever approach it. He just loses temper and swears."

"Then how ought I to approach the situation, Camilla? Would you be better pleased if I worked it out mathematically?"

"Don't you say anything against mathematics!" she breathed. "In higher mathematics, for those who can understand, it's the most romantic and imaginative conception ever dreamed of! Mathematics gave us the space age ..."

"Hooray."

"... and other things reactionaries sneer at because they hate progress! In the higher mathematics ... not that I ever got that far myself. ..."

"Well, I never even got within shouting distance. To me mathematics means the activities of those mischievous lunatics A, B, and C. In my time they were always starting two trains at high speed from distant points to see where the trains would collide somewhere between. Which as the man said in the story, is a hell of a way to run a railroad."

"Does the problem anywhere say the two trains are on the same track? It doesn't; you know it doesn't; you won't stop to think! All the book wants to know is where those two trains will pass!"

"And why does the book want to know that? So one engineer can wave to the other or give him a raspberry in passing? Why?"

"Somebody's got to ask, you know, Camilla, it suggests a legal ruling once said to have been made in Arkansas about a disputed right-of-way. 'When two trains approach this intersection, both shall stop; and neither shall proceed until the other has passed.'"

"That's just about as sensible as most decisions now handed down by the United States Supreme Court. But we haven't finished with your tireless friends A, B, and C. When the silly dopes weren't wrecking trains or computing the ages of two of their children without seeming to know how old the brats were, two of 'em had a passion for pumping water out of a tank while the third poor mug pumped water into it. Camilla, how many afternoons do you spend doing that?"[5]

What Alan does not realize, in the above example, and what Camilla mentions but does not emphasize is the fun and recreational aspect of mathematics. The fact that each branch of mathematics is a different game—like chess, checkers, monopoly—is the important idea to a pure mathematician. If someone were to set up a chessboard with a few pieces on it and challenge you to checkmate in five moves, you are not expected to have to explain why you want to solve the problem—it's just fun. If you ask why a knight combines the move of a castle and bishop by

moving straight one place and diagonally the next, the question has no meaning because it is one of the rules of the game. If you change the rules then the game is a different game. In the same way each branch of mathematics has different rules. Each branch of mathematics starts with a few axioms or assumed rules that are unsupported by proof and with a few terms that are not defined. Some terms are not defined because the mathematician does not want a circular definition. In plane geometry, for example, words like "point," and "line" are not defined. From the few assumed statements, axioms, the whole system or branch is built just as chess has a relatively few rules on which a whole game with various shades of playing pleasure can be managed.

You may recall from an earlier chapter that mathematical truth differs from the usual definition of truth because in mathematics there is not a necessary correspondence to "reality" just as the game of "Life" does not necessarily correspond to real life. This feature is often overlooked by individuals and even scientists today and was certainly not a part of the belief system of the Renaissance where mathematics was regarded as a true model of reality. Mathematics is and was used as a language and as a model but it is sometimes forgotten that the model can never be the "real" thing and that it is true only so far as it corresponds to reality.

Models are, for the most part, caricatures of reality, but if they are good, then, like good caricatures, they portray, though perhaps in distorted manner, some of the features of the real world.

The main role of models is not so much to explain and to predict—though ultimately these are the main functions of science—as to polarize thinking and to pose sharp questions. Above all, they are fun to invent and to play with, and they have a peculiar life of their own. The 'survival of the fittest' applies to models even more than it does to living creatures. They should not, however, be allowed to multiply indiscriminately without real necessity or real purpose.[6]

There are several examples that can be cited to show that mathematics is not uniformly a model for reality.

The lecture at our PTA meeting was devoted to the new math. One point covered in detail was: No matter in what sequence numbers are added, the answer is the same. Thus, 2 plus 3 plus 4 is the same as 4 plus 3 plus 2.

Touring the building after the meeting, I spotted a blackboard on which a shapely female figure had been drawn. A doubting convert to new math had written beside it: 36-24-36, or 24-36-36, or 36-36-24 does not add up to the same figure.[7]

Another similar example, "she married and had a baby" is not the same as "she had a baby and was married." In other words, there are real situations where a particular branch of mathematics can not be used for a model or for communication. There may be another branch of mathematics that does fit or there are situations which can not be covered by any now known branch of mathematics.

The systems of mathematics are frequently divided into three or four large areas. Where there are three, these are algebras, geometries, and analysis and when there are four probability and statistics are added. The following list, while not complete, should give some idea of the variety of courses in mathematics. There is, of course, arithmetic which is a part of one form of algebra and plane, solid, spherical and analytical geometries; differential and intregal calculus; probability theory, theory of differential equations, principles of analysis, functional analysis, abstract algebra, linear algebra, theory of numbers, concepts of geometry, projective geometry, differential geometry, topology, algebraic topology, mathematical logic, set theory, advanced calculus, linear analysis, multilinear analysis (tensor analysis), boundary value problems of differential equations, optimization problems, qualitative theory of differential equations, partial differential equations, complex analysis, operational calculus, functions of a complex variable, elements of stochastic processes, square summable power series, entire functions, theory of approximation, mathematical methods in games and programming, topics in random walk and Markov chains, abstract structures, series of functions, theory of Fourier series, integral equations, functional analysis, linear partial differential equations, commutative algebra, theory of rings and algebras, group theory, modern differential geometry, algebraic geometry, algebraic topology, algebraic closure spaces, axiomatic set theory, and statistical methods.

The fact that should be strikingly strange is that there should be any connection at all between any system of mathematics and reality. It's like expecting a relationship between business and Monopoly or War and Stratego. Why should there be a connection between a mental fantasy and reality?[8]

Periodically someone feels that there is a correspondence between a mathematical system and something that is going on in the world and applies or misapplies mathematics to it.

Some uses of mathematics could include the beginning of probability which is attributed to Cardan during the Renaissance and applied to gambling; or the application of matrix algebra by Steinmetz to electrical load distribution; or its application to many complex problems of today in the fields of economics, sociology, industrial management, and modern psychology.

One might also cite a number of abuses of mathematics. Historically, man has always sought means of predicting future events. One of these methods makes use of mathematics. Called numerology, it was used by the Egyptian priests, by the Pythagoreans, by cabalist rabbi and many others. Sometimes this technique applies the changing of letters to number values, (e.g., the number of the beast in *Revelation*) following which one could also obtain geometric figures from these numbers. According to one explanation there are twenty-two Hebrew letters and also twenty-two regular polygons which can be described within a circle. Shades of Kepler—at least we can see that he and his mystical works are a part of a continuum which lasts even today. From the number of the beast and applying numbers to names, it is possible to find out when the world will end and the anti-Christ has been various people including Nero, Hitler, Luther, and some of the Popes. It's all a matter of making numbers do what you want them to do.

Another method abusing mathematics is to start with a great building such as the Eiffel Tower or the Pyramids and use some of their measurements.

> ... Georges Barberin in his *Secret de la Grande Pyramide,* discovered, in a rather surprizing fashion and in the measurements of the great gallery, the date of the beginning of the First World War. The length of the great gallery's flooring is 1,884-1/3 inches. Now if we add to this number of 1,884-1/3 inches (taking an inch for each year) to 5th April, A.D. 30, the date of the death of Christ (which was the beginning of the Christian era) we get the date 4th-5th August, 1914.

> By a similar method applied to the Eiffel Tower, if we subtract from the number 1,927 (that of the steps and landings from the base to the summit) the number of those who partook of the last supper, we get 1914, the date of the beginning of the First World War. If we add 12 (the number of the Apostles) to 1,927 we get 1939, the date of the outbreak of the Second World War [if we subtract 3 (the Holy Trinity) from 1,927 we get 1924, the year in which I was born.][9]

It is not only the numerologist who abuses mathematics but also many writers. This abuse may be conceptual in nature. (e.g., "Consider the famous example of the chimpanzees left in a room with ten typewriters. It is mathematically certain that, given enough time, the chimpanzees would type out *Hamlet.*")[10] While it is true that, from the standpoint of probability, the example is strongly probable it is by no means mathematically certain and it is only strongly probable within the system of mathematics designated as probability. Descartes in his *Medita-*

tions states: "For whether I am awake or asleep, two and three together always form five . . ." Today, we would have to regard this statement as being utter nonsense because the operation is dependant on the mathematical system being used and the number base used must contain "2," "3," and "5" in order that the statement be true. In addition, all of the numbers must be cardinal and not ordinal or a mixture of cardinal and ordinal, in other words the numbers must be of the same kind.

Some of you may have noticed while driving on country roads that every once-in-a-while the road turns sharply through 90° and back through 90° again. This effect is brought on through the application of the wrong mathematical model to the real world. Our country is laid out in squares and, at least for most of the East, the basic unit of government is the township. The roads with the sharp turns are laid out along township lines and a township is about six miles square. In the process of surveying the country into little squares—you may have wondered why there are so many squares in this country—a basic difficulty was encountered. You may see the problem easily by taking an orange and a ball point pen. Draw an equator on the orange then draw lines to quarter it. Take one of the quarters and begin to draw squares of the same size which will uniformly divide the quarter you have chosen. The problem that presents itself is the application of a form of mathematics designed for a plane to a surface, the earth, that is spherical.

Most beginning physics books have projectiles being fired and most of the books make a statement to the effect that they are going to ignore the effects of the air—in other words their cannons are fired in a vacuum. We don't live in a vacuum so how much difference is there between the mathematical model that assumes that we live in a vacuum and the real world?

> A simple example shows the importance of air resistance in ballistics. A body projected from the earth at 2700 feet per second at an angle of 30° to the horizontal would have a horizontal range of about 65,000 yards if there were no atmosphere. A well-designed shell, fired with the same initial velocity and elevation, hardly achieves 30,000 yards range in still air.[11]

The difference, as shown by the example, is 100% and more.

As we can see from these examples, there is not a necessary correspondence between the mathematical model and reality. This little understood fact creates chaos for the person who is unacquainted with mathematics. He is under the delusion that mathematics is all the same thing—a homogenous mass—which can be used to prove something. As we have shown, there are many systems of mathematics which, like games, have different rules and sometimes are a model of portions of reality but really prove nothing in and of themselves. We have also seen how mathematics can be used and abused. Mathematics can and has been used as support, in the same way that a drunk uses a lamppost, rather than for illumination as it should be. Or expressed differently—figures do not lie but liars often figure.

While we will not be working directly with mathematics, for the most part, we do use some figures and we will be talking about mathematics because of its increasing importance, not only in science but in many other fields as well.

There are three qualities, ill defined, that are frequently measured in the physical sciences and in ordinary living. These are: extension (including length, width, and depth), weight (physicists usually use the term mass which we shall not), and time.

In the Middle Ages, and even later in some countries, every area, and sometimes every town, had its own units of measurement. As countries centralized they tended to have national standards, (e.g., the English system) but science was and is international and as the use of figures in science grew so did the need for a uniform system. The French, following their revolution, wanted to do away with everything that was old—changing the names of the months, the days of the week, etc.—and part of this was the establishment of the metric system.

The metric system is a decimal one (i.e. based on multiples of ten) so that units can be converted by simply moving the decimal point. This system was readily adapted by persons working in science all over the world fairly quickly and easily so that they could communicate results of what they were doing much better.

One of the major problems with the metric system is that while it is supposed to be a system of multiples of ten, some of the units are not used and this makes their use somewhat difficult. The unit for extension is the meter, that for weight the gram, and for volume the liter. To each unit there is supplied a series of prefixes to indicate what part of the unit is being spoken of.

> milli- one-thousandths of the unit
> centi- one-hundredth of the unit
> deci- one-tenth of the unit
> the unit
> deca- ten times the unit
> hecto- one hundred times the unit
> kilo- one thousand times the unit

Of these prefixes: deci-, deca-, and hecto- are almost never used. Our money system is partly metric and we use the mil as the basis of real estate tax or one dollar for a thousand dollars of valuation. We use cent to mean a hundredth part of a dollar but the others we don't use.

A millimeter is about as wide as a pencil lead and a centimeter about the width of a finger, while a meter is slightly longer than a yardstick.

For very far distances (astronomical) two units are used. These are the parsec which is the distance that a star must be from the earth in order to have a parallax of 1 second and is equal to about 19 trillion miles; and the light year which is the distance that light travels in a year (light travels at about 186,000 miles/sec. or 3×10^8 m/sec.) or about 6-trillion miles.

In the preceding paragraph reference was made to a kind of shorthand which is often used with very large or very small numbers. In the instance above, 3×10^8, stands for 300,000,000. The 10^8 means that 10 is multiplied by itself 8 times. However, an easy way to remember this is that the superscript gives the number of places the decimal point is moved. On the other hand, very small numbers such as 0.1 and 0.01 are written as 10^{-1} and 10^{-2} which means that they are the same as   and   . An easy way to remember this is that the total negative superscript gives the number of places to the right of the decimal point (e.g., 10^{-8} would be the same as 0.00000001).

Occasionally, we will weight objects and again we will use the metric system. There are 1,000 grams (g.) in a kilogram. Most of the weighing we will do will be in grams, tenths of grams or smaller subdivisions of grams.

The measurement of time is the same, at least for the second, minute, and hour, over the world. Why should this be the case? What is different about the way the units of time were established from the way units were established for extension and weight? Questions like these are hard to answer even though they may be of interest to the historian. Since we are more concerned here with measurements we only raise the questions with passing interest.

When a direct measurement is made using a measuring device (meter stick, scales for weighing, venier calipers, thermometers, watch) error is introduced. The amount of error is dependent on the accuracy of the instrument and on the training and ability of the person making the measurement, in addition of course, to the inherent errors in perception. So that in any direct measurement there will be some difference between the measurement and the "real" object or phenomena. The only numerical values that are completely accurate are those which are

set by definition (such as 10mm. = 1 cm.) Keep this in mind when making calculations. It is not necessary to carry an answer to twenty places because of the lack of accuracy of the direct measurement. The value of pi, since it is a defined relationship between the diameter and circumference of a circle, has been carried out to over 100,000 places. You should note the degree of accuracy of your measurements to determine how far to carry an answer.

An indirect system of measurement is used for weights and extension where it is not possible to measure directly. (The distance across a lake, e.g.) One of the first large scale direct measurements was made by Eratosthenes (c. 275 B.C.), a Greek astronomer and geographer, who wanted to measure the circumference of the earth. (He also wrote literary criticism and poetry and in mathematics invented the sieve method for finding prime numbers.) He observed that at Cyrene (near present day Aswan) on a particular day that the sun at noon could be seen from the bottom of the well. He also observed (or had a friend observe) that on the same day the sun was 7-1/2° from being directly overhead at Alexandria, somewhat north of Cyrene. He had the distance between the two cities paced and obtained a value of 5,000 stades. So he set up the relationship by ratios that 5,000: x as 7-1/2°: 360° and obtained the value of 240,000 stades for the circumference of the earth. We don't know how accurate he was since we do not

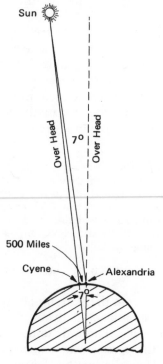

Figure IX-1. Circumference of the Earth.

know the value for a stade. Some have assumed that there were 10 stades to the Roman mile and if so this would give a value of about 24,000 miles for his calculations, which is not a bad approximation, about a 5% error. The distance between the two cities gave him what is called a base line and is used today and is referred to as triangulation. A similar method is used for range finders.

In his work, Eratosthenes made a number of assumptions, as indeed one must do in any indirect measuring. Some of these were:

(1) That Euclidian geometry would give a true model.

(2) That the earth was spherical.

(3) That light goes in straight lines.

(4) That Alexandria is directly north of Cyrene.

(5) That both cities are at sea level.

(6) That the sun's rays are parallel.

(7) That his measurements are accurate.

(8) That the base line is straight, not curved.

As we have seen the methods for measurements, both direct and indirect, had been at hand for many years. The Arabs had made use of measurements in calculating the densities of various substances hundreds of years before the Renaissance and the applications of mathematics to science in the west.

It is the Neoplatonic influence placing an emphasis on geometry and the successes in applying mathematics to motion that is the beginning of making science more quantitative with an emphasis on exactness.

STUDY AND DISCUSSION QUESTIONS

1. Look up definitions for extension, weight, mass, and time. Can you define any of these terms better?

2. Write the number six trillion in scientific notation (i.e., in powers of ten).

3. Examine each of Eratosthenes assumptions. Can you think of any others? What difference would it make if any or all assumptions were wrong?

4. Alexandria is 3° apart from Cyrene. How much difference does this make? In miles? In 1% of error?

5. What was the value of arithmetic, according to Caxton?

6. What is mathematics?

7. Does mathematics correspond to reality?

8. What can be proved by mathematics?

9. What are some of the abuses of mathematics?

10. What are some of the uses for mathematics?

11. How did the metric system spread into general use over Europe?

12. Is it true that the more places to which an answer is carried the better the answer? Why or why not?

FOOTNOTES

1. Caxton's, *Mirrour Of The World, op. cit.,* pp. 36-38.

2. Eugene P. Wigner, "The Unreasonable Effectiveness of Mathematics in the Natural Sciences" in *Communications On Pure and Applied Mathematics,* 13:1-14 (1960), p. 2.

3. These quotes are from the 1966 edition of the *Americana,* Vol. 18, p. 433.

4. Richard Courant and Herbert Robbins, *What Is Mathematics?* (Oxford University Press, 1941), p. 107.

5. John Dickson Carr, *Dark Of The Moon,* (Harper and Row, 1967), pp. 60-62.

6. Mark Koe, "Some Mathematical Models in Science" in *Science* 166:695-699 (Nov. 7, 1969), p. 699.

7. Submitted by Warren C. West of Seattle, Washington and printed in the *Reader's Digest* under "Life In These United States."

8. Wigner, *op. cit.* has an interesting discussion.

9. Maurice Bouissen, Magic: *Its History and Principal Rites,* Appendix 2, (E. P. Dutton & Co., 1961). Also see "The Great Pyramid, The Golden Section and Pi" in Willy Ley's, *Another Look At Atlantis,* (Doubleday, 1969).

10. Dunean Walters, "Towards a Definition of Art" in *The Undergraduate Journal of Philosophy,* 1:1-6, p. 2 (1969).

11. O. G. Sutton, *Mathematics in Action* (Harper & Brothers, 1969), p. 71.

Further General Reading:

Morris Kline, *Mathematics in Western Culture,* Oxford University Press, 1953.

Mathematics In The Modern World, W. H. Freeman and Company. (Reprints from *Scientific American*).

X

On the Shoulders of Giants

As we have seen in the last chapter, pure mathematics consists of a number of games but, at least until recent years, mathematical models are pressed into the service of man very soon. Mathematics had its origins in magic and religion. The Hindu-Arabic numbers were developed to calculate religious feasts and we have seen the importance of mathematics in calculating Easter. Frequently, then, the needs of the political, mercantile, or religious portions of the community will dictate the form that the models of mathematics will take.

Aristotle had divided motions into natural and unnatural (or violent). Natural motions were subdivided into—celestial and sub-lunar. Celestial motions were circular and were around the earth while the sublunar (motions below the moon) were directly toward the center of the earth, in the case of earthy and heavy objects, or directly away from the center of the earth, in the case of ethereal or fiery substance. Unnatural motions included projectiles of various sorts that were forced into motion.

We have already discussed the revolution associated with celestial motions and how they became elliptical, moving about the sun instead of the earth. It should not be assumed that the questions about unnatural motions arose during the Renaissance anymore than to assume that Copernicus was the first to invent the heliocentric system. There is a long and continuing stream of persons who were not satisfied with some aspect of Aristotle's system on motion as we shall indicate, but here we shall be more concerned with the works on motions of Galileo Galilei (1564–1642).

His works with the telescope and his writings, in the vernacular had considerable influence in completing one revolution and his works on motion were, in a sense, the beginnings of one.

While the cannon and gunpowder had been applied during what is classed as the Middle Ages, it was refined during the Renaissance because of the many small wars between petty states and principalities. It may certainly be considered as a major factor in the revival of studies in motion, although not the

Figure X-1. Galileo.

only one.[1] (e.g. "It was left to Philip III in 1590 to order a lectureship in mathematics to be reinstituted at Salamanca, because of a lack of experts on artillery.")[2]

Men of genius do not, as a rule, create from nothing. They are a product of their age and draw heavily on the works of their predecessors. Galileo is no exception. The problem of motion had been studied by many who were unhappy with Aristotle's explanation of unnatural motions. In fact, by the sixth century John Philoponos of Alexandria had challenged some of the more obvious fallacies. There were also criticisms by Alexander of Aphrodisias and Themistius. After the rise of Islam the arguments were kept going and embellished by the Arabs particularly Avicenna (Ibn Sina), Averroes (Ibn Rusd), and his students—Siger of Brabant and John of Iandun—Avempace, and Al-Farabi. The site of the arguments moved into Europe after the 12th century when Arabic works of the classics were translated. During the late Middle Ages contributions were made by learned scholastic churchmen including William of Ockham, Francis de Marchia, Nicolaus of Cusa, Albertus Magnus, St. Thomas Aquinas, Duns Scotus, Thomas Bradwardine and his student Richard Swineshead (or Swseth) who made the first serious attempt at quantifying changes, John (Jean) Buridan, whose idea of impetus was useful, Albert of Saxony, Nicholas Oresme, who developed a geometric method of explaining motion, Marsilius of Inghen, Roger Bacon, and Petrus Johannes Olivi. In Italy itself there was the work of Giovanni Benedetti and Simon Stevin, who had dropped unequal weights from a tower and noted that the weights struck at the same moment.[3]

This sketches briefly the background prior to Galileo. He is noted for inventing the law of falling bodies, the principle of inertia, and the resolution and composition of independent motions. The first is the most celebrated.

In deriving his laws there are several influences to consider: the technology and artisans, Neoplatonism, mathematics, and experimentation.

The artisans and technology provided, not only the equipment needs for verifying his thought experiments, but also a common sense background of observations of natural phenomena to which logic could be applied and a pure idea extracted. This leads to the influence of Neoplatonism in removing the accidental qualities from an event and considering only those concerned with the event itself. The qualities that he considered associated with an event were those that were measurable or which had the quality of number (e.g., weight, time, length). He would exclude those which could not (e.g., taste, smell, color). (This led others into the realm of philosophy discussing primary and secondary qualities—qualities in the body itself or those in the mind of the observer.)

Underlying all his thought was the assumption that nature is a simple, orderly system whose every proceeding is thoroughly regular and inexorably necessary. 'Nature doth not that by many which may be done by few.' Furthermore, his thinking was dominated by the conviction that to understand the pattern of nature we must express observations of it in quantitative terms—nature is the domain of mathematics: 'Philosophy is written in that great book which ever lies before our eyes—I mean the universe—but we cannot understand it if we do not first learn the language and grasp the symbols in which it is written. This book is written in mathematical language.'[4]

In his experimentation he used two kinds—mental and physical—but the physical was merely to verify what the mental or thought experiment had shown him. He realized that motion had to be slowed in

some way in order that it might be studied and this realization led him to make a number of devices.

The devices that Galileo uses to study the motion of falling bodies are an inclined plane, a pendulum, and a water dispenser. The water came out in a thin stream which was collected in a glass and weighed. The time, then, for the descent along an inclined plane or the period of the pendulum could be obtained in terms of the weight of water collected during the event. He realized that a freely falling body would move too fast for direct measurement so he "diluted" gravity by using a brass ball moving down an inclined plane.

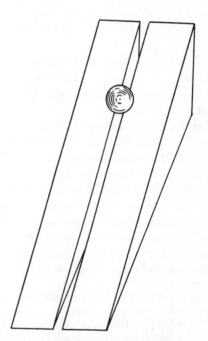

Figure X-2. Inclined Plane.

His approach to the law of falling bodies was not experimental, but rather by use of deductive reasoning and mathematics. He assumes that a body in falling will be accelerated uniformly. From this assumption he considers two possibilities: the first that the speed of a falling body would be proportional to the distance through which it fell or secondly, that it was proportional to the time that the body was falling. Because he saw a contradiction in the first, he selected the second. Today, we would express his statement as V for velocity or speed is equal to a, the acceleration times t, the time or $V=at$. At that time, and for hundreds of years thereafter, algebra was not used because geometry was mathematics and only by using the proofs of geometry could relationships be shown. We are using algebraic notation because it is easier to write. From this Galileo derived the relationship that the distance, s, that a body would fall would equal one-half the acceleration times the square of the time or $s=1/2at^2$. As a check he performed experiments with a ball on a grooved inclined plane. He measured the angles, the distances, and the times.

> In 'experiments near an hundred times repeated,' Galileo found that the times agreed with the law, with no differences *'worth mentioning.'* His conclusion that the differences were not 'worth mentioning' only shows how firmly *he had made up his mind beforehand,* for the rough conditions of the experiment would never have yielded an exact law. Actually the discrepancies were so great that a contemporary worker, Père Mersenne, could not reproduce the results described by Galileo, and even doubted that he had ever made the experiment.[5]

In other words, the experimental facts did not support Galileo any more than the physical facts supported Copernicus when he put forth his heliocentric system.

He develops the concept of inertia which later becomes Newton's first law of motion: a body remains stationary or in uniform motion in a straight line unless it is acted upon by an external force. This means that uniform motion in a straight line is the same as a body at rest which is contrary to the older physics. He states and demonstrates that the path of a projectile, such as a cannon ball, is that of a parabola. (An idea used by Newton in the formulation of his universal law of gravitation.)

> According to the Galilean analysis, a projectile has two independant components of motion, horizontal and vertical, like the ball on the inclined plane. If fired horizontally from a gun, it moves foreward the same distance in every second if we disregard the small factor of air resistance [sic]. As it emerges from the barrel it also begins to fall toward the earth. During the first second it will fall 16 feet; during the second second, 48 feet; during the third, 80 feet, and so on. [If you plot the time and distance fallen on a graph you will find that] ... the path of the shell will be parabola. Here was a brand new discovery that was of the utmost practical importance in the new science of artillery-ranging.[6]

We are prone to think of Galileo as one of the first men of modern science. As such, we may think of him as not being typical of a man of the late Renaissance period in which he lived. He was a contemporary of Shakespeare and was born in the year Michelangelo died. His father, a composer and one of the founders of opera, taught him to play the organ and lute. In later life, he told friends that, had circumstances permitted, he would have become a painter. Cilogi, the painter, gave Galileo credit for

having taught him what he knew about the art of perspective. Galileo demonstrated his appreciation for music and art in a letter to the painter.

> Will we not admire a musician who moves us to sympathy with a lover by representing his sorrows and passions in song much more than if he were to do it by sobs? And this we do because song is a medium not only different from but opposite to the expression of pain while tears and sobs are very similar to it. And we should admire him even much more if he were to do it silently, with an instrument only, by means of dissonances and passionate musical accents; for the inanimate strings are of themselves less capable of awakening the hidden passions of our soul than is the voice that narrates them.[7]

During the Renaissance, the educated man was inspired to become expert in many diverse fields. Leonardo da Vinci, for instance, became eminent as both a painter and a scientist. Galileo, in addition to being a scientist, possessed outstanding literary talents and had a great interest in the arts. In the present age, it is possible to be an expert in only a very narrow field and we think of the artist as having a basic nature opposite that of the scientist.

The Renaissance was a rebirth of the culture of ancient Greece and Rome that occurred in Europe during the fourteenth, fifteenth and sixteenth centuries. It was characterized by a flowering of painting, sculpture, and music. The medieval emphasis on life after death had shifted to an emphasis on the enjoyment of life on earth, however, religious convictions still were deeply imbedded.

Since Galileo was accustomed to symbolism in the arts, he also found it in the *Bible.* In a letter to Father Castelli he said, "There are in Scripture words which, taken in the strict literal meaning, look as if they differed from the truth."[8] He noted that when one is in the hold of a moving ship he speaks as if the ship were at rest although he realizes that it is not. Galileo felt that statements in the Scripture about the movements of the sun should be taken in the same way.

Among the works of science, Galileo's *Dialogue*[9] may be considered one of the outstanding works of art. Since the *Dialogue* was intended to persuade the average literate layman of the validity of the Copernican theory, literary excellence was important. In achieving it, he used a dialogue format, which had been used by Plato.

In many instances, Galileo was far ahead of his period in the understanding of physical phenomena. He formulated mathematical laws for the pendulum and for falling bodies. He made telescopic observations of the mountains on the moon, the satellites of Jupiter, the phases of Venus, and the spots on the sun. He interpreted these observations and this led him to accept the Copernican theory.

In other instances, he adhered to the ideas of the past. Although Galileo disagreed with Aristotle about the center of the universe and the nature of celestial material, he considered these to be minor points. He felt that basically the Copernican theory and his own laws of falling bodies supported Aristotle's ideas. Galileo saw a universe with Aristotelian order in which bodies moved toward their proper place with a uniform acceleration. Circular motion he also considered natural because it could be associated with a particular place, the center of the circle.

By his work on the motion of bodies Galileo was certainly one of the giants that Newton will speak of as a forerunner of his system. But in another sense Galileo is just another strand in the warp and woof of man's achievement. Perhaps it could be argued that a man is no more than his press agent or Boswell is able to make him out to be. Certainly a person's stature from a historical point of view depends on what he and others have said of him and the documentation that is available.

> Galileo's greatest general contribution was the idea that mathematics was the language of motion, and that change was to be described mathematically, in a way that would express both its complete generality and necessity, as well as its universality and applicability to the real world of experience.[10]

Still another interpreter of Galileo sees him in a far more general light and in a broader context.

> By the example of his life and work Galileo bequeathed to posterity three more intellectual ideals. These are (i) his recognition of the dynamic nature of science; (ii) his faith in the unity of all creation, which led him to look for consistency in all valid knowledge or human experience; and (iii) his passionate antagonism to any kind of dogma based on human authority.[11]

One might question the degree of Galileo's convictions in that although he had written in defense of the Copernican system he did not teach it in his class at the University but retained the Ptolemaic system. Further one might question his virtue because when he is questioned by the inquisition he recants.[12]

But at any rate, today he is regarded as an eminent scientist. He also produced a classic literary work and had an avid interest in the arts. Furthermore, he was deeply religious and was an adherent of Aristotelian philosophy. In all these respects, he was a man of the Renaissance.

STUDY AND DISCUSSION QUESTIONS

1. How does Galileo's description of motion differ from that of Aristotle?
2. Why does Galileo idealize his laws of motion?
3. In what ways was Galileo a man of his times?
4. What were Galileo's major accomplishments?
5. How important are experiments to Galileo?
6. What is Galileo's method of approaching a problem?
7. What factors played an influence in Galileo's works?

FOOTNOTES

1. For a number of other explanations see E. A. Burtt, *The Metaphysical Foundations of Modern Science* (Doubleday & Co., 1954).
2. Henry Kamen, *The Spanish Inquisition,* (New American Library, 1966). p. 300.
3. For a more elaborate study of the background to motion see the following:
   E. J. Kijksterhuis, *The Mechanization of the World Picture* (Oxford University Press, 1961).
   George Sarton, *A History of Science,* Vol. I, (Harvard University Press, 1952), pp. 509 & 515.
   Siegfried Gredion, *Mechanization Takes Command* (Oxford University Press, 1948), pp. 14-17.
   James A. Weisheipl, *The Development of Physical Theory in the Middle Ages* (Sheed and Ward, 1959). This is particularly good.
   Cecil J. Schneer, *The Evolution of Physical Science* (Grove Press, 1960), pp. 70-73.
4. R. E. Givson, "Our Heritage from Galileo Galilei" in *Science* 145:1274 (1964).
5. I. Bernard Cohen, "Galileo" in *Scientific American* 81:45. Italics are mine. Aug. 1949.
6. *Ibid.*
7. Erwin Ponofsky, *Galileo as a Critic of the Arts,* (Martiners Nejhoff, 1954), pp. 36-37.
8. Georgio de Santilla, *The Crime of Galileo* (University of Chicago Press, 1955), p. 46.
9. Galilei Galileo, *Dialogue Concerning The Two Chief World Systems,* (University of California Press, 1962).
10. Cohen, *op. cit.,* p. 46.
11. Gibson, *op. cit.,* p. 1275.
12. Owsei Temkin, "Historical Reflections on The Scientist's Virtue" *Isis* 60:437 (1969).

Bibliography

Clagett, Marshall (ed.). *Nicole Oresme and the Medieval Geometry of Qualities and Motions.* University of Wisconsin Press, 1968.

Drake, Stillman and I. E. Drabkin (eds. & Trans.). *Mechanics in Sixteenth-Century Italy.* University of Wisconsin Press, 1969.

Seeger, Raymond J. *Galileo: His Life and His Works.* Pergamon Press, 1966.

XI

Newton and the World He Built

NEWTON, THE MAN

Read by Mr. Geoffrey Keynes at the Newton Tercentenary Celebrations at Trinity College, Cambridge, on 27 July, 1946. Reprinted by permission of the publisher, Horizon Press, from *Essays In Biography* by John Maynard Keynes. Copyright 1951.

... I believe that Newton was different from the conventional picture of him. But I do not believe he was less great. He was less ordinary, more extraordinary, than the nineteenth century cared to make him out. Geniuses are very peculiar. Let no one here suppose that my object ... is to lessen, by describing, Cambridge's greatest son. I am trying rather to see him as his own friends and contemporaries saw him and they without exception regarded him as one of the greatest of men.

In the eighteenth century and since, Newton came to be thought of as the first and greatest of the modern age of scientists, a rationalist, one who taught us to think on the lines of cold and untinctured reason.

I do not see him in this light. I do not think that anyone who has pored over the contents of that box which he packed up when he finally left Cambridge in 1696 and which, though partly dispersed, have come down to us, can see him like that. Newton was not the first of the age of reason. He was the last of the magicians, the last of the Babylonians and Sumerians, the last great mind which looked out on the visible and intellectual world with the same eyes as those who began to build our intellectual inheritance rather less than 10,000 years ago. Isaac Newton, a posthumous child born with no father on Christmas Day, 1642, was the last wonder child to whom the Magi could do sincere and appropriate homage.

Had there been time, I should have liked to read to you the contemporary record of the child Newton. For though it is well known to his biographers, it has never been published in *extenso,* without comment just as it stands. Here, indeed, is the makings of a legend of the young magician, a most joyous picture of the opening mind of genius free from the uneasiness, the melancholy and nervous agitation of the young man and student.

Figure XI-1. Newton.

For in vulgar modern terms Newton was profoundly neurotic of a not unfamiliar type, but—I should say from the records—a most extreme example. His deepest instincts were occult, esoteric, semantic—with profound shrinking from the world, a paralyzing fear of exposing his thoughts, his beliefs, his discoveries in all nakedness to the inspection and criticism of the world. "Of the most fearful, cautious and suspicious temper that I ever knew," said Whiston, his successor in the Lucasian Chair. The too well-known conflicts and ignoble quarrels with Hooke, Flamsteed, Leibnitz are only too clear an evidence of this. Like all his type he was wholly aloof from women. He parted with and published nothing except under the extreme pressure of friends. Until the second phase of his life, he was a wrapt, consecrated solitary, pursuing his studies by intense introspection with a mental endurance perhaps never equalled . . . I believe that the clue to his mind is to be found in his unusual powers of continuous concentrated introspection. A case can be made out, as it also can with Descartes, for regarding him as an accomplished experimentalist. Nothing can be more charming than the tales of his mechanical contrivances when he was a boy. There are his telescopes and his optical experiments. These were essential accomplishments, part of his unequalled all-round technique, but not, I am sure, his *peculiar* gift, especially amongst his contemporaries. His *peculiar* gift was the power of holding continuously in his mind, a purely mental problem until he had seen straight through it. I fancy his pre-eminence is due to his muscles of intuition being the strongest and most enduring with which a man has ever been gifted. Anyone who had ever attempted pure scientific or philosophical thought knows how one can hold a problem momentarily in one's mind and apply all one's powers of concentration to piercing through it, and how it will dissolve and escape and you find that what you are surveying is a blank. I believe that Newton could hold a problem in his mind for hours and days and weeks until it surrendered to him its secret. Then being a supreme mathematical technician, he could dress it up, how you will, for purposes of exposition, but it was his intuition which was pre-eminently extraordinary—"so happy in his conjectures," said de Morgan, "as to seem to know more than he could possibly have any means of proving." The proofs, for what they are worth, were, as I have said, dressed up afterwards—they were not the instrument of discovery. There is the story of how he informed Halley of one of his most fundamental discoveries of planetary motion. "Yes," replied Halley, "but how do you know that?" "Have you proved it?" Newton was taken aback—"Why, I've known for years," he replied. "If you'll give me a few days, I'll certainly find you a proof of it"—as in due course he did.

Again, there is some evidence that Newton in preparing the *Principia* was held up almost to the last moment by lack of proof that you could treat a solid sphere as though all its mass was concentrated at the center, and only hit on the proof a year before publication. But this was a truth which he had known for certain and had always assumed for many years.

Certainly there can be no doubt that the peculiar geometrical form in which the exposition of the *Principia* is dressed up bears no resemblance at all to the mental processes by which Newton actually arrived at his conclusions.

His experiments were always, I suspect, a means, not of discovery, but always of verifying what he knew already.

Why do I call him a magician? Because he looked on the whole universe and all that is in it *as a riddle,* as a secret which could be read by applying pure thought to certain evidence, certain mystic clues which God had laid about the world to allow a sort of philosopher's treasure hunt to the esoteric brotherhood. He believed that these clues were to be found partly in the evidence of the heavens and in the constitution of elements (and that is what gives the false suggestion of his being an experimental natural philosopher), but also partly in certain papers and traditions handed down by the brethren in an unbroken chain back to the original cryptic revelation in Babylonia. He regarded the universe as a cryptogram set by the Almighty—just as he himself wrapt the discovery of the calculus in a cryptogram when he communicated with Leibnitz. By pure thought, by concentration of mind, the riddle, he believed, would be revealed to the initiate.

He *did* read the riddle of the heavens. And he believed that by the same powers of his introspective imagination he would read the riddle of the Godhead, the riddle of past and future events divinely foreordained, the riddle of the elements and their constitution from an original undifferentiated first matter, the riddle of health and of immortality. All would be revealed to him if only he could persevere to the end, uninterrupted, by himself, no one coming into the room, reading, copying, testing—all by himself, no interruption for God's sake, no disclosure, no discordant breakings in or criticism, with fear and shrinking as he assailed these half-ordained, half-forbidden things, creeping back into the bosom of the Godhead as into his mother's womb. "Voyaging through strange seas of thought alone," not as Charles Lamb, "a fellow who believed nothing unless it was as clear as the three sides of a triangle."

And so he continued for some twenty-five years. In 1687, when he was forty-five years old, the *Principia* was published.

• • •

During these . . . [years at Trinity] of intense study, mathematics and astronomy were only a part, and perhaps not the most absorbing, of his occupations. Our record of these is almost wholly confined to the papers which he kept and put in his box when he left Trinity for London.

Let me give some brief indications of their subject. They are enormously voluminous—I should say that upwards of 1,000,000 words in his handwriting still survive. They have, beyond doubt, no substantial value whatever except as a fascinating sidelight on the mind of our greatest genius.

Let me not exaggerate through reaction against the other Newton myth which has been so sedulously created for the last two-hundred years. There was extreme method in his madness. All his unpublished words on esoteric and theological matters are marked by careful learning, accurate method and extreme sobriety of statement. They are just as *sane* as the *Principia,* if their whole matter and purpose were not magical. They were nearly all composed during the same twenty-five years of his mathematical studies. They fall into several groups.

Very early in life Newton abandoned orthodox belief in the Trinity. At this time, the Socinians were an important Arian Sect amongst intellectual circles. It may be that Newton fell under Socinian influences, but I think not. He was rather a Judaic monotheist of the school of Maimonides. He arrived at this conclusion, not on so-to-speak rational or sceptical grounds, but entirely on the interpretation of ancient authority. He was persuaded that the revealed documents give no support to the Trinitarian doctrines which were due to late falsifications. They revealed God was one God.

But this was a dreadful secret which Newton was at desperate pains to conceal all his life. It was the reason why he refused Holy Orders, and therefore had to obtain a special dispensation to hold his Fellowship and Lucasian Chair and could not be Master of Trinity. Even the Toleration Act of 1689 excepted anti-Trinitarians. Some rumors there were, but not at the dangerous dates when he was a young Fellow of Trinity. In the main the secret died with him. But it was revealed in many writings in his big box. After his death Bishop Horsley was asked to inspect the box with a view to publication. He saw the contents with horror and slammed the lid. A hundred years later, Sir David Brewster looked into the box. He covered up the traces with carefully selected extracts and some straight fibbing. His latest biographer, Mr. More, has been more candid. Newton's extensive anti-Trinitarian pamphlets are, in my judgment, the most interesting of his unpublished papers. Apart from his more serious affirmation of belief, I have a completed pamphlet showing up what Newton thought of the extreme dishonesty and falsification of records for which St. Athanasius was responsible, in particular for his putting about the false calumny that Arius died in a privy. The victory of the Trinitarians in England in the latter half of the seventeenth century was not only as complete, but also as extraordinary, as St. Athanasius's original triumph. There is good reason for thinking that Locke was a Unitarian. I have seen it argued that Milton was. It is a blot on Newton's record that he did not murmur a word when Whiston, his successor in the Lucasian Chair, was thrown out of his professorship and out of the University for publicly avowing opinions which Newton himself had secretly held upwards for fifty years past.

That he held this heresy was a further aggravation of his silence and secrecy and inwardness of disposition.

Another large section is concerned with all branches of apocalyptic writings from which he sought to deduce secret truths of the Universe—the measurements of Solomon's Temple, the Book of Daniel, the Book of Revelations, an enormous volume of work of which some part was published in his later days. Along with this are hundreds of pages of Church History and the like, designed to discover the truth of tradition.

A large section, judging by the handwriting amongst the earliest, relates to alchemy—transmutation, the philosopher's stone, the elixir of life. The scope and character of these papers have been hushed up, or at least minimized, by nearly all those who have inspected them. About 1630, there was a considerable group in London, round the publisher Cooper, who during the next twenty years revived interest not only in the English Alchemists of the fifteenth century, but also in translations of the medieval and post-medieval alchemists. There is an unusual number of manuscripts of the early English alchemists in the libraries of Cambridge. It may be that there was some continuous esoteric tradition within the University which sprang into activity again the twenty years from 1650 to 1670. At any rate, Newton was clearly an unbridled addict. It is this with which he was occupied "about 6 weeks at spring and 6 at the fall when the fire in the elaboratory scarcely went out," at the very years when he was composing the *Principia*—and about this he told Humphrey Newton not a word. Moreover, he was almost entirely concerned, not in serious experiment, but in trying to read the riddle of tradition, to find meaning in cryptic verse, to imitate the alleged but largely imaginary experiments of the initiates of past centuries. Newton has left behind him a vast mass of records of these studies. I believe that the greater part are translations and copies made by him of existing books and manuscripts. But there are also extensive records of experiments. I have glanced through a great quantity of this—at least 100,000 words, I should say. It is utterly impossible to deny that it is wholly magical and wholly devoid of scientific value; and also impossible not to admit that Newton devoted years of work to it. Sometime it might be interesting, but not useful, for some student better equipped and more idle than I to work out Newton's exact relationship to the tradition and MMS. of his time.

In these mixed and extraordinary studies, with one foot in the Middle Ages, and one foot treading a path for modern science, Newton spent the first phase of his life, the period of life in the Trinity when he did all his real work. Now let me pass to the second phase.

After the publication of the *Principia* there is a complete change in his habit and way of life. I believe that his friends, above all Halifax, came to the conclusion that he must be rooted out of the life he was leading at Trinity which must soon lead to decay of mind and health. Broadly speaking, of his own motion or under persuasion, he abandons his studies. He takes up University business, represents the University in Parliament; his friends are busy trying to get a dignified and remunerative job for him—the Provostship of King's, the Mastership of Charterhouse, the Controllership of the Mint.

Newton could not be Master of Trinity because he was a Unitarian and so not in Holy Orders. He was rejected as Provost of King's for the more prosaic reason that he was not an Etonian. Newton took this rejection very ill and prepared a long legalistic brief, which I possess, giving reasons why it was not unlawful for him to be accepted as Provost. But, as ill-luck had it, Newton's nomination for the Provostship came at the moment when King's had decided to fight against the right of Crown nomination, a struggle in which the College was successful.

Newton was well qualified for any of these offices. It must not be inferred from his introspection, his absentmindedness, his secrecy and his solitude that he lacked aptitude for affairs when he chose to exercise it. There are many records to prove his very great capacity. Read, for example, his correspondence with Dr. Covell, the Vice-Chancellor, when, as the University's representative in Parliament, he had to deal with the delicate question of the oaths after the revolution of 1688. With Pepys and Lowndes he became one of the greatest and most efficient of our civil servants. He was a very successful investor of funds, surmounting the crisis of the South Sea Bubble, and died a rich man. He possessed in exceptional degree almost every kind of intellectual aptitude—lawyer, historian, theologian, not less than mathematician, physicist, astronomer.

And when the turn of his life came and he put his books of magic back into the box, it was easy for him to drop the seventeenth century behind him and to evolve into the eighteenth-century figure which is the traditional Newton.

Nevertheless, the move on the part of his friends to change his life came almost too late. In 1689, his mother, to whom he was deeply attached, died. Somewhere about his fiftieth birthday on Christmas day, 1692, he suffered what we should now term a severe nervous breakdown. Melancholia, sleeplessness, fears of persecution—he writes to Pepys and to Locke and no doubt to others, letters which lead them to think his mind is deranged. He lost, in his own words, the "former consistency of his mind." He never again concentrated after the old fashion or did any fresh work. The breakdown probably lasted nearly two years, and from it emerged, slightly "gaga," but still, no doubt, with one of the most powerful minds of England, the Sir Isaac Newton of tradition.

In 1696, his friends were finally successful in digging him out of Cambridge, and for more than another twenty years he reigned in London as the most famous man of his age, of Europe, and—as his powers gradually waned and his affability increased—perhaps of all time, so it seemed to his contemporaries.

He set up house with his niece Catherine Barton, who was beyond reasonable doubt the mistress of his old and loyal friend Charles Montague, Earl of Halifax and Chancellor of the Exchequer, who had been one of Newton's intimate friends when he was an undergraduate at Trinity. Catherine was reputed to be one of the most brilliant and charming women in the London of Congreve, Swift and Pope. She is celebrated not least for the broadness of her stories, in Swift's *Journal To Stella*. Newton puts on rather too much weight for his moderate height. "When he rode in his coach, one arm would be out of his coach on one side and the other on the other." His pink face, beneath a mass of snow-white hair, "when his peruke was off was a venerable sight," is increasingly both benevolent and majestic. One night in Trinity after Hall, he is knighted by Queen Anne. For nearly twenty-four years he reigns as President of the Royal Society. He becomes one of the principal sights of London for all visiting intellectual foreigners, whom he entertains handsomely. He liked to have clever young men about him to edit new editions of the *Principia*—and sometimes merely plausible ones as in the case of Fatio de Duillier.

Magic was quite forgotten. He had become the Sage and Monarch of the Age of Reason. The Sir Isaac Newton of orthodox tradition—the eighteenth-century Sir Isaac, so remote from the child magician born in the first half of the seventeenth century—was being built up. Voltaire returning from his trip to London was able to report of Sir Isaac—" 'twas his peculiar felicity, not only to be born in a country of liberty, but in an Age when all scholastic impertinences were banished from the World. Reason alone was cultivated and Mankind could only be his Pupil, not his Enemy." Newton whose secret heresies and scholastic superstitions it had been the study of a lifetime to conceal!

But he never concentrated, never recovered "the former consistency of his mind." "He spoke very little in company." "He had something rather languid in his look and manner."

And he looked very seldom, I expect, into the chest where, when he left Cambridge, he had packed all the evidence of what had occupied and so absorbed his intense and flaming spirit in his rooms and his garden and his elaboratory between the Great Gate and Chapel. But he did not destroy them. They remained in the box to shock profoundly any eighteenth- or nineteenth-century prying eyes . . .

In 1888, the mathematical portion was given to the University Library at Cambridge. They have not been indexed, but they have been edited. The rest, a very large collection, were dispersed in the auction room in 1936 by Catherine Barton's descendant . . .

As one broods over these queer collections, it seems easier to understand—with an understanding which it not, I hope, distorted in any other direction—this strange spirit, who was tempted by the Devil to believe, at the time when within these walls he was solving so much, that he could reach *all* secrets of God and Nature by the pure power of mind—Copernicus and Faustus in one.

STUDY AND DISCUSSION QUESTIONS

1. What is Cohen's view of Newton?
2. How does Cohen's view differ from the one presented by Keynes?
3. What was Newton's physical appearance?
4. Why does Keynes call Newton the last of the magicians?
5. What was Newton's peculiar gift?
6. In what way was Newton a Neo-Platonist?
7. What is an anti-Trinitarian?
8. What was Newton's dreadful secret?
9. What difference does it make where Arius died?
10. How does Newton's life change after the publication of the *Principia?*
11. Who were his contemporaries?
12. What were Newton's contributions?
13. How does Newton's childhood account for his characteristics?

NEWTON'S WORLD SYSTEM

The Copernican system had not only fractured the total Aristotelian system but had displaced man as a central figure in the universal design. As more persons came to believe in the Copernican system they too were aware of man's lost favor in the Divine scheme. Among those aware of this problem was René Descartes (1596–1650) and his whole philosophy was an attempt to reestablish man in a central position. He cannot replace man in a physically central position but he can make man a central part of the Divine scheme.

Since he could not affect man's physical position he turned inwardly and began by doubting all knowledge that he possessed. He finally reached the ultimate position that he could not doubt that he thought. From this he could then deduce the state of his existence or briefly stated, "I think, therefore I am." He gradually evolves a dualism consisting of the dead world of matter, which has motion and extension, and a world of the soul and mind. Although the two worlds are totally separate they are kept in a relationship by the will of God.

God is a prime mover and caused all motion to be. The motion is maintained and transmitted through an invisible substance called "ether." Space is not nor can it be empty but must be filled with the transmitting material. These moving regions of ether are in closed systems like whirlpools or a vortex and the contact between vortices carry the visible planets and stars. This basic idea was extended to explain chemistry by the revival of the Greek concept of atom and even light by the use of a corpuscular theory. The total range of experience in the physical world was explained by this system. Heat, color, motion, growth and decay can all be explained in terms of this matter in motion.

Man was reestablished because he was the only one of God's creations that partook of both the world of things and the world of mind and soul. This meant that, save for man, the whole of physical creation, the material world, was a giant machine.

The machine (animal) plus the soul was a man. Since animals were machines lacking souls, they could feel nothing and the squeaks they might emit were like the creaking and groaning of a Flemish mill—mere mechanical camouflage. For a period there were Cartesian [those that followed the ideas of Descartes] philosophers who would vivisect an animal before their classes to marvel at the precise mimicry of suffering that the machine displayed. For if one were not protected by the armor of deductive logic and did not know that the beast lacked a soul, one would have sworn that the animal was in pain.[1]

Descartes' philosophy was so successful that it remained popular for over a century. It was this system which was the accepted theory over most of Europe that Newton's works had to overcome.

Newton was very aware of Descartes' theory and mathematics as well as the planetary laws of Kepler and the works of Galileo, Huygens, Halley, Hooke and others. He recognized that his achievements were the culmination of the works of others. In a letter to Robert Hooke (Dtd. Feb. 5, 1675/6) he wrote, "If I have seen further (than you and Descartes) it is by standing upon ye shoulders of Giants (Kepler and Galileo e.g.)." Ideas about gravity had been proposed by a number of persons and the crux of the important part of the law of universal gravitation was current in Newton's contemporaries.

Newton could not accept the ideas of Descartes in view of Galileo's proof for the law of inertia. He felt that the proper question was to ask what deflected the planets from a straight line course rather than to ask what kept the planets in motion. He assumed that the same force caused an apple to fall must also act on other bodies. He extended the idea of projectile motion from the top of a very high mountain (one too large to exist on earth) to find out the force needed to keep the moon from falling into the earth.

There were two sticky problems related to the question of gravity. The first was by how much does the force of gravity decrease as one body separates from another and the second was where can the distance be measured. That is, can the distance be measured from the edge of the body, from the center, or from a point midway between the center and edge?

Copernicus has an idea of gravity because he said that weight is a certain material appetite which God gave pieces of matter so that they would go together to form a globe. Borelli, an Italian astronomer, in 1666 had speculated on the prospect of some central forces. Both of these, however, were guesses and unsubstantiated by evidence. Gilbert had proposed a sort of magnetic effuvia to hold the planets on their course and Galileo had been working with a kind of circular inertia to explain gravity.

The idea that the sun must hold the planets as the inverse square of the distance had been proposed by

Bullialdus about 1646 and by 1684, the date that Newton returned to his studies of gravitation several persons were discussing the idea. Sir Christopher Wren, an architect who helped rebuild London after the Great Fire; Robert Hooke, who helped Wren and also contributed to the discovery of a number of physical laws; and Edmund Halley, later Astronomer Royal, were contemporaries of Newton and were all convinced that the key to universal gravitation was the inverse square relationship. They felt that the relationship was implied in the second and third laws of Kepler, as indeed it was but they couldn't prove it.

What were the facts that were available at the time Newton was working on the motion problems? It was known that the moon described an elliptical orbit around the earth that was nearly circular. This was not a fact of direct observation, of course, but was an interpretation of observed facts that no one doubted. The radius of the moon's orbit was known to be about sixty times the earth's radius. The other facts consisted of the slightly more than twenty-seven days of the moon's period of revolution and that an object dropped near the earth's surface will fall about sixteen feet per second if the friction of the air is ignored.

In order to study motion he also needed to propose methods of measuring forces. Several of these are almost stated in the works of Galileo but Newton is usually given credit for them. The laws of motion as they would be stated today run as follows: The first law—Every body remains at rest or moves with a steady speed in a straight line unless it is forced to change its speed by some force acting on it. The second law states that when an outside force acts on a body that it will change direction and speed in the direction of the applied force—it goes where and how it is pushed. The third law is original with Newton and states that to every action there is an equal and opposite reaction.

We shall examine a reconstruction of Newton's solution of the moon "falling" towards the earth as an example. We will not consider his proofs for the inverse square law or that the mass may be considered as being at the center of a body. These are essential but the geometric reconstruction of his proofs would be tedious.

In the diagram:

L = position of the moon
L' = position of the moon one minute after L
S = position of the moon if it traveled for one minute in a straight line
C = earth's position

The other letters are for puposes of geometric construction and discussion.

The circumference of the moon's orbit was found to be 7,603,200,000 feet and the time it took for one orbit was 27 days, 7 hours, 43 minutes or 39,393 minutes. By dividing the distance by the time the speed of the moon was calculated at 193,250 feet/minute. Thus in one minute the moon traveled 193,250 feet—the distance from L to L'.

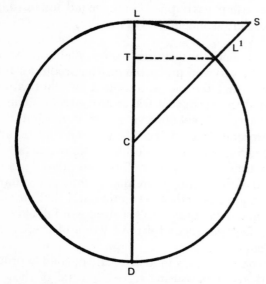

Figure XI-5. Moon Falling.

From geometry theorems, the distance that the moon falls (S to L') can be calculated:

$$\frac{LT}{TL'} = \frac{TL'}{TD}$$

Cross multiplying this gives,

$$(TL')^2 = (TD)(LT)$$

Using near approximations and substituting terms we get,

$$(LL')^2 = (LD)(SL')$$

Dividing both sides by LD,

$$\frac{(LL')^2}{LD} = SL'$$

or

$$\frac{(193,250 \text{ feet})^2}{7,603,200,000 \text{ feet}} = SL'$$
$$\frac{}{pi}$$

SL' comes out to be 15.43 feet. Newton had calculated 16.1 feet.

He proved geometrically that all the force that a body exerts may be measured as if all the mass of the object were at the center of the object. From these and other considerations, Newton is able to derive his law of universal gravitation: every particle of matter in the universe attracts every other particle of matter in the universe with a force which is directly proportional to the product of the masses and inversely proportional to the square of the distance between their centers. All of the proofs in Newton's work are geometrical, but today his law is represented by the formula:

$$F = G \frac{m_1 m_2}{R^2}$$

where F stands for force, m for mass, R for distance, and G for gravitational constant which was not determined until 1798. All of this and his laws of motion was presented in his book, *Mathematical Principles of Natural Philosophy,* usually called the *Principia* published in Latin in 1687.

All that Newton needed to do to deduce Kepler's third law was to combine his laws of motion with the universal law of gravitation. A circular orbit is used in the following to avoid using calculus (although the same thing would be shown) and algebra rather than geometry is used to make it less messy. The distance between the planet and the sun is R; the planet is m_2; the sun is m_1; the time for the planet to go around the sun is T; the velocity or speed of the planet is v; the circumference of a circle is 2 pi R. v is also distance divided by time but the distance here is orbit of the planet or the circumference and the time is the time it takes to go aorund the sun or T so that v is equal to 2 pi R divided by T. The force of the planet in motion is equal to $\frac{m_2 v^2}{R}$; thus this must be equal to the force of gravity or $G \frac{m_1 m_2}{} = \frac{m_1 v^2}{}$

(1). Since $v = \frac{2piR}{T}$

we can substitute it in the right hand side of equation

(1) in place of v and get: $G \frac{m_1 m_2}{} = \frac{m_2 (2piR/T)^2}{R}$

Then squaring and clearing we get:

$$G \frac{m_1 m_2}{R^2} = \frac{4 pi^2 m_2 R^2}{T^2 T}$$

Then putting all the T's and R's on one side we get:

$$\frac{R^3}{T^2} = \frac{G m_1}{4pi^2}$$

The left hand equation is Kepler's Third Law and the fact that it shows up in this manner is a strong support for Newton's universal law of gravitation.

Descartes' theory of gravity, sometimes called, "the most beautiful romance of physics—too bad it isn't true," slowly died before Newton's law of universal gravitation which would predict and which brought understandable order to the cosmos.

Read how Henry Cavendish (1731–1810) determined the value of G in Morris H. Shamos (ed.), *Great Experiments in Physics,* Henry Holt, New York, 1959.

STUDY AND DISCUSSION QUESTIONS

1. Why is Descartes' explanation more satisfactory and better accepted on the continent than is Newton's at first?
2. What is the relationship of the laws of motion to the universal law of gravitation?
3. Why must all phenomena whose effects are noted at a distance be governed by the inverse square law?
4. What is Newton's universal law of gravitation?
5. Why was Newton's law accepted if it were lacking experimental proof?
6. What factors did Cavendish consider when making his apparatus?
7. What experimental errors or factors does Cavendish try to exclude? How?
8. What density value for the earth does Cavendish obtain? How does this compare with the value accepted today?
9. How may the mass or weight of the earth be computed by knowing the density? Is the density uniform?

FOOTNOTES

1. Cecil J. Schner, *The Evolution of Physical Science,* (Grove Press, 1960), p. 99.

THIS IS THE WORLD THAT NEWTON BUILT

The direct application of Newton's Universal Law of Gravitation was limited. It did permit the rather precise prediction of the daily positions of the sun, moon, and planets which did assist navigation; and it explained tides for the first time. At a later date its application made possible the discovery of the planets Neptune (1845) and Pluto (1930). Of considerably more importance were the indirect effects, not only on various disciplines of science but also, on practically every other field.

What is sometimes called the Newtonian World Machine is really a combination of Newtonianism and Mechanism. Mechanism, the belief that everything can be explained in purely mechanical terms, has its beginning with Harvey and progresses in an almost unbroken stream through Galileo, Borelli, Malphigi, and Bellini from the 17th Century through the 18th and into the 19th Century.

William Harvey (1578–1657) in 1628 published a book describing the circulation of the blood and the function of the heart. His explanation was a mechanical one in which the heart was likened to a pump and the circulation as one would describe through pipes. Galileo, as we have already seen, studied the solar system and motion in mechanical terms. Borelli, Malphigi and Bellini were biological mechanists. Descartes' system is a fully mechanistic one as is Boyle's explanation of the behavior of gasses and chemical actions.

The blending of science and mathematics had already been successful in a number of areas. In 1643, Torricelli, Galileo's pupil, invented the barometer and weighed the atmosphere. Pascal and Boyle expanded these studies into other fluids. Facts about the atmosphere and pressure were applied by the development of the Newcomen steam engine within twenty years. In the science of optics, which was originated by Kepler and Descartes, there were systematic developments by Christian Huygens and Newton. Before the end of the century, 1695, Olaus Roemer (1644–1710) was able to measure the speed of light.

Newton's work is both the capstone and the beacon in the application of mathematics to physical phenomena. With one law he explains the spinning planets, the motion of the cosmos, and the falling of a sparrow's feather. God may know when it falls but Newton describes its motion.

In compiling his work, Newton uses mathematics and a few rules of reasoning. These are (1) to reduce the number of causes of natural events to those that are true and sufficient to explain them; (2) to assign to the same kinds of events the same causes; (3) to assign the same causes to similar events everywhere since properties of matter are the same everywhere; (4) to assign causes based on experiment since it is only through experiment that we can know the properties of matter. This is the methodology of what becomes the physico-mathematical method.

Science is to be founded, like geometry, upon a few positively true axioms. From these axioms it would be possible to deduce every law of nature. An axiom is something that we know to be true intuitively, with absolute certainty, that requires no proof. Examples of these from Euclidian geometry are: the whole is equal to the sum of its parts, and a straight line is the shortest distance between two points. It was felt that these axioms existed not only in science but in every field of human endeavor—morals, politics, ethics, religion. These axioms were suggested to man if he looked at the natural behavior of any phenomena and once formulated into natural law it was possible to then deduce and predict any phenomena.

The idea of seeking natural laws spread quickly to many areas and what was called natural history (botany, zoology, geology, and chemistry) was avidly studied by all from kings to commoners. They were in the collecting stage which preceeds classification, order and the application of mathematics. Naturalists such as Georges Louis Leclerc comte de Buffon (1707–88) who wrote his great Natural History which covered all animals in 44 vols.; Jean Baptiste Pierre Antoine de Monet chevalier de Lamarck (1744–1829), who classified invertebrates, worked with some plants, and developed a theory of evolution; Baron Georges Leopold Chretien Frederic Dagobert Cuvier (1769–1832) worked with zoology and comparative anatomy; Karl von Linne (Linneaus) (1707–78) the great systematizer of botany; were all devoted to collecting and classifying specimens. Geologists including James Hutton (1726–97), Abraham Gottlob Werner (1750–1817), and William Smith (1769–1839) mapped and classified rocks, formulated theories, and prepared the way for Charles Lyell in the next age. In the chapter on chemistry it will be noted that a number of elements are collected in the 18th Century including work by Cavendish, Priestley and Lavoisier—again the collection and classification of items. The classification at this time only included element, compound and mixtures and these were not totally clear until the next century with Dalton's work.

In philosophy, the Newtonian World Machine raised several basic questions again. How do we know? In Newtonian thought, we know by experiment—in other words through the senses or extensions of them. But since the senses are defective then our knowledge must be defective. This gives rise to a certain amount of skepticism. If knowledge is gained by experiment only then this leads to the concept that the real world is the material world. This is known as the doctrine of materialism—i.e. that the only reality is matter reacting in accord with fixed mathematical laws.

Materialism leads naturally to its corollary—determinism. If there is only matter behaving in accord to fixed mathematical laws, then the future is already established. Applied to man, as Voltaire was quick to see, there is no free will since the human will is determined by external and internal physical causes. Voltaire stated: "It would be very singular that all nature, all the planets, should obey eternal laws, and that there should be a little animal, five feet high, who, in contempt of these laws, could act as he pleased, solely according to his caprice."

If all matter is subject to immutable laws of nature then as one enthusiastic follower stated:

> If we knew truly the relative position of all phenomena which constitute the real universe, we could at this moment with a certainty equal to that of the astronomers tell the day, the hour, the minute at which, say England will evacuate India, when Europe will have burned the last lump of coal, when such a criminal, still to be born will assassinate his father, when such a poem, yet to be conceived will be composed. The future is contained in the present as all the properties of a triangle are contained in its definition.[1]

This consequence of mechanism had been foreseen and summarized by Laplace in his *Philosophical Essays on Probability:*

> An intelligence that, at a given instant, knew all the forces by which Nature is moved and the respective situation of the entities of which it is composed, would, if besides it were great enough to submit these data to analysis, embrace in a single formula the movement of the largest planets in the universe and those of the lightest atom. For this intelligence nothing would be uncertain, and the future as well as the past, would be present in its eyes.[2]

Newtonian mechanics showed that the planets and other bodies moved according to the same laws of motion that held on earth. Galileo had already shown that the moon and other planets seemed to be of the same material as the earth. The mystery of the sky was removed and the crystalline spheres that Copernicus had shattered were transparent. There was no God nor angels in the bright light that could be seen by the vision of reason.

If nature is uniform and acts through immutable laws there can be no miracles. God, if there is a God, is the great mathematician and prime mover. God had been called the great mathematician by Kepler, Newton, Descartes, Leibnitz and others. For these men, however, God was the creator and the "deus ex machina" or rectifier when necessary to His creation, (e.g., to correct the motion of mercury). To men of this type, and they still exist, the wonder and nature of God can be found by examining the nature and wonder of His creation. Most men of the 18th century are led through reason to Deism.

Deists range in their beliefs but most of them reject established religions. In the mild state they are seeking a return to the pure religion of God undefiled by corruption of man's additions—in a sense an extension of the Protestant movement. The stronger position which borders on atheism was stated by Laplace in answer to a question put to him by Napoleon as to why he made no use of God in his book on celestial mechanics, "I have no need for such a hypothesis."

The true believer, as pointed out by Eric Hoffer,[3] is intolerate. Our forefathers came here for religious freedom—their own. This freedom which they wanted they did not extend to others. Probably more men have been killed because someone believed himself to be right than for any single cause. As Montaigne describes it: "It is setting a high value on one's opinions to roast men on account of them." It was because the age of reason refused to accept beliefs on blind faith alone and because of the multiplicity of Protestant sects that a measure of toleration grows.

The weakening or denial of religion in addition to giving toleration also revived the issues over moral law. Since most morals prior to the 18th Century was founded in religion and hence in the concept of avoidance of sin. The concept of sin, largely a Christian concept, is founded in Adam's fall. This also explains the presence of evil in the world. However, the growth of materialism and determinism not only eliminate sin, and since man is not a free agent, he is not responsible for his actions. Sin is often equated to pleasure and although it doesn't stop man from his actions it keeps him from enjoying it. The whole question of virtue vs. vice is explored with some reaching the conclusion that vice is nice, and that virtue is its own reward—its only reward. This conclusion is reached by the Earl of Shaftesbury, La Mattrie; and perhaps reached its summit, or depths depending on your viewpoint, in *Justine* (1791) by Donatien Alphonse François comte de Sade (1740–1814). The problem is also explored in *The Bridge of San Luis Rey* (1927) by Thornton Niven Wilder (1897–   ).

It should not be assumed from this discussion that science is responsible for weakening the position of religion by itself or even that it was the main factor, but certainly it contributed. For an examination of that particular problem see Kenneth Hamilton's book.[4]

If the moral code is not to be dependent on religion then where is it to come from? Several sources were suggested. Locke claimed that it was possible to give a mathematical demonstration of morality. Some claimed that there was a sense of right and

wrong within us—a natural sense in the same way that one can choose the beautiful from the ugly. A third way, and one which was to become dominant in the 18th Century and was to form the beginning of sociology and anthropology, was to examine the behavior of men in a natural state. Thus the ultimate appeal was to cultures which were undeveloped and uncorrupted by the manners and rituals of civilization. The noble savage was exalted and imitated to seek those natural laws governing the behavior of man. Once these laws were found man could do away with the cultural garbage of the centuries which only obscured the actions which were right and proper to man. Further, man would obey these laws easier because they were natural to him.

The same approach could be seen in the economic sphere. Remove the interferring laws of man and the natural economic laws will become apparent and

Figure XI-3. de Sade.

effective. Under the leadership of François Quesnay (1694–1774) the physiocrats sought to have the government restrict itself to the enforcement of the necessary laws of human society. These simple and axiomatic laws were three: property, security, and liberty. The 18th Century idea of liberty was the right to own property. The policy of the government should be *Laissez-faire*—no interference with business. There should be no restrictions on manufacturing and no tariffs. Other than the protection of property, the only function of government was to promote business—(e.g., building roads and canals). Adam Smith (1723–90) pushed much the same view in England in his book *An Inquiry into the Nature and Causes of the Wealth of Nations* (1776).

The early throes of industrialism in England were already apparent but the poverty and misery of a large group of the population was easily explained from "Natural Laws." Thomas Robert Malthus (1766–1834) wrote a book in 1798, *An Essay on the Principle of Population,* in which he contended that such a situation is unavoidable. The increase of population is a geometric one while the means of subsistence is only arithmetic. If you try to reduce the misery and poverty by increasing wages, charity, or legislation you only increase the number of children. Thus the factory owner could argue that by holding wages down he was reducing misery and poverty by decreasing the number of the miserable and poverty-stricken. The rich get richer and the poor have children—all according to a natural law.

The founders of our country were influenced by three strands: the thought and logic of the ancient Greeks, the political writings and events of the 17th Century England, and the Newtonian system. A few quotations from the Declaration of Independence will suffice to show the latter point. ". . . the separate and equal station to which the *Laws of Nature* and of *Nature's* God entitle them . . . We hold these truths to be self-evident, (axiomatic) that all men are created equal . . ." The very founding documents of our country are written as a mathematical treatise—a reflection that natural laws can be found in all fields of human endeavor by the application of reason.

Attempts, too, were made to examine the past and to ascertain the laws that govern nations and peoples over a long period of time. One of the earliest was that of Edward Gibbon (1737–94) who composed the *Decline and Fall of the Roman Empire.* It appeared in six volumes between 1776–83 and tried to show, in a rational manner the reasons leading to the fall of Rome. In the same manner, Karl Marx (1818–83) took the economic laws of the 18th Century and blended them with historical laws to show the stages of social development which would culminate in pure communism. This was published as *Das Kapital* Vol. I (1867) and Vols. II and III (1885–94).

In order to make language more precise, like mathematics, it was necessary that the meaning of words be clear and spelled the same. (Prior to the 18th Century, one of the marks of an educated man was the variety which he could introduce in his spelling.) The standardization of English was begun by Samuel Johnson (1709–84) in *A Dictionary of the English Language* (1755). The number of scientists in the 18th Century who were also Puritan were a factor towards advocating plain and unadorned speech.

Figure XI-4. Gibbon.

Not only was the language itself to be changed but also the style. Whereas even science had been written in the form of poems or dialogues before the 18th Century it then changed to the form of the plain essay or satire. Poetry was dealt a mortal blow from which she has never fully recovered.

The 17th Century poets either ignored science or mathematics while the 18th Century lauded Newton, mathematics and science. By the 19th Century many writers felt that there was something more to a rainbow than the description of the refracted wave lengths of light, and a reaction against a purely mechanical world had set in.

Even in the fields of architecture, landscaping and art the influence of Newton was to be seen. Emphasis was placed on geometric balanced structures and subjects which would stimulate the mind rather than the emotions. The rules for beauty in art were to be found either by the observation of nature or by the establishment of *á priori* laws of aesthetics in the manner of mathematical systems. Here, too, a later reaction would form against the mechanical reproduction of beauty.

The overall results which we can still find in many fields was the ideal of seeking for and finding those natural laws which are mathematical in nature and which will describe and predict phenomena in all areas. That these laws applied with reason will overcome all problems to the progress of man. It also contributed an overwhelming feeling of optimism that a materialistic man can deal effectively with his environment.

STUDY AND DISCUSSION QUESTIONS

1. What areas of science were quantified?
2. What change of approach was indicated to physiologists and psychologists?
3. What is the mechanistic view?
4. What was the chief characteristic of this new approach to knowledge?
5. What is an axiom? What axioms are used in physics?
6. What is the doctrine of materialism?
7. In materialism what happens to the soul and mind?
8. What is determinism? Is determinism a necessary corollary to materialism?
9. What is free will and is it subject to determinism?
10. Today the phrase "God is dead" is heard. How did the discovery of natural and mathematical laws contribute to this statement?
11. How does revelation fit into a world of reason?
12. Can miracles exist in a world of natural laws?
13. What becomes the basis for moral law and ethics without religion?
14. In what three ways did science influence the words in the English language?
15. In what ways was English prose reformed? Relation to Puritanism?
16. How were the discoveries of Torricelli, Pascal, and Boyle applied?
17. What is the importance of the universal law of gravitation?
18. How were the ideas of Newton spread?
19. What is the method of Newtonian science?
20. In what ways is there conflict between the rational and experimental methods?

FOOTNOTES

1. Richard L. Schanck, *The Permanent Revolution in Science,* (Philosophical Library, 1954), pp. 17-18.
2. *Ibid.* p. 16.
3. Eric Hoffer, *The True Believer* (Harper & Row, 1951).
4. Kenneth Hamilton, *God is Dead* (1966), especially pp. 24-26.

Bibliography

Kline, Morris. *Mathematics in Western Culture.* (Oxford University Press, New York, 1953.) Chapters 16-18. Also see J. H. Randall, Jr., "The Newtonian World Machine" in A. B. Arona and A. M. Bork (eds.) *Science and Ideas.* Prentice-Hall, 1964. pp. 138-165.

Mill, John Stuart. *Auguste Comte and Positivism.* Ann Arbor Paperback, 1961. pp. 6-9, 12, 14-15.

Randall, John Herman, Jr. *The Making of the Modern Mind.* Houghton-Mifflin Co., 1940. pp. 253-386.

Whitehead, A. N. "The Romantic Reaction" in *Science and the Modern World.* Macmillan Company, 1925. pp. 109-138.

XII

Thermodynamics and Heat Engines

One of the greatest innovations of man was to devise machines to use energy other than the muscles of animals, including man. It has made the machine the servant of man or, as some claim, man the servant of the machine. It is true that man had used the wind to drive his ships for thousands of years and occasionally waterwheels had been used in antiquity, but the great flowering of windmills and waterwheels happens toward the end of the Middle Ages. They are used to drain swamps, mines, to grind grain, and saw wood. They do not disappear suddenly. In this country in the last century gristmills and sawmills were frequently water powered. There are remains of these mills to be found in most older cities in the United States.

The application of heat engines to these, and other machines, presents an interesting interaction among pure science, technology, and society. There is no simple relationship. We might expect, as we frequently do today, that a discovery in pure science will find an application in technology which, in turn, will have an influence on society. This is not the case, at least, in this particular problem. A number of technological advances were made in the heat engine before the nature of heat was understood.

Let us examine the nature of heat first. Heat is one form of energy in a condition of transition. It may be produced in a number of ways: mechanically (e.g., friction), chemically (e.g., burning of coal), electrically, nuclear (e.g., conversion of matter to energy). It may be transmitted by conduction, convection, or radiation.

It is a common mistake to confuse the quantity of heat with the temperature. To use an illustration here, if we have water at 212°F and steam at the same temperature, the burn you would receive would be more severe from the steam because it contains more heat than does the water. The temperature is an indication of the speed of the atoms. Almost all substances can exist in one of five states (solid, liquid, gas, colloidal, plasma) depending, in part, on the temperature. Temperature is measured with a thermometer devised by Galileo about 1638.

Between 1592 and 1742, the thermometer developed into its present form. In 1779, there are nineteen different scales for thermometers which have since boiled down to three in common use. The scale which we and other English speaking people use is the fahrenheit, named for its German inventor Gabriel Daniel Fahrenheit (c. 1704). The scale is based on two erroneous assumptions: (1) that the temperature of a normal person is 96° and that the zero point is the lowest temperature possible with a mixture of salt and water. Scientists all over the world and the French use the centigrade scale devised by a Swede named Celsius (sometimes called the Celsius scale) in 1742 with the zero at the freezing point of water the boiling point at 100°. R. A. Ferchault de Reaumur, a Frenchman, devised a scale named for him in 1731, used principally in Germany, with the freezing point of water at its zero and the boiling point at 80°. The two standard points for calibrating thermometers today is the freezing and boiling points of pure water. Why do you suppose water is frequently the standard —(e.g., density, thermometers, etc.?)

Joseph Black (1728–1799) of the Universities of Glasgow and Edinburgh was the first person to make the distinction between temperature and the quantity of heat. It is interesting to note here that Watt was the scientific-instrument maker at the University of Glasgow and in spite of the fact that they were both at the same institution and are supposed to have had some connection with each other about the steam engine, neither realized that the steam engine was merely a special device for converting heat into mechanical energy could be changed one into the other and had a proportional relationship to each other.

Nearly fifty years before the idea of equivalence of heat and work was recognized, Benjamin Thompson (1753–1814), better known as Count Rumford, in 1798 performed a numerical experiment relating work and heat. He was supervising the boring of a brass cannon and he noticed that the amounts of heat being produced were disportionately large to be accounted for by heat theory in which caloric was sup-

Figure XII-1. Black.

Figure XII-2. Joule.

posed to be squeezed out of the metal. He measured and found that the quantity of heat was the same in the shavings as in the brass gun. He then performed an experiment in which he weighed a quantity of water and determined how much the temperature changed in a specific time using one horse to turn the boring equipment. (Hence, one horsepower.) He then calculated the mechanical equivalent of heat. However, he did not recognize the significance of what he had done, and the value he obtained was quite crude—1034 ft-lbs./B.T.U. (Today's accepted value is 778 ft-lbs./B.T.U.)

In 1842, using data which had long been known about the amount of heat needed to keep expanding air at the same temperature, Julius Robert Von Mayer (1814–1878), an obscure German physician, published an article in which he gave the relation between heat and work at 725 ft-lbs./B.T.U. In the same article he stated what has since become the law of conservation of energy—i.e. that energy cannot be destroyed, but can only change its form. He was treated as a freak or a madman and at one time became so disturbed that he tried to commit suicide. His work was not recognized nor appreciated until he was awarded the Copley medal in 1871.

He was eventually given his due but, the credit for determining the value for the mechanical equivalent of heat was given to James Prescott Joule (1818–1889). Joule was a brewer in Manchester, England and a scientist by avocation. He made the determination a number of times and to show the extent to which he would go he took a thermometer with him on his honeymoon in Switzerland to measure the temperature at the top and bottom of an eight-hundred foot waterfall that he and his bride were going to. He stated his value of 772 ft-lbs./B.T.U. and that the amount of heat produced in friction is proportional to the work expended. He also made a statement on the conservation of energy, in a paper in 1843.

"I shall lose no time in repeating and extending these experiments being satisfied that the grand agents of nature are, by the Creator's fiat, indestructible; and that wherever mechanical force is expended, an exact equivalent of heat is always obtained."[1]

Earlier than the work of Joule and Mayer was that of Nicolas Leornard Sadi Carnot, (1796–1832) a French engineer who in 1824 noted in his *Reflections On The Motive Power of Heat* that in heat engines the heat always goes from one body where the temperature is elevated to another where it is lower. In other words the efficiency of an engine is dependent on the incoming temperature of the steam and the temperature to which the exhaust can be lowered. From a practical standpoint then, the steam should be introduced at a very high temperature but the only way in which this can be done is to increase the pressure since the temperature of steam is dependent on the pressure at which it is formed. Of equal importance is to lower the exhaust temperature as low as possible, usually done by using water

Figure XII-3. Carnot.

to condense the steam. This was the advantage that Watt obtains by using a separate condenser.

Carnot's work remained relatively unnoticed until 1849 when William Thomson (1824–1907) published a paper, "Carnot's Theory of the Motive Power of Heat with Numerical Results Deduced from Regnault's Experiments on Steam." Thomson was no mere press agent. He had graduated from the University of Glasgow when he was 10 and was appointed Professor of Natural Philosophy when he was 22. He knew Joule and Henri Victo Regnault (a French Chemist). He was later, 1866, knighted Lord Kelvin.

Figure XII-4. Lord Kelvin.

Thomson could not reconcile the work of Joule and Carnot because he was hung up on the old Caloric theory of heat. He finally found the solution but before he could publish his ideas Rudolf Julius Emmanuel Clausius (1822–1888) showed the idea in his "On the Moving Force of Heat and the Laws of Heat which may be Deduced Therefrom." Clausius was a German mathematical physicist who developed the term "entropy" which is a measure of statistical disorder applied to what is called the Clausius statement of the Second Law of Thermodynamics. It states, "It is impossible for a self-acting machine, unaided by any external agency, to convey heat from one body to another at a higher temperature." This denies the possibility of a perpetual motion machine. The Kelvin statement of the Second Law of Thermodynamics (which is equivalent to that of Clausius') states, "It is impossible, by means of inanimate material agency, to derive mechanical effect from any portion of matter by cooling it below the temperature of the coldest of the surrounding objects."

The First Law of Thermodynamics, sometimes called the law of conservation of energy, was stated by Kelvin as, "when equal quantities of mechanical effect are produced by any means whatever from purely thermal sources, or lost in purely thermal effects, equal quantities of heat are put out of existence or are generated."

There is a third law of thermodynamics, but as it is more related to chemistry, it will not be considered.

THE DEVELOPMENT OF THE STEAM ENGINE

No one knows who first confined air or water in a vessel and applied heat; but that person was in for a big surprise. The development of the heat engine must be dated as beginning with the first faltering step to confine a heated fluid. Any development passes through certain distinct phases. The first phase is usually as a toy, a gimmick, or magic. The second is a competitive one in which the new must overcome and replace the old. The third phase is one of supremacy in which the device is applied everywhere in many ways. While the fourth phase is one of replacement and decline as a new device begins to replace it.

"Great inventions are never, and great discoveries are seldom, the work of any one mind. Every great invention is really either an aggregation of minor inventions, or the final step of a progression."[2] This statement is certainly true of the steam engine as it is hoped that this section will show. It should be noted that our discussion here is of a scanty nature.

No attempt has been made to even list all of those who contributed to the development of the steam engine nor to describe the numerous intermediary steps in the development. Our purpose here is to very briefly sketch some of the developments and to relate these developments to the laws of thermodynamics.

The first written evidence of heat producing motion in fluids is to be found in Heron's manuscript, *Spiritalla seu Pneumatica,* occurring sometime between the second century B.C. to the third century A.D. in Alexandria, Egypt in the heyday of what is referred to as the Hellenistic Age or what might be called the Greeks abroad. A number of devices are described but it is not known Heron invented them himself or, as is suspected, that he was simply describing devices well known to the time and invented by others. We will examine two of these devices, as examples, in some detail.

The Aeolipile or Ball of Aeolus was a toy. It was a sphere with a number of small tubes bent to 90 degree angles. Steam was fed into the sphere through hollow supporting tubes and jetted out through the bent tubes which caused the sphere to spin.

The other is what might be called a Rube Goldberg device, whose purpose was to mystify the temple-attending populus. In front of the temple was a hollow altar. When a fire was built on the altar, supposedly as an offering, the heat forced the air

from inside the altar into a sphere containing water. The hot air as it expanded forced the water through a siphon tube into a bucket. The weight of the bucket and water pulled on two ropes attached to and wrapped around two cylinders causing them to rotate. The cylinders were fastened to the temple doors and opened them. When the fire was extinguished, the air in the altar cooled and contracted. The air pressure over the bucket forced the water out of the bucket through the siphon tube back into the sphere. As the bucket became lighter another weight caused the cylinders to rotate in the opposite direction which caused the temple doors to close. All the people saw was when the fire was lighted the temple doors would open, apparently by themselves, and when the fire was put out the temple doors would close.

The first known applications of steam are toys and magic. When Justinian closed Plato's Academy in 529 A.D., Greek learning became inaccessible to the West for some seven hundred years. The term "Dark Ages" means that the light provided by Greek learning had gone out. This term is not applicable to technology for during the Middle Ages there is a steady stream of technological innovations, inventions, and applications.

With the revival of Greek learning, which begins in the twelfth century due in part to the Arab-West conflict, the works of Heron and Archimedes are translated into Latin and are again available in the West.

The demands made for armour and weapons, for bells and horseshoes, for coining money—all increased during the period 1100–1300. These demands exhausted existing surface mines and forced the mines deeper. As mines went deeper, water seepage and ground water increased the need for more efficient means of removal. All of the traditional sources of power were used—wind, water, muscle. Of these the most dependable and mobile source was muscle. The muscles of the ox, horse, ass and man were applied in various combinations.

In the 1450's the invention of movable type was applied to printing and by the sixteenth century many books were being printed in the vernacular. Among these were the writings of Heron. During the sixteenth century many speculations were made by men of practically every country in Europe as to the way steam could be used for various purposes including the raising of water.

The first English reference to steam appears in a patent granted to David Ramseye by Charles I dated January 21, 1630; but this device, as well as others, that are described by various workers of this period

Figure XII-5. Hero's Steam Engine.

Figure XII-6. Edward Somerset.

Figure XII-7. Thomas Savory.

were never built. The first working device that was built is described by Edward Somerset (1601–1667), second Marquis of Worcester, in a work titled, "A Century of the Names and Scantlings of Inventions by me Already Practiced."

Essentially the device sprayed a stream of water into the air by means of steam pressure. It had been used by others for ornamental fountains but was applied by Somerset to do useful work. While the device worked, its use did not spread and Somerset, like many inventors, died a poor man.

The first person to use a piston in a steam engine was Denis Papin (1647–1712), a French physician. He borrowed the idea from a gunpowder engine devised by Christian Huyghens which was also the forerunner of internal combustion engines since the fuel was consumed in the cylinder. The problem with the gunpowder engine, which Papin realized, was the lack of power for the return stroke of the piston. Too much gas remained in the cylinder after combustion and Papin felt, rightly, that the steam in condensing would give a more perfect vacuum which would permit the atmospheric pressure to force the piston back.

Papin was aware of the air pump which von Guericke had invented and he worked with Boyle on gas pressures during which he invented the double air-pump and the air gun. He had also noted that when steam was prevented from escaping, it raised the temperature of its container above the boiling point of water. "He found that the higher temperatures cooked many things more rapidly than did ordinary boiling. He devised special containers for this purpose, which he called 'Digesters.' "[3]

His device used a small quantity of water placed in the cylinder which was then heated. The steam produced forced the piston up to the top of the cylinder where it was caught by a latch. The fire was then removed and the cylinder allowed to cool. The latch was then released and atmospheric pressure pushed the piston back to the bottom of the cylinder. The cylinder was 2-1/2 inches in diameter and would lift 60 pounds once a minute. (This would be about 7 gallons of water.) About 1695 he describes a steam boiler which made enough steam to operate his device, which he called a digester, at a rate of four strokes a minute. He died without seeing any of his inventions a practical success.

Edward Somerset's engine is made a commercial success by Thomas Savory (1650–1715). He was granted a patent by King William III, July 25, 1698. The Savory engine raised 3,000 gallons per hour, filling the receiver four times a minute, and required a bushel of coal per day. The cost was 50 pounds. (50 pounds was considered an adequate annual income for a gentlewoman in the 18th century.) To operate in a mine separate engines had to be set up every 60 or 80 feet.

The high pressures required in Savory's engines to pump from great depths made it unsafe. It was uneconomical because of the great heat loss in the cold forcing cylinders; it was slow in operation and unreliable as well as its high intitial cost and high repair costs.

The genius of Thomas Newcomen (1663–1729), as with other great men of genius in other fields, was the ability to synthesize from the ideas and devices already present a new steam engine. Newcomen was an iron-monger and blacksmith by trade. In combination with John Calley and Thomas Savory, who held a patent on part of the idea necessary to the new engine, he took out a patent.

The atmospheric engine of Newcomen was a combination of the ideas of Huyghens, a cylinder and piston operated by gunpowder, Papin's improvement of substituting steam for gunpowder, and the method of condensation used in the Savory engine. It was modified to use for pumping water from mines by the use of an overhead beam with the piston suspended from one end and the pump rod from the other. It was further improved by injecting water directly into the cylinder which gave a faster return stroke.

A piston two feet in diameter could lift 12,240 gallons of water an hour to a height of 162 feet. The horsepower was about 8. As can be seen, this is about four times as much water as the Savory engine lifted twice as far. Since it operated at about 10 pounds per square inch, it was a safer engine. Here we might insert an example of the relative effectiveness of the Newcomen engine. In 1739 one was used for de-watering a French mine. The engine did in 48 hours what had been done in a week using 50 men and 20 horses working around the clock in shifts.

Newcomen's basic patents expired in 1733 and the door was open for many innovators to try their hand at improving the device. One of the more successful of these was John Smeaton (1724–1792) who made an experimental model in 1769 and performed some 130 experiments showing the stroke length, number of strokes per minute, cylinder diameter, quality of boiler feed water, type of fire, and size of boiler. He tabulated his results from which he was able to make the Newcomen engine much more efficient.

He was commissioned in 1773 by Catherine II of Russia to construct a pumping engine to drain a great dry dock in St. Petersburg. The dry dock was begun by Peter the Great and was large enough to hold ten ships. Previously it had been drained by two windmills 100 ft. high but a year was required for the operation. The new pump did the job in two weeks.

Smeaton is a new breed of man—the first to call himself a civil engineer since he was a civilian rather than being a military engineer. (I had often wondered why some engineers were called civil since there is little in their attitude to indicate such a title.) The first engineering school is in France and was established in 1747. (Ecole des Ponts et Chaussees)

As will be noted, the steam engine has a rather extensive background and it was well developed before James Watt (1736–1819), usually credited

Figure XII-8. James Watt.

with inventing the steam engine, appears in the picture.

Glasgow University had a Newcomen engine which didn't work right and in 1764 Watt was asked to repair it. Instead of doing this job, he examined the engine itself. He found it necessary to determine several physical constants experimentally, including the specific heats of certain materials and the latent heats of vaporization and condensation. He then calculated that the amount of steam needed for the engine was a sum consisting of that amount needed to fill the cylinder, plus the amount needed to heat the cooled cylinder walls, the piston, the water remaining in the cylinder and the sealing water. In other words, he found that only 1/3 of the total steam in the Newcomen engine was actually being used.

These defects he corrected by using a separate condenser for the steam and supplying steam for both motions of the piston to make two power strokes. So much did the improvements save that Watt and Boulton could sell their engine on a contract that called for the periodic payment of 1/3 of the fuel saving from the replacement of Newcomen engines. So that the purchaser got a free engine and saved money at the same time.

Watt also standardized the meaning of horsepower. He found that the average horse could lift 1 cwt (112 lbs.) 196 feet high in 1 minute and could work at this steady rate. This would be 365.86 ft-lb/sec but he increased this to 550 ft-lb/sec (about 50%) so that purchasers of his engine should have no complaint.

The next major development was the application of high pressure to the steam engine. Two men are outstanding in this application—an American, Oliver Evans (1755–1819), and an Englishman, Richard Trevithick (1771–1833).

Evans was a self-made man and, although he went to country school until he was 14, was largely self-educated. His ideas about steam arose when he observed some boys playing about a blacksmith's fire. They had an old gun barrel in which they had placed some water and a wad. They then placed the breech in the fire and the resulting reaction sounded like gunpowder. This convinced him that low pressure engines like Watt's were not the answer.

In 1786, he tried to get the state legislatures of Pennsylvania and Maryland to give him a monopoly on steam engines and steam carriages in those states but they considered the idea too ridiculous to even consider. He had no prestige so it was in England that the steam locomotive was to first develop.

He did, however, make stationary engines. In 1807, in Philadelphia, he founded the Mars Iron Works which produced his steam engines. The beam which translated rotary motion to an up and down motion had a peculiar loping motion and his engine became known as the grasshopper type. His engines operated as far west as Ohio and were described as being able to grind 300 bushels of grain or 12 tons of plaster in a 24-hour period.

We might here highlight the application of steam to transportation:

John Fitch in 1787 had a paddle-wheel steamboat operating on the Delaware—the framers of the Constitution went down to see it. It could not compete in payload so the first commercially successful steamboat was made by Robert Fulton and placed in operation on the Hudson River in August, 1807.

Early trains were horse drawn. Their use had been made in mines and for hauling passengers before the application of steam. Such trains had speeds of three to six miles per hour. The first steam locomotive was built in 1769 by Nicholas Joseph Cugnot for the purpose of drawing artillery. Richard Trevithick (1771–1833) designed and built a locomotive that in 1804 was in operation on a Welsh coal road. It pulled five cars at five miles per hour.

There were others that worked on steam locomotion, but the man credited with being the inventor and founder of the railways was George Stephenson (1781–1848) another self-educated man. In competition at Liverpool in October, 1829 Stephenson's *Rocket* won. It ran 12 miles in 53 minutes.

THE WORKINGS OF THE STEAM ENGINE

The Newcomen engine was also called an atmospheric engine because the pressure of the atmosphere returned the piston to its original position. See Figure XII-9. In this engine steam was produced in the boiler (1) with valve A open and valve B closed. The steam entered the cylinder (2) and pushed the piston (3) upward which pushed the right hand side of the walking beam up and the left hand side down. The pump rod (6) was also pushed down which caused the pump piston to be pushed down. Valve A was then closed and valve B opened which allowed cold water from the water reservoir (8) to enter the cylinder (2). The cold water cooled the cylinder causing the steam to condense which produced a partial vacuum. The air pressure pushed the piston down causing the piston rod to go down which caused the right hand side of the walking beam to go down and the left hand side to go up. This action also pulled the pump rod (6) up and caused water to be raised from the mine.

In the Watt engine, pictured in Figure XII-10, there are several improvements made which made the steam engine much more efficient. No longer was cold water injected into the cylinder but instead there was a separate condenser into which the steam could go for condensation without losing heat. Also it was possible for steam to be used to return the piston which meant that the engine could operate at a higher temperature and pressure than the Newcomen engine. Steam was produced in the boiler and valves A and D were opened. The steam entered the cylinder (2) and pushed up the piston (3) which forced any spent steam through valve D to the condenser. The piston rods changed reciprocal motion into circular motion. Then valves A and D were closed and valves B and C were opened. This permitted the steam to enter (2) the cylinder and force the piston (3) back down while pushing the spent steam through valve C to the condenser.

STUDY AND DISCUSSION QUESTIONS

1. What is the nature of heat? How may it be produced and transmitted?
2. What is the difference between heat and temperature?
3. Why is water frequently used as a standard?
4. What is a calorie? a B.T.U.?
5. What are the first two laws of thermodynamics?
6. How does a steam engine work?
7. How do the improvements in the steam engine reflect the laws of thermodynamics?

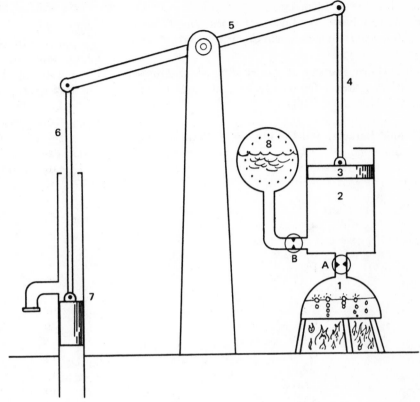

A and B Valves

1. Boiler
2. Cylinder
3. Piston
4. Piston Rod

5. Walking Beam
6. Pump Rod
7. Pump
8. Water Reservoir

Figure XII-9. Newcomen Engine.

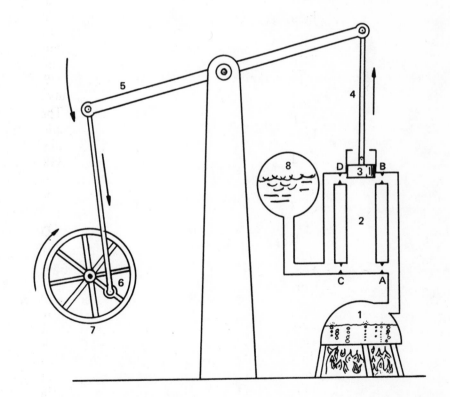

A, B, C, and D Valves

1. Boiler
2. Cylinder
3. Piston
4. Piston Rod

5. Beam
6. Crank
7. Wheel
8. Condenser

Figure XII-10. Watt Engine.

FOOTNOTES

1. As cited in Lloyd W. Taylor's *Physics: The Pioneer Science* (Houghton-Mifflin Company, 1941), p. 300.

2. Robert Henry Thurston, *A History of the Steam-Engine* (Kegan, Trench, Trubner & Co., London, 1895), pp. 2-3.

3. Edmund Berkeley and Dorothy Smith Berkeley, *The Reverend John Clayton* (The University Press of Virginia, 1965), p. xxi.

Suggested Readings:

John U. Nef, *Cultural Foundations of Industrial Civilization,* Cambridge University Press, 1958.

John F. Sandfort, *Heat Engines.* Doubleday and Co., Inc., 1962. pp. 1-89.

Charles Singer, *et. al.* (eds.), *A History of Technology,* Vol. IV. Oxford University Press, 1958. pp. 148-198.

Holland Thompson, *The Age of Inventions,* Yale University Press, 1921. pp. 53-83.

Abbott Payson Usher, *A History of Mechanical Inventions,* Oxford University Press, 1929. pp. 332-357.

XIII

The Industrial Saltation

It has been argued quite reasonably that the Industrial Revolution or saltation[1] would have occurred without the invention of the steam engine. The industrial saltation is characterized by the use of power machines in a factory system. The power, wind or water, had been applied to machines for a long period of time prior to the 1760's when there was an acceleration of rate of change. It is this rapid change that has given rise to the use of words like "saltation" and "revolution" to what is really an evolutionary process. The word "revolution" implies a dam to change and violence and while there was some opposition to the factory system and a few cases of violence, there was not enough to warrant the use of this term.

In the Eighteenth Century in England there were a number of factors—economic, social, and political —which favored the rapid change that occurred.

England had enjoyed a rapid growth of its economy due in large measure to the wool trade which was stable and well established with government and guild regulations. So much was England dependent on wool that even today there is a wool sack in Parliament to remind its members of its importance. The commerce of the Seventeenth Century had led to an organized system of banking and finance—so much so that it became possible to borrow from the future, (e.g., the Seven Years War (what we call the French and Indian War) was financed on credit). Thus by the end of the Eighteenth Century, there was a fluid source of investment capital. Another economic factor was the changes in agriculture, which had seen few changes since the Middle Ages, in the Eighteenth Century. The development of scientific stock-breeding, introduction of the four-field system, the use of nitrogen-restoring crops and a variety of new vegetables—all contributed to producing more food for a growing population with fewer persons required for agricultural work. The enclosure of the open fields and commons by acts of Parliament[2] gave rise to more scientific farming and larger farms while at the same time pushing many people off their lands. These agricultural changes meant that a moveable labor pool was created and that it was larger, while at the same time there was a larger overall population and a surplus of food. The larger population created by the food surplus and by colonization increased the demand for textiles both of wool and of the new cotton.

Among the social factors was the fact that oligarchy control was open to talent and, hence, there was a degree of social fluidity. A poor man with talent and/or education could rise above his class. Another way to rise in the class structure was money. Thus, there was a reward for making money. The church non-conformists (persons who were non-Anglicans— e.g. the Puritans) had limited citizenship, but many of these persons had the Calvin idea of deadness to the world but recognized that one sign of salvation was earthly success. Thus the acquisition of money and property, while it could not be used for finery, horse racing or pleasure, was a sign of divine favor and encouraged them to invest their money and to make more money. A number of the men who fought their way up the economic ladder were cut-throats, pirates and had little regard for laws or moral rights in the battle for economic survival. A socio-religious factor was the spread of Christianity to warm climated countries. For with Christianity came the doctrine against the naked body which fit nicely the needs of a growing textile industry.

Political developments of the Eighteenth Century aid an industrial growth. There gradually grows a tacit unexamined *Laissez-faire* in which Parliament does nothing to restrict commerce and permits the development of free trade internally—the largest such area in Europe at that time. At the same time, Parliament passes a number of acts favorable to trade. Enclosure acts passed by Parliament increase all during the Eighteenth Century, and numerous acts are passed to permit toll roads and canals to be built. The growth and linkage of roads and canals are necessary for any industrial development as the cargos are too heavy for pack mules and horses. The growth of British imperial control is an important factor, not only from the standpoint of larger markets for manufactured goods, but also because the British gain control of India which is the source of cotton.

There are, of course, obstacles to industrial change which may be classed as technological, traditional, and economic.

To be sure, there had been some slight mechanization of textile production but most of the process was still done by hand. The sheep were sheared or the cotton picked then the wool was carded and combed and the cotton was deseeded and combed—all hand operations. The spinning process was one of making the fibers hang together in a long single strand. The older process made use of a spindle, made of wood about nine to fifteen inches long tapered and rounded at both ends with a notch near one end to catch the yarn during the twisting process and toward the other end was an attached disk of clay or stone to give momentum to the rotating spindle. The fibers were attached to a piece of wood held under the left arm of the operator and the spindle set in motion by rolling with one hand against the leg. When the rotating spindle had pulled out a full length, the yarn was rolled onto the spindle and the process repeated until the spindle was filled. By the Fourteenth Century in India and somewhat later in Europe a wheel was used to give motion to the spindle—hence a spinning wheel. The next step was the weaving of the yarn.

Essentially no improvement had been made in the loom since the beginning of the Christian era. The process of weaving is performed by interlacing at right angles, two or more strands of flexible materials, the vertical or those fibers extending from and toward the operator are called the warp and those that go crosswise in front of the operator are called the woof. Basically the warp is made by attaching a series of yarns to two sets of rollers; there is a set of rods for lifting alternate layers of warp or some pattern, and a shuttle that carries the woof across. The shuttle is sort of tossed across, carrying the yarn of the woof, then the rod is changed so that the layers of the warp criss-cross, after which the shuttle is again tossed over. This process is repeated forming a piece of cloth. There are additional steps in textile production that we shall leave out for simplicity.

In short, there was no mechanism available for increasing the output beyond what could be done by increasing the number of workers. The domestic system had no large investments in machinery because the persons who worked supplied their own spinning wheels and looms.

In addition to the lack of machines, there was the opposition that was traditional—if it was good enough for father, it's good enough for me. The wool

trade was hampered by many government and guild regulations. Input could not be increased because of the rules governing the training of men—apprentice for seven years, journeyman and finally master—and those which regulated quality and the quantity that might be produced during a working day, etc. (Much like the productive restrictions of unions today.) There were also those who had a vested interest in textiles as they were then being produced and they wanted no mechanical competition.

Cotton production had the economic deterrents of competition with the wool trade and the fact that the East India Co. had investments in India to produce cotton cloth.

The revolution begins in the cotton industry because it is not hampered by guild regulations and because it must meet the strong competition and opposition from the East India Co. The cotton manufacturers begin a bootleg process to compete with the import, from India, of Calico cottons which are lightweight and cheap. They even make an unsuccessful attempt to grow cotton in England. The East India Co. makes strong protests to Parliament which responds with some laws which are ignored by the cotton manufacturers. Both the East India Co. and the ancient wool industry have objections to the home manufacture of cotton. The fact that cotton is a bootleg operation and not restricted by traditional guild regulations provides incentive to get inventions. Cotton also does not have a vested interest in existing machines or techniques.

Figure XIII-1 shows a graphical representation of the state of the process at the beginning of the industrial revolution in textiles. As you can see, this is a very large market but only the amount that could be woven could be sold. This created a weaving bottleneck.

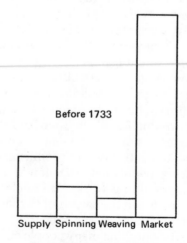

Figure XIII-1. Process Graph.

In 1733 John Kay (1704–1764) developed a flying shuttle which increased the speed of weaving as it freed one hand to manipulate the batton. It also permitted wider cloth with less labor as before this invention two weavers were needed for cloth over 30 inches wide. This invention merely moved the bottleneck from weaving to spinning as shown in Figure XIII-2.

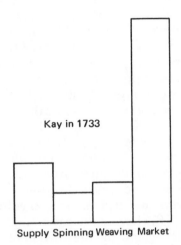

Kay in 1733

Supply Spinning Weaving Market

Figure XIII-2. Process Graph.

To give a spinning summary, John Wyatt and Lewis Paul developed two new concepts: (1) the application of rollers to the manipulation of the yarn and (2) mounting the spindles on a movable carriage to duplicate the drawing out of the yarn by the hand process. Using the work of Wyatt and Paul, as well as their equipment (they went bankrupt) Richard Arkwright (1732–92) made a water frame in 1767 which could produce a coarse yarn. James Hargreave (d. 1778) developed in 1765 what he called a spinning jenny that enabled one man to spin as much as eight had spun. The water frame and the jenny were placed on the market in 1768. Samuel Crompton combined the best features of both, hence a mule (cross between a horse and a jenny). Power weaving began to come in about 1769–1779 and these two improvements caused a bottleneck at the supply point. This is shown in Figure XIII-5.[3]

The supply of cotton was a function of the number of acres that could be planted and cultivated; as well as the number of hands that could be brought to bear on the deseeding and combing processes. The deseeding of the cotton bolls required a considerable number of hands for a considerable period—in other words, cheap labor. The cheapest, at least so many Southerners believed, was the slave. In 1794, Eli Whitney (1765–1825) invented the cotton gin which consisted of a sieve and wooden cylinder to remove

Figure XIII-3. Richard Arkwright.

the seeds. The first primitive model could do the work of two men for a day in one hour. Each gin required a backup of 50 field hands. In the early Seventeenth Century, slavery had nearly died out or was in the dying process. The invention of the cotton gin was to perpetuate the peculiar institution for a growing money crop.

The location of the early cotton mills was dictated by two factors in this country (three in England)—water was necessary for power and this meant a hilly terrain, secondly cotton thread needs a high humidity for weaving for if the thread gets too dry, it breaks easily. In England there was the additional requirement that the mills be located in a sparsely settled area far from the central government for like stills in this country the manufacture of cotton was a bootleg industry.

Figure XIII-4. Eli Whitney.

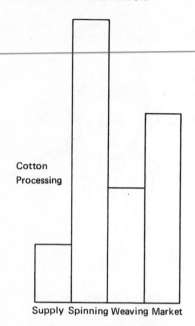

Figure XIII-5. Process Graph.

Cotton stimulates a variety of other industries. In the Middle Ages most machinery that existed was wooden, but with the faster machines possible with steam power the machines had to be made of iron. The iron mines and the coal mines, as they were dug deeper, needed the steam engine to pump the water seepage from them so that the mines acted as a stimulus for the steam engine which in turn acted as a stimulus for the coal and iron industries. Britain was geologically fortunate in that it was rich in coal and iron and they were located close together.

The factories themselves had to be built and the towns were enlarged to provide for the large numbers of workers that went to the factories. Construction work grew by leaps and bounds. If outsiders didn't come in and build the workers housing, then the factory owners had houses built and only those persons who worked in the factories could live in the houses.

In the factory system, the early plants were built on streams in hilly terrain to provide the power. There was a large capital investment required for the mill and the attendant machinery. Since the worker could no longer afford his equipment if he wanted to work he had to move within walking distance of the factory. After the advent of steam engines, plants were located near systems of transport such as the canals, sea ports, and rail lines.

The factory system gave England a head start and a virtual monopoly. She became the very image of a successful empire, and this was brought about in large measure by entrepreneurs who dug the canals, mined coal, revolutionized the textile and other in-dustries, erected factories powered with steam and laid the basis for over a century of industrial domi-nance.

Some of the individuals who did most to bring about the revolution were themselves least re-warded. The inventors and innovators were attacked from two sides—on the one hand from the entre-preneurs who wanted to use their invention but didn't want to pay royalties and who fought in court and on the other, by workers who saw the machines depriving them of their livelihood. In the first in-stance many inventors went broke trying to win their case in court (justice is only for the rich) and in the second, their property was destroyed and some-times they were physically attacked.

Two cases will serve to illustrate the latter. In 1753, a mob broke into Kay's house and destroyed it and he was forced to flee for his life. He was also ruined by the legal expenses of a series of suits which he had to bring to keep manufacturers from using his device. Hargreaves suffered much the same fate with the manufacturers and had his machines smashed by workers in 1767. Laws were passed under the reign of George III against the destruction of mills, facto-ries and machines and in the early part of the Nine-teenth Century a group of men were hanged for such a riot. At that time, at least and throughout most of human history, property rights had a higher value than human life. But as we shall see in the next chap-ter perhaps they were right for life, for most people, had little value—it was a period of suffering.

STUDY AND DISCUSSION QUESTIONS

1. What is the difference between a revolution and a saltation?
2. In what sense is the industrial revolution a revo-lution?
3. What were the social, economic, and political factors present in England which were favorable to the revolution? Those which opposed?
4. Explain spinning. Weaving.
5. Discuss the bottleneck explanation of develop-ment. How did solving one bottleneck create an-other?
6. Explain the domestic system.
7. Discuss how the factory system differed from the domestic system.
8. Why did the revolution in the textile industry start with cotton?
9. Explain why Whitney's invention contributed to and perpetuated slavery.
10. Discuss the locating of factories.

FOOTNOTES

1. Saltation means a leap as applied to a dancer and has been applied to the sudden leap in industrial activity by W. H. Cowley of Stanford.

2. Gave rise to this often quoted poem of Goldsmith's, "The Deserted Village"

> Sweet smiling village, loveliest of the lawn,
> Thy sports are fled, and all thy charms withdrawn;
> Amidst thy bowers the tyrant's hand is seen
> And desolation saddens all thy green.
> One only master grasps the whole domain,
> And half a tillage stints thy smiling plain . . .
> And trembling, shrinking from the spoiler's hand,
> Far, far away, thy children leave the land . . .
> Where then, ah! where shall poverty reside
> To 'scape the pressure of continuous pride?
> If to some common's fenceless limits strayed,
> He drives his flock to pick the scanty blade,
> Those fenceless fields the sons of wealth divide,
> And e'en the bare-worn common is denied . . .
> Ye friends to truth, ye statesmen who survey
> The rich man's joys increase, the poor's decay,
> 'Tis yours to judge how wide the limits stand
> Between a splendid and a happy land . . .
> Ill fares the land, to hastening ills a prey,
> Where wealth accumulates, and men decay;
> Princes and lords may flourish, or may fade;
> A breath can make them, as a breath has made,
> But a bold peasantry, their country's pride,
> When once destroyed, can never be supplied.

3. I am indebted to Professor Wilcox of the History Department of the University of Michigan for the "bottleneck" thesis.

Additional Suggested Reading

Paul Mantoux, *The Industrial Revolution in the Eighteenth Century,* Harper and Row, 1961.

T. S. Ashton, *Industrial Revolution, 1760–1830,* Oxford University Press, 1948.

Arnold Toynbee, *The Industrial Revolution,* Beacon Paperback, 1956.

Melvin Kranzberg and Carroll W. Pursell, Jr. (eds.), *Technology in Western Civilization,* Oxford University Press, 1967. Vol. I, pp. 217-245.

Eric E. Lampard, *Industrial Revolution: Interpretations and Perspectives,* American Historical Association, 1957.

XIV

Man Makes Like a Machine

With the gradual enthronement of the natural, agrarian man and the romanticised ideal of living close to nature as opposed to the artificial and basically evil urban setting, many historians and others have come to regard the industrial revolution and mechanization, in general, as an evil. Most have, at best, argued that it was a necessary evil and that it is impossible to return to the glorious, unfettered agrarian past. This theme has been repeated in many plays, movies and television performances. The pure simple farm person going to the city and being corrupted by the various evil influences or, at least, suffering exploitation at the hands of the city slicker is a far more frequent theme than that of a city boy going to the country and being exploited as a share-cropper or being corrupted by the evils of farm life.

In fact, many of the economic and social evils associated with the factory system and the industrial revolution have been shown to predate the steam engine and the factory. Throughout most of the history and prehistory of man over most of the globe the large mass of the working class have and still do suffer at the hands of these "well born" or the "well-to-do." There were, of course, some differences, but keep in mind that for the poor the difference was only of degree.

> Instead of working in their homes they were obliged to work in mills; and instead of being comparatively their own masters . . . they were under masters who made them work for what wages they chose to give, and during what hours they chose to dictate.

> . . . water [steam or wind] could now be employed to do the harder part of the work formerly done by the men . . . and machinery was invented which children could manage . . .

> In this way a demand for child-labour was created, and the supply was not deficient. But it was effected in a manner which scarcely seems creditable to the humanity of to-day [1833]; large bodies of children were drafted from the workhouses of London . . . and other great cities, and placed in the mills as 'apprentices,' where . . . they were worked unmercifully, and treated with . . . brutality . . .

> As early as 1796, voices were raised in protest against the cruel wrongs inflicted on these poor children . . . but . . . those voices were unheeded. Meantime, the condition of these unfortunate children was growing from bad to worse, until at last the cruelty of the system under which they were held was hardly paralleled by the abominations of negro slavery. A horrible traffic had sprung up; child-jobbers scoured the country for the purpose of purchasing children to sell them again into the bondage of factory slaves. The waste of human life in the manufactories to which the children were consigned was simply frightful. Day and night the machinery was kept going; one gang of children working it by day, and another set by night, while, in times of pressure, the same children were kept working day and night by remorseless task-masters.

> Under the 'Apprentice System' bargins were made between the churchwardens and overseers of parishes and the owners of factories, and the pauper children—some as young as five years old—were bound to serve until they were twenty-one.

> In some cases alluring baits were held out to them; they were told they would be well-clothed and fed, have plenty of money, and learn a trade. These deceptions were practiced in order to make the children wish to go and thus give an opportunity to the traffickers to say that they went as volunteers, and not under compulsion. Generally, the spell was broken when, like livestock, these children were packed in waggons, and sent . . . wherever their destination might be. If the illusion did not vanish then, it did when the gates of the 'Pentice House' closed upon them, and they were checked off, according to invoice, and consigned to the sleeping berths allotted to them, reeking with the foul oil with which the bedding of the older hands was saturated.

> Their first labours generally consisted in picking up loose cotton from the floor. This was done amidst the burring din of machinery, in an average heat of 70° to 90° Fahrenheit, and in the fumes of the oil with which the axles of twenty thousand wheels and spindles were bathed.

> Sick, with aching backs and inflamed ankles from the constant stooping, with fingers lacerated from scraping the floors; parched and suffocated by the dust and flue —the little slaves toiled from morning till night. If they paused, the brutal overlooker who was responsible for a certain amount of work being performed by each child under him, urged them on by kicks and blows.

> When the dinner-time came, after six hours' labour, it was only to rest for forty minutes, and to partake of black bread and porridge, or, occasionally, some coarse Irish bacon.

In process of time more important employment was given to them, involving longer hours and harder work. Lost time had to be made up by overwork—they were required every other day to stop at the mill during the dinner-hour to clean the frames, and there was scarcely a moment of relaxation for them until Sunday came, when their one thought was to rest. Stage by stage they sank into the profoundest depths of wretchedness. In weariness they often fell upon the machinery, and almost every factory child was more or less injured; through hunger, neglect, over-fatigue, and poisonous air, they died in terrible numbers, swept off by contagious fevers.

There was no redress of any kind. The isolation of the mills aided the cruelties practised in them. The children could not escape, as rewards were offered for their capture and were eagerly sought; they could not complain when the visiting magistrate came, for they were in abject fear of their task-masters, and, moreover, on those days the house was swept and garnished for the anticipated visit, and appearances would have given the lie to complaints; if they perished in the machinery, it was a rare thing for a coroner's inquest to be held . . . And when the time came that their indentures expired, after years of toil, averaging fourteen hours a day, with their bodies scarred with the wounds inflicted by the overlookers—with their minds dwarfed and vacant, with their constitutions, in many instances, hopelessly injured; in profound ignorance that there was even the semblance of law for their protection—these unfortunate apprentices, arrived at manhood, found that they had never been taught the trade they should have learned, and that they had no resource whatever but to enter again upon the hateful life from which they were legally freed.[1]

The above is merely typical of the many descriptions of working conditions that existed (and still do exist in many places) in England's early industrial phase.[2]

While the machine age had caused some of these evils, it was also the machine age that moved to end them. The application of power to printing made possible more books and newspapers at cheaper prices. The growth of many cheap items which could be sold for a profit created a larger middle-class. Middle-class morality opposed the exploitation of women and children and caused the passage of acts against mill operators. The first acts passed gave slight relief even where enforced and their enforcement depended on the Justice of the Peace who was the mill owner.

The importance of the Act of 1847 becomes very apparent, when we remember that out of 544,876 persons employed (according to the returns of that year) in the textile industries, no less than 363,796 were young persons and women, whom the act directly affected; the time of their labour being limited, (from the date of passage) . . . to eleven hours a day or sixty-three hours weekly, and from May the 1st, 1848, to ten hours a day or fifty-eight hours weekly.[3]

We have seen legislation finally affecting the work week of men as well as women and children as the hours have decreased to a forty-hour week and projected to a thirty-hour week for the future. Overworking the poor was but one part of the gigantic new industrial society but probably contributed strongly to a continuing stream of anti-industrialism.

Early in the century the Romantic movement sounded a major theme of literature thereafter—a revulsion against industralism. Writers attacked not the machine itself so much as its by-products: the rule of getting and spending, the utilitarian philosophy, the defilement of the landscape, and the horrors of the industrial town's and their factories. Blake's 'dark Satanic mills.' In France, Balzac and Flaubert concentrated on a devastating analysis of the rising bourgeois and their values. The Victorians deplored the complacence over a purely material progress. 'Our true Deity is mechanism,' wrote Carlyle, and he summed up the progress: 'We have more riches than any Nation ever had before; we have less good of them than any Nation ever had before.' Ruskin added: 'There is no Wealth but Life.' He also attacked the ugliness of the products of industrialism, which even some men dazzled by the Great Exposition had noticed; the taste displayed in design and ornament was about the worst in the history of craftsmanship. Matthew Arnold, the apostle of culture, saw the chief menace as the 'faith in machinery' of a civilization grown 'mechanical and external.' Zola, Hauptmann, William Morris, and others spoke out on behalf of the workers, who had been degraded by industrial progress. In general, the nineteenth century created a great literature, more varied and abundant than that of any century before it, in a way suited to the extraordinary energy and creativity it was exhibiting in technology; yet, writers kept growing more critical of their society, often hostile to it, than writers had ever been before. William Morris wrote that his leading passion was 'hatred of modern civilization.'[4]

The use of steam instead of water made the machines go too fast to be made of wood any longer so that the age of steam led to the fulfillment of the age of iron and both ages left their imprint on man and his works.

Their color spread everywhere, from grey to black: the black boots, the black stove-pipe hat, the black coach or carriage, the black iron frame of the hearth, the black cooking pots and pans and stoves.

Iron became the universal material. One went to sleep in an iron bed and washed one's face in the morning in an iron washbowl: one practiced gymnastics with the aid of iron dumb-bells or other iron weight-lifting apparatus; one played billiards on an iron billiard table . . . ; one sat behind an iron locomotive and drove to the city on iron rails, passing over an iron bridge and arriving at an iron covered railroad station . . .[5]

Even as early as Elizabethan England c. 1500, acts had been passed to prevent the burning of soft coal in the hearth's of London to prevent smog. With the increasing use of the steam engines in transport and in industry the heavy black smoke spread over and darkened the industrial cities. Soot settled everywhere and increased the cost of keeping clean. Within a day of a snowfall the snow had turned a dirty gray. The tannery and the rendering works had always smelled but impurities in the coal as well as many chemical industries produced fumes that could be deadly as well as unpleasant.

The slums and other problems associated with large numbers of people living together was as old as the pyramids. Rome had had to build aqueducts because human sewage had made the Tiber unusable. With the spread of industrial cities these problems were magnified and man has continued to make one body of water after another unusable for man or beast until today we are approaching a crisis. The concentration of large numbers of persons, at any time, had made conditions ripe for diseases and their spread. Typhoid and other intestinal diseases from the improper disposal of sewage; typhus and diseases carried by vermin due to poor construction and sewage; diseases due to lack of sunlight and nutritional deficiencies—all of these conditions while not due to industrialization were caused by the needs of the factory system for concentrations of people.

But there were other types of environmental degradation besides these forms of pollution. Foremost among these was that resulting from the regional specialization of industry . . . Thus England . . . turned all its resources and energy and manpower into mechanical industry and permitted agriculture to languish: similarly, within the new industrial complex, one locality specialized in steel and another on cotton, with no attempt at diversification of manufacture. The result was a poor and constricted social life and a precarious industry. By reason of specialization a variety of regional opportunities were neglected, and the amount of wasteful cross-haulage in commodities that could be produced with equal efficiency in any locality was increased; while the shutting down of the single industry meant the collapse of the entire local community. Above all, the psychological and social stimulus derived from cultivating numerous different occupations and different modes of thought and living disappeared. Result: an insecure industry, a lopsided social life, and impoverishment of intellectual resources, and often a physically depleted environment. This intensive regional specialization at first brought huge pecuniary profits to the masters of industry; but the price it extracted was too high. Even in terms of mechanical efficiency the process was a doubtful one, because it was the barrier against that borrowing from foreign processes which is one of the principal means of effecting new inventions and creating industries. While when one considers the environment as an element in human ecology, the sacrifice of its varied potentialities to mechanical industries alone was highly inimical to human welfare: the usurpation of park sites and bathing sites by the new steel works and coke-ovens, the reckless

placement of railroad yards with no respect to any fact except cheapness and convenience for the railroad itself, the destruction of forests and the building up of solid masses of brick and paving stone without regard for the special qualities of site and soil—all these were forms of environmental destruction and waste.[6]

The foremost center of industrialism and typifying it was the city of Manchester. Alexis de Tocqueville described it: "From this foul drain the greatest stream of human industry flows out to fertilize the whole world. From this filthy sewer pure gold flows. Here humanity attains its most complete development and its most brutish, here civilization works its miracles and civilized man is turned almost into a savage."

The smoke-filled towns were the drabbest, grimiest, and ugliest in all history. As they went about 'making a new Heaven and a new Earth—both black,' they made little or no provision for sanitation and recreation, parks and playgrounds, or any open space where people could gather and relax. The industrial slums that started growing up in the old cities too made drabness, filthiness, and ugliness seem still more like a natural, normal condition of industrial progress. Respectable people were not shocked by the foul surroundings in which men were learning to live, and could hope to live with any contentment only by virtue of deadened senses, with the help of gin.[7]

The major drive then as now for the production of slums was profit. In the nineteenth century landlords in New York made a comfortable 40% a year on their investment with no income tax to pay.

Movements of population toward industrial centers, slums, and factory work were part and parcel of the same picture.

The gradual enclosure of the common lands, the improvement of agriculture with fewer farm laborers needed and the attendant population growth forced large numbers to the cities and factory work. They were held there by a natural disinclination to move, their own poverty, and the parish poor laws.

On the other hand the skilled handicraft worker was gradually forced into competition with the machine as the work he was doing was gradually replaced by a factory. The factory has always moved in the direction of unskilled labor the very mechanization of which destroys craftsmanship and pride of work and was replaced by the bounds of starvation, ignorance, and fear which held the worker as a part of the pulse of the machine. Even if by some quirk of fate the factory hand had an opportunity to move toward the frontier of the new world the specialization and partitioned functions required of the machine ill suited him as preparation for life as a farmer or pioneer.

The factory also demanded discipline and subordination to the machine. Women and children were more docile than men and easier to handle. Children, not paupers or orphans, were forced into the factories by blackmailing the parent—"if you want a job then you must bring your child." Combinations of workmen or unions were illegal. The law was used as well as the blackball, violence, and "scabs." As if these were not enough, sometimes industries would move away from labor trouble (as they still do) and sometimes invention of technological improvement was the factory owner's answer. Some have regarded strikes to be more productive of good than evil since they stimulated invention. " 'In the case of many of our most potent self-acting tools and machines, manufacturers could not be induced to adopt them until compelled to do so by strike. This was the case of the self-acting mule, the wool-combing machine, the planing-machine, the slotting-machine, Nasmyth's steam-arm, and many others.' "[8]

Factory work was quite different from the routine toil and drudgery that poor people had always been accustomed to. Instead of setting their own pace, workers had to learn the 'new discipline,' conformity to 'the regular celerity of the machine.' They were always under the eye of a supervisor, often a severe taskmaster who kept them working at high speed through the long day, and at best they were no longer free to pause, chat, or idle now and then. All became factory 'hands'—a suitable name for machine-tenders who were no longer doing man-sized jobs but were turning out parts instead of finished products, and who served as interchangeable cogs in the system. As Karl Marx put it, the factory 'transforms the worker into a cripple, a monster, by forcing him to develop some highly specialized dexterity at the cost of a world of productive impulses and faculties— much as in Argentina they slaughter a whole beast simply in order to get his hide or tallow.' Work had given both peasants and artisans a definite social status and function, however humble, but most factory work was humanly meaningless except as a necessary means of making a living. Workers could not express themselves in their mechanical work—they could only deny themselves. If or when higher wages resigned them to machine-tending, it could only be at some sacrifice of their natural interests and capacities.[9]

The pace of the machine was probably best illustrated in the film "Modern Times" (1936) written, acted and directed by Charlie Chaplin. It is a satire on the mechanized civilization of the twentieth century in which the factory owner tries out a worker feeding machine so that the worker will not have to stop working in order to eat. Charlie is the worker on whom it is tried and in the process of trying it out the machine goes amuck and so does he. The factory leaves no time for leisurely eating or a restful nap after.

Factory workers had to acquire habits of regularity and punctuality that at first were as strange as their barrack-like surroundings. Throughout the year they had to get up at fixed hours, often in darkness, just as through the working day they had to forego their natural impulse to rest when they felt like it. Neither could they take vacations, nor ever enjoy the long spells of relative leisure peasants had known after harvesting their crops.[10]

Evidence of this change is present in the "twelve days of Christmas" of the carols that disappear in the early industrial revolution and reappear in the emaciated form as one day. Capital investments can not stand idle and even those visionaries that try to get stores to close on Sunday do not attempt to force Sunday closings on the factories—they know where the butter and their bread comes from.

The factory system provided its poisons both at work and at home for the worker for the owners felt no responsibility for his wage slave or the consumer.

Industrial diseases naturally flourished in this environment: the use of lead glaze in the potteries, phosphorus in the match-making industry, the failure to use protective masks in the numerous grinding operations . . . increased to enormous proportions the fatal forms of industrial poisoning or injury: mass consumption of china, matches, and cutlery resulted in a steady destruction of life. As the pace of production increased in certain trades, the dangers to health and safety in the industrial process itself increased: in glass-making, for example, the lungs were over-taxed, in other industries the increased fatigue resulted in careless motions and the maceration of a hand or the amputation of a leg might follow.[11]

Under the stress of competition (and lacking guild regulations) adulterants in food become a commonplace . . . flour was supplemented with plaster, pepper with wood, rancid bacon was treated with boric acid, milk was kept from souring with embalming fluid . . . Stale and rancid food degraded the sense of taste and upset the digestion: gin, rum, whiskey, strong tobacco made the palate less sensitive and befuddled the senses: but drink still remained the 'quickest way of getting out of Manchester.'[12]

Of course drink has always been the helpmate of the worker and his wife and children. It is cheaper than food and takes away the hunger pangs. In England and this country the "do gooders" of the middle class have removed this pain killer for the masses under the title of uplifting the lower born. Children of working mothers were often given a little opium to keep them quiet while the mother was away at work. The pains of the system could be lessened.

The factory system made a wage slave of the apprentice and changed the relationship that had existed between master and worker. In the Middle Ages the apprentice had lived with and worked with the owner and master but in the industrial system the owner did not even know the people that worked for him. The skilled and semi-skilled sank on the economic ladder and joined the unskilled poor.

Economic fluctuations were increased by the factory system. Machines that worked night and day producing cheap goods to be dumped on a created market of colonialism produced too well and too much. The market was glutted and the workers were paid too poorly to be consumers for much of the goods. When goods could not be sold the machines stopped and the factories closed. The workers were the victims of their product. In the domestic system the skilled or semi-skilled worker had a little garden, a pig or cow and they could subsist repeated depressions but with the factory workers who lived in slums they were too specialized for farm work and they had no space for gardens.

According to some the condition of overproduction can only be solved by wars which use up the production of a nation in a manner that people will accept out of patriotism. Since all material in a war is expendable it is thrown away or used up and creates a constant demand for new goods.

Certainly it can be argued that a good deal of the empire building at the end of the nineteenth and the beginning of the twentieth century was motivated by an urge to obtain markets for factory production and to divert the worker's mind from his own misery.

The United States followed England into industrialism about one hundred years later. The industrial revolution began before the Civil War but the greatest rate of growth was in the generation following the War. It was the same men who had supplied the Union Army with rotten beef and shoddy cloth who were well supplied with capital from their connivance to give this country the greatest period of growth she has ever known.

Yet the heroes also distinguished themselves by exploitation, plunder, and fraud, on a colossal scale unknown in Britain. They made this generation the most flagrant in the nation's history for routine corruption, in both government and business; they thoroughly earned another name for themselves, the 'robber barons.' With the help of the Republican Party, which ruled the country in their interests to the end of the century, they also made America the most backward of the industrial countries in social legislation to protect workers against the abuses of private enterprise and the hazards of industry—just as the interests of business had helped to make it the last 'Christian' nation to abolish slavery. The economic freedom they prized, no less because the authors of the Constitution had neglected to mention it, was a freedom only for themselves, the men on top; they fought bitterly the enterprise of workers who sought to achieve more freedom for themselves by organizing in labor unions. ... The magnate who said 'The public be damned' was only putting crudely the freedom they defended most zealously—the right to be socially irresponsible.[13]

But there is another side to the story of industrialization. The industrial revolution produced an abundance of cheap goods whose main market was the common man. It was, as C. P. Snow has reminded us, the only hope of the poor who, constituting the majority of mankind, had always gotten the short end of the economic stick throughout most of history. When some owners finally realized that the workers were a potential market and paid them enough to buy it made possible a new age for the common man. This is, of course, what the Marxist or communist cannot see or understand that the road to material betterment lies in evolution not revolution with an incentive program. The better material goods that the worker has gained has lessened the range between king and commoner, between millionaire and worker.

With cheap books, record player, television the common man can command performers in all fields of the past and present to enter his living room and play for him. This is something that not even kings could do a few years back. In the industrial west the common man lives cleaner, healthier, longer and more comfortable than ever before. Many of the evils of the factory system have disappeared and others are under attack.

As for children, indignation over their ordeals in factories and mines was a novel as tardy, for the children of the poor had always been worked hard, often treated callously. ... It was only with the spread of free public education in the industrial countries that the idea began taking hold that youngsters were entitled to a childhood of school and play, freedom from adult responsibilities. ... The age of Materialism produced far more humanitarian legislation than had any previous age, or the most religious societies, in part just because of the materialism. It spread the idea that poverty and physical suffering were evils, not really good for the soul.[14]

One of the means by which ideas were spread and materials moved was the application of steam to motion in the form of the locomotive and the steam boats. Most of the common man has always been forced to walk at a speed of 3-4 miles per hour. Royalty and the rich had horses or could afford boat passage but the poor walked. Even the rich were not traveling much faster than the pedestrian but they didn't have to walk and carry.

Some were afraid that they would not be able to breath at the pace that the train went. Others feared explosions and the noises of the train. But gradually railroads were the bands that pulled nations together and aided nationalism. Whereas before towns had been near navigable streams and canals with the railway towns fought to have rail transport and settlements were made in advance of the rails.

Just as the factory was the pulse of the town so the railways forced nations towards a standard time which was achieved in the United States in 1883. Previous to the railroad each town had kept what amounted to sun time with noon when the sun was overhead. The factory and the railroad forced man to accept the tyranny of time. One of the stages in the mechanization of man—he eats because its time to eat, goes to bed because its bedtime. The mainspring of modern man is his watch.

The development of a rail system made it possible to settle the central portion of this country at a rapid pace. Hordes of immigrants lured by the cheap rail rates moved to fill the population gap. The railroads needed people to produce and to have produce which could be shipped by rail. It destroyed the canal boats and the river boats in the competition of the late nineteenth century. The railroads moved the bulk products of cheap mass production cheaply and relatively quickly.

A warning note should be sounded that technology does not hold the answers to all of man's problems nor are good intentions always fulfilled.

In coal mining one of the problems that developed early aside from the water level that was solved or aided by a pump worked by a steam engine was the explosive conditions created by finely divided coal and the release of various gases. Sir Humphry Davy visited Newcastle (a large coal mining region in England) with the purpose of alleviating the condition and making life safer for the miner. He developed a lamp which would not cause an explosion. He only wanted to serve humanity and didn't even patent a device which would have given him an income of twenty-five to fifty thousands of dollars per year. Unfortunately, the Davy lamp made life harder and more dangerous for the miner because he could then work deeper and more dangerous seams of coal and the number of accidents actually increased. Even if the innovation is intended as a blessing or good for man it may not be used in that manner.

STUDY AND DISCUSSION QUESTIONS

1. Why do most states have compulsory school attendance laws?
2. Compare the domestic system to the factory system.
3. What proportion of young persons and women were employed in the textile industry in 1848?
4. What were the chief objections to industrialism by the Romantic writers?
5. Why is the technology of steam and iron related?
6. Relate the factory system to the economy and the political scene.
7. Should factories close on Sunday if stores do?
8. How has industrialism made the western world better for the common man? For everyone?
9. Why did the working class drink?
10. How did the railroads force acceptance of Standard Time?
11. What has been the influence of railroads?
12. If pollution is as old as Rome why are we uptight about it today?
13. Why do slums exist?
14. How can technology be used to solve labor disputes?
15. What are the arguments for and against featherbedding?

FOOTNOTES

1. Edwin Hodder, *The Life and Work of the Seventh Earl of Shaftesbury,* (K. G. Cassell & Co., Ltd., 1893), pp. 75-78.
2. For other descriptions see Paul Mantoux, *The Industrial Revolution in the Eighteenth Century* (Harper Torchbooks, 1961), pp. 399-439.
   Jurgen Kuczynski, *Labor Conditions in Great Britain* (International Publishers, 1946).
   G. D. H. Cole and A. W. Tilson, *British Working Class Movements Selected Documents 1789–1875* (St. Martin's Press, 1965).
   Eric E. Lampord "The Social Impact of the Industrial Revolution" in Melvin Kranzberg and Carroll W. Pursell (eds.) *Technology in Western Civilization,* Vol. I (Oxford University Press), pp. 302-321.
   Norman Ware, *The Industrial Worker* 1844–1860 (Quadrangle Books, 1964).
3. Hodder, *op. cit.,* p. 369
4. Herbert J. Muller, *The Children of Frankenstein* (Indiana University Press, 1970), p. 63.
5. Lewis Mumford, *Technics and Civilization* (Harcourt, Brace and World, 1963), pp. 163-164
6. Ibid., pp. 171-172
7. Muller, *op. cit.,* p. 55
8. Mumford, *op. cit.,* p. 175
9. Muller, *op. cit.,* p. 59
10. *Ibid.,* p. 60
11. Mumford, *op. cit.,* p. 175
12. *Ibid.,* p. 179
13. Muller, *op. cit.,* pp. 56-57
14. *Ibid.,* pp. 72-73

XV

Bird or Bag of Hot Air?

LIGHTER-THAN-AIR CRAFT

Greek, Assyrian, Egyptian and Oriental mythology are filled with references to man's desire to fly and most involve imitation to soaring like a bird. The legend of Icarus, who was supposed to have used wax and feathers to make wings and fly but his vanity made him fly too near the sun where the heat melted the wax and he fell to earth, testifies to the antiquity of the dream of man to fly like a bird. However, our story of flight has a rather wet beginning.

ARCHIMEDES PRINCIPLE

King Hiero II ordered a crown made of pure gold. When he received the crown, he suspected that the artist had added too much alloy. The question was, how could he be sure?

He took the problem to Archimedes (b. 287-212 B.C.) at Syracuse, Sicily. No chemical tests had been devised for determining purity, but the idea of density was known. The crown was irregular in shape and did not lend itself to a simple solution.

While taking a bath, Archimedes noticed that the water seemed to buoy him up—just as you have noticed that you seem to be pushed up by the liquid in a bath (unless you always take a shower and do not swim). The idea struck rapidly and, so excited was he, that he jumped from the bath and ran naked through the streets shouting, "Eureka!" (this is Greek and means, I have found it!) Not only was he able to prove that the king's crown had an alloy with it, he also formulated what has come to be called Archimedes Principle.

ARCHIMEDES PRINCIPLE

A body surrounded by a fluid (water, oil, air) loses as much weight as the weight of an equal volume of the fluid. The fluid is pushed aside (no two objects can occupy the same space at the same time) to the extent of the volume of the body.[1]

Evangelista Torricelli (1608–1647), a student of Galileo, in 1643 invented a barometer. It was a glass tube with one end sealed, filled with mercury, and

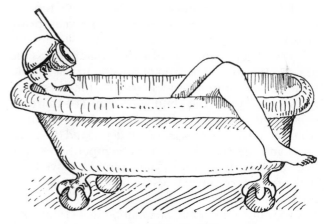

Figure XV-1. The Bath.

inverted in a cup of mercury. The pressure of the air would support a column of mercury about 30 inches high leaving a vacuum in the tube above 30 inches. This demonstrated that air must be a something since it could support a column of mercury. In taking it up a mountain side, it was found that the mercury column gradually became less which showed that the amount of the something decreased as altitude increased.

Francesco de Lana, a Jesuit monk, in 1670 reasoned that "no air" is lighter than some air and therefore a container from which air was removed would be buoyed up like wood or cork floating on water. He proposed that a metal container which weighed less than the air which it would normally contain should give him a vacuum balloon.

The reason why such an idea was not then practical was the weight of any metal container is heavier than the air contained in its volume if it is to be strong enough to withstand the air pressure.

Joseph Galien, (1699–1782) a Dominican friar, using the theory that air near the top of the atmosphere would be lighter than that at the bottom, in 1755 suggested that an airship with a volume of 1,000,000 (1×10^6) cu. ft. of this upper atmosphere would have a lifting force of 7,500,000,000 (7.5×10^9) pounds. The question might be asked as to how one was to get high enough to get the air of the upper atmosphere?

The first successful lighter-than-air craft in the Western world, was developed in late 1782 by two French brothers, Joseph Michel (1740–1810) and Jacques Etienne Montgolfier (1745–1799). They used a balloon made of silk, open at the bottom, and when a flaming paper was held at the opening, the bag filled with hot air. Since hot air is less dense than cold air, the balloon rose.

The Montgolfiers may have been aware of Archimedes principle but they got their idea from careful observations of chimney smoke, a practical rather than a theoretical approach. It is also an illustration of cross fertilization—an idea from one area producing results in another.

After additional tests their first public launching was on June 5, 1783, with a balloon of 23,430 ft³ volume and 105 ft. in circumference made at Annoney, France.

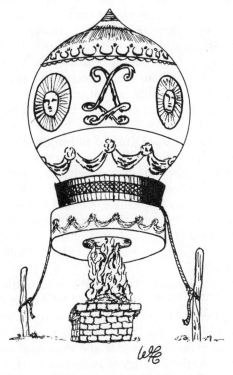

Figure XV-2. Hot Air Balloon.

On September 19, 1783, they employed a balloon made of linen and paper and using smoke from a fire of straw and wood the balloon went up 1,500 ft. and flew 1-1/2 miles. The flight lasted eight minutes carrying the first passengers—a sheep, a duck, and a cock.

Henry Cavendish (1731–1810) had proved in 1766 that hydrogen was seven times lighter than air. It occurred to Joseph Black (the same Black that had

ideas about the heat engine but who failed to apply them as Watt did) that a thin-skin container filled with hydrogen could be lifted upward. As in the case of the steam engine he had the idea but lacked the practicality to do anything about it. Actually the first recorded suggestion for the use of a gas filled balloon was made by Roger Bacon (c. 1214–1294) about 1250.

Tiberio Cavalio (1749–1809), an Italian physicist in England inflated soap bubbles with hydrogen and caused them to rise.

Again, as in the case of the Early Renaissance, the merger of the intellectual or the man of ideas with the craftsman is the most successful combination for material progress. Professor Jacques Alexandre Cesar Charles (1746–1823), a French physicist, with the help of the brothers Charles and M. N. Roberts, on August 27, 1783, using a spherical balloon made of lustring silk, 13 ft. in diameter, and coated with a varnish made of rubber, inflated with hydrogen made a flight. The flight took place in the heart of Paris and lasted 45 minutes covering a distance of 15 miles.

It was destroyed on the ground by a mob of terrified French peasants.

The first man-made aerial flight took place on November 21, 1783, and was made by Jean François Pilatre de Rozier (1756–1785) and the Marquis Fraçois Laurent d' Arlandes (1742–1809) in a hot air balloon over Paris. The flight lasted 20 minutes and covered a distance of five miles reaching a height of 3,000 ft.

At this point the balloon is somewhat similar to the development of early boats—no means of propelling or steering. It also gives rise to a sport which becomes quite popular in the 18th and early 19th centuries—ballooning. We tend to apply what has been used with success in other fields. So using the boat analogy the first attempts to control the flight were oars, sails (twin parasols), propellers, and paddle wheels.

Using paddle wheels, Francios Blanchard (1753–1809), a Frenchman, reputed to be the inventor of the parachute (1785), carried a passenger Dr. John Jeffries (1744–1819), an American, across the English Channel on January 7, 1785. They left Dover and arrived in Calais, a distance of about twenty miles, three hours later.

While we do not usually think of the early national period as being the air age, it was in 1793, that Blanchard received special permission from George Washington and flew his balloon from Philadelphia, then the capital of the United States, to New Jersey, just east of Woodbury (about 5 miles). The flight lasted 45 minutes and was viewed by Washington, his cabinet, members of Congress and an astonished population.

APPLICATION OF POWER

The Dirigible—a cigar shaped balloon with power.

Henri Giffard (1825–1882) made the first directed flight from the Paris Hippodrome September 24, 1852, using a three-horsepower steam engine which weighed 351 pounds and operated a propeller which drove a 88,000 ft^3 hydrogen aircraft.

Many other persons of many nationalities continued work in the field of lighter-than-air craft. (e.g., Paul Haelein (1835–1905), a German engineer, in 1872, flew the first dirigible using a four-cylinder gasoline internal combustion engine.)

The Zeppelin—similar to the dirigible but having a rigid structure.

The basic principles of a rigid aircraft were patented by Joseph Spiess (1839–1917), a French engineer in 1873. But the credit for the invention and construction of the first zeppelin is credited to Count Ferdinand von Zeppelin (1838–1917), a German military man, who served in the balloon corps of the Union Army during the U.S. Civil War. He formed a public company for their construction in 1898. The first zeppelin made its trial flight on July 2, 1900. Between 1900 and 1920 they built 150 commercial and military craft.

The German Airship Transportation Company formed in 1910 carried 34,228 passengers and traveled 107,180 miles without accident during a four-year period.

After World War I, on September 18, 1928, the *Graf Zeppelin* made its maiden trip. In 1929, it flew around the world from Friedrickshafen, Germany via Tokyo, Los Angeles, and Lakehurst in 20 days, 15 hours and 17 minutes.

The Hindenberg 7 x 10^6 ft^3 in volume and built at a cost of $2,600,000 was launched in March, 1936. In 14 months it made 54 flights, 36 of these were trans-Atlantic. On May 6, 1937, the Hindenburg was making a landing at Lakehurst, N.J., when it burst into flames. Thirteen passengers were killed. This was the effective end of rigid lighter-than-air craft.

HEAVIER-THAN-AIR CRAFT

Leonardo da Vinci (1452–1519) in 1505 was one of the first to state some of the rational principles of flight by mechanical imitation of nature. e.g., birds, bats, flying squirrels, flying fish, seeds of certain trees and plants that are carried by air currents. He kept his writings secret by reverse script that could only be read by using a mirror, thus his concepts had little effect during the Renaissance and became known to the rest of the world about 1797.

The Glider—similar to an airplane but lacking power except gravity and natural air current. Early models used take-offs from high points of land or running tows behind humans or horses, whereas today airplanes or jeeps are used for take-offs.

It is interesting to note that almost a century was to pass after da Vinci's works became known before the first practical experimental works were done on heavier-than-air craft. This might be explained away as due to cultural lag, but I favor the explanation that it was more due to the successes of the lighter-than-air craft which sidetracked other efforts. At any rate, no connection seems to exist between da Vinci's works and the brothers Lilienthal—Otto and Gustav.

A group of Englishmen, Sir George Cayley (1773–1857), William Samuel Henson (1805–1888), and John Stringfellow (1700–1883), made theoretical investigations and model flight studies in the first half of the 19th century, but it remained for Otto Lilienthal (1848–1896) and his brother, Gustav (1849–1933) to make gliding practical.

He, like da Vinci, gathered data on flight by observing the flight of birds. They began their experiments in 1867, but Otto did not build his first man carrying glider until 1891. It was a frame constructed of peeled willow rods over which a tough cotton fabric was stretched. The wings were curved instead of flat planes—this from the birds and his own experiments. He put his arms through padded rubber tubes and grasped the cross bar. Control of the glider's flight was by body movement. Otto was killed in his experiments.

Octave Chanute (1832–1910), a structural engineer, who in 1896 at the age of 64 began to make gliding flights. He tried five sets of wings, then three, and finally two. Further, he elaborated control by means of a vertical rudder and movable wings. His bi-plane weighed 23 lbs. or including the pilot hanging beneath it 178 lbs. So well did he build that his 2,000 flights were without an accident.

John J. Montgomery (1859–1911) used gliding as a means of aero-dynamic study. Between 1883–1894 he experimented with a bi-plane which had a seat, pulley operated horizontal tail, and curved wings. Side motion was by body movement. His first flight, without running, achieved a 600 ft. glide in an 8 to 12 mile an hour wind.

In 1903, he stretched a cable between two mountains and used various models to find the proper form for a wing surface. He tested the result by a manned flight a year later to observe in-flight air flow effects.

His largest glider, 45 lbs., was launched April 29, 1905, from a hot air balloon at an altitude of 4,000 ft. The flight lasted 20 minutes. (The 1952 gliding record was 56 hours, 15 minutes.)

FROM GLIDERS TO POWER FLIGHT

The brothers Wright operated a successful bicycle manufacturing and repair shop in Dayton, Ohio. In 1899, they wrote the Smithsonian Institute in Washington, D.C., for a list of reading material on aeronautics. That institution recommended some articles by Lilienthal and Chanute which the brothers avidly read. They also wrote to Chanute and began an exchange of technical data which lasted over a period of time.

Their first idea was to build a kite capable of carrying a man. They wrote the United States Weather Bureau to try to find a suitable location of up-drafts and other conditions favorable to their undertakings. The Bureau recommended two locations—one on the west coast and the other at Kitty Hawk, N.C. The brothers selected the North Carolina location because it was closer to Dayton.

In the spring of 1900, when the bicycle manufacture season slacked off, they made a model based on the writings of Lilienthal and Chanute. In the fall they took it to Kitty Hawk and flew it as a kite controlled by two ropes. They used a 50 lb. payload of chains.

Since it seemed to fly well, the Wright brothers built a larger man carrying glider which they tried at Kitty Hawk in the fall of 1901 with Chanute watching. The glider proved unsuccessful in flight and convinced the brothers that there were flaws and errors in the data which had been supplied by Lilienthal and Chanute.

They, thus, began their own experimental studies which was to lead to successful powered flight. In the winter of 1902–03, they built a wind tunnel from an old wooden starch box which confirmed the idea that the data they had been using was wrong. They then built a more efficient wind tunnel powered by a single cylinder gasoline engine. They experimented with some 200 different shaped airfoils. (One of the early wind tunnels is part of a display in Greenfield Village near Dearborn, Michigan.) From the data thus secured, they constructed a new glider which was quite different from those of previous gliders in design.

In the fall of 1902, they made over a thousand flights at Kitty Hawk. Some into winds of 36 miles/hour and in some glides went over 600 ft.

They then decided to put power to gliders. Their analysis of the problem showed that steam engines which had been tried were too heavy for powered flight. Their specification called for an 8 H.P. engine that would give 1 H.P. for every 20 lbs. of engine weight. They couldn't find a company willing to undertake the manufacture of such a motor. Since they had designed (and had their mechanic, Charles Taylor, build the motor for) the wind tunnel they decided to build their own engine for the aircraft.

Between the fall of 1902 and that of 1903 the bicycle business was neglected while they tackled the problem of engine building. Their engine had direct fuel injection into its cylinders and was made of aluminum.

Their success at bicycle building may be gauged from the facts that they could afford to be absent from the plant for long periods of time; that the cost of transporting models to Kitty Hawk and staying there for a period of time cannot have been inexpensive and that they were proposing to use a metal which was then relatively rare and that cost far more than iron.

The Wright engine weighed 170 lbs. and gave 16 H.P. for the first 15 seconds at 1,200 r.p.m. They then applied their knowledge gained from bicycle manufacture (the bicycle in many ways was the forerunner of both the automobile and the airplane) to rig two chain drives to transmit power to the two pusher type propellers.

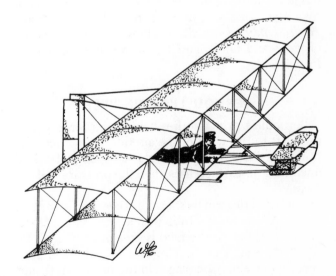

Figure XV-3. Early Airplane.

They did not arrive at Kitty Hawk until September, 1903, where a combination of bad weather and mechanical problems delayed their flight until December 14, 1903. They tossed a coin and Wilbur (1867–1912) won the prize of the first flight. He was up 3-1/2 seconds, climbed too steeply, and crashed.

After two days to repair the damage it was Orville's (1871–1948) turn. He was up 12 seconds and landed 120 ft. from the take-off point successfully. They made three more flights that afternoon. The longest flight lasted 59 seconds and covered a distance of 852 ft.

The press was informed of the event, but only three papers in the United States carried the story. Most of the press was skeptical and did not believe the story. The first complete account of the Wright's flights was reported in the March, 1904, issue of the magazine *Gleanings in Bee Culture.* Man had flown with power in a heavier-than-air craft, but few believed that it was possible.

THE BASIS OF HEAVIER-THAN-AIR CRAFT

As had been mentioned previously, air is a fluid so that the theoretical aspects of aerodynamics draws heavily on the field of hydrodynamics. We will examine some of the more pertinent principles that apply to the ideas of flight by a comparison of aerodynamics and hydrodynamics.

Daniel Bernoulli (1700–82) came from a family that had fled from religious persecutions in Antwerp (what is now Belgium) to Switzerland. The family had many members who have contributed greatly to the sum of man's knowledge, especially in the various fields of mathematics. Daniel's main interest was hydrodynamics and in 1738, he published a book, *Hydrodynamics* which develops the kinetic theory of gases and fluids; and also formulates Bernoulli's Principle.

BERNOULLI'S PRINCIPLE

Bernoulli's principle lies behind the force which gives lift to the wing of the aircraft. In essence, the principle states that as the velocity (speed) of a fluid (including gases) increases, then the pressure will decrease.

In a liquid this can be shown by the use of a Venturi meter. A Venturi is a tube with a more slender portion. The speed of the fluid is greater at B (see Fig. XV-4) than at A or C. To make this into a meter, vertical tubes may be placed in the tube at A, B, and C. The pressure in the horizontal tube is shown by the height to which the fluid rises in the vertical

tubes. At h_2 where the fluid speed is greatest, the pressure is least.

Why should this be? As we have already seen, and as most of you already knew, you can't get something for nothing (the physicist calls this the law of conservation of energy). In this case, the pressure (P) times the velocity must be the same everyplace in the tube. $P_1 V_1 = P_2 V_2$. So that at point B where the velocity is increasing in order to keep the amount of energy the same, the pressure must decrease. If B is 1/2 the cross-sectional area of A then the velocity in B is

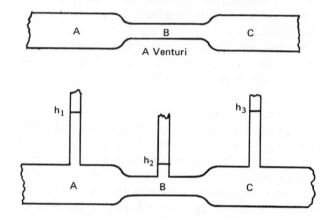

Figure XV-4. Bernoulli's Principle.

doubled and drops the pressure to 1/4 in B as compared to A.

Let us see how this applies to the wing of an aircraft. (See Fig. XV-5.) By the shape of the wing sections more air is forced over the top of the wing at A than at B. This has the same effect as forcing a fluid through a smaller opening so the velocity of the air at point A is greater than at point B. Since the velocity of the air at point A is greater, the pressure must decrease. The pressure at point B is then greater than at point A so this provides the lift for the wing. (Bernoulli's Principle also explains how atomizers work; why a baseball curves up, down, right or left; why roofs are lifted off in high winds; why a ping pong ball remains in a vertical jet of water.)

In addition to Bernoulli's principle, there is also the effects from the air particles which impart a force as they strike the wing surface. This is described in Newton's Third Law of Motion—that for every force there is an equal and opposite force and while the energy of one air molecule is not much this must be multiplied by the number of air molecules.

In order for a plane to fly, there are a number of other factors which we will not consider in detail. These are thrust, weight, center of gravity, induced drag, friction drag, downwash, etc.

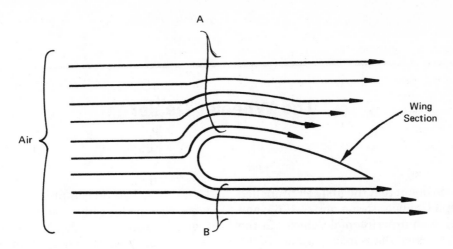

Figure XV-5. Wing Cross Section.

STUDY AND DISCUSSION QUESTIONS

1. Explain the principle by which lighter-than-air-craft rise.
2. Why was the development of a lighter-than-air-craft so late? When was it developed in the East?
3. Why were hydrogen and hot air used?
4. What is the most successful combination for material progress? Why?
5. When did the air age begin in this country? Why isn't more said of this in history?
6. What are the differences among a balloon, dirigible, and zeppelin?
7. How does lighter-than-air-craft control develop?
8. What were some of the natural phenomena that suggested flight to man?
9. Why didn't the Wright Brothers' work suffer the same fate as Mendel's?
10. What are the basis of heavier-than-air-craft? Explain.
11. Describe the Wright Brothers' approach to the study of flight.

FOOTNOTES

1. T. L. Heath, *The Works of Archimedes* (Cambridge, 1897). or *Great Books of the Western World, Vol. II.*

Suggested Readings

Auguste Piccard, *Earth, Sky and Sea.* (Oxford University Press, 1956). pp. 1-18. General, readable and recent. A personal account.

J. L. Nayler and E. Ower, *Aviation: Its Technical Development.* (Dufour, 1965). Illustrated, readable on both balloons and heavier-than-air craft.

Oliver Stewart, *Aviation: The Creative Ideas.* (Praeger, 1966). Readable history of heavier-than-air craft.

Ascher H. Shapiro, *Shape and Flow* (Doubleday Anchor, 1961). Especially pp. 138-142. Well illustrated and fairly readable.

J. H. Means, *James Means and the Problem of Manflight* (Smithsonian, 1964). Detailed on early period 1882–1920 and related more to one man. Less readable, but good illustrations.

Charles H. Gibbs-Smith, *The Invention of the Aeroplane* (Taplinger, 1966). Detailed, but readable.

XVI

The Wheeler Dealer

While there is no doubt that the first self propelled vehicle was Joseph Cugnot's steam one of 1765, and there were other such experimental vehicles in the early nineteenth century, this is mainly a story of the internal combustion engine applied to free self-moving vehicles. To a certain extent the early airplane and the automobile stories overlap in the development of engines.

Both the airplane and automobile are descendants of the bicycle, to a certain extent, for it was the rise in popularity of the bicycle in the 1860's that gave the impetus for a self moving vehicle. The bicycle, which would propel a man at 10 to 12 miles an hour, liberated men for short trips and for pleasure that he could not afford with a horse. There were other factors at the start of our Civil War that contributed to the development of the automobile. These revolve about the solution of three major problems—fuel, machining technology, and metallurgy.

There were some in the seventeenth century with the idea of an internal combustion engine using gunpowder, Hughens and Papin, for example. Others thought that it might be possible to use various 'inflammable airs' such as vaporized turpentine, e.g., R. Street, George Cayley, while working with his ornithopter experiments, in 1808 actually built a gunpowder engine and it was the first example of hot-tube ignition. Samuel Brown in 1825 followed an idea of the early steam engine and constructed and sold an engine which used the combustion of hydrogen sulfide to create a vacuum after which atmospheric pressure drove the piston. Although the engine had a bore of 12 inches and a stroke of 24, it only developed 4 horse power. Brown's engine used 1,000 cubic feet of gas per horse-power-hour and thus was not economical.

The solution to the fuel problem for the engine was in the developing illuminating gas industry. Every place that set up a city light plant suggested the idea to some of power as well as light. Joseph Etienne Lenoir (1822–1900) was the first to make use of such an engine about 1860 which modeled after a steam engine in valves and in double acting were used in factories. It had no compression and unless the fuel were also in the compressed form, had a disadvantage of the cost of a rather long pipe to the source of supply. In 1862, he modified an engine to run on liquid fuel and put it on a vehicle. It took two to three hours for a six mile trip.

Liquid fuel could be supplied by a rising petroleum industry (the first oil well was in 1859) but gasoline was not used until the carburetor was perfected about 1890. Lubricating oils could also replace vegetable oils and animal fats, although caster oil was the preferred lubricant for racing until 1936.

While the theories of thermodynamics had not been developed in time to help with the steam engine the work done by Mayer in the 1840's and by Joule in the 1850's was of material aid to those engineers constructing automobiles. Then too the theoretical calculations of the fuel heats from these beginning ideas in chemical calorimetry were of an aid in the design of engines. There were timely technical innovations as well.

Both the steam engine and the internal combustion engines suffered from the lack of accuracy that was present in cutting the bore for the cylinder and the piston as well as the proper transmission of power through well cut gears. These had to wait for industrial development and perfection. In the early engines, including internal combustion engines, after the piston was fitted it had to wear into position by running and some were so crudely built that a person was sent with the machine to make adjustments in the fit. One of the crucial machines for the advancement of material progress, the lathe, had by 1850 developed into a machine tool of precision and great capacity.[1] By the same period production gear-cutting machines had been developed[2] and before the turn of the century, 1897, production grinding machines for automobile crankshafts were available.[3]

Nikolaus August Otto (1832–1891), a traveling salesman, developed an engine which worked on the atmospheric principle in 1863. He formed a partnership with Eugene Langen (1833–1895) who provided the capital and business acumen. This man, like Boulton with Watt, and his type are often overlooked in the fame that goes with the development process. The role of capital and the capitalist in implementing a change for profit is an important one. I would not care to speculate on the number of inventions that have not been developed for lack of capital.

In 1862, Alphonse Beau de Rochas (1815–1891) published an analysis of the cycle of operations for a successful gas engine—the four cycle arrangement used today. Ten years later Gottlieb Daimler (1834–1900) who was then employed by Otto and Langen made the first commercially successful engine using the idea of de Rochas. Later Daimler made his own engines—one mounted on a bicycle in 1885 and his first car in 1886.

> In the original Otto machine a charge of gas and air was drawn into the cylinder and then fired to raise an exceedingly heavily weighted piston and rod; on the piston descending at the end of this stroke motion was given to a shaft and flywheel by rack, pinion and automatic clutch —the actual work on the flywheel shaft being, therefore, performed by gravity. A suitable arrangement of links operated the valves and the slide for the ignition flame. This crude and unlovely contrivance, which made a noise like a brontosaurus with hiccups throwing a set of fire-irons downstairs, was more mechanically efficient than Lenoir's (used half the fuel and was twice as fast) but was very erratic in action.[4]

Together Daimler and Otto produced a gas engine using the four stroke principle and patented in 1876. It was marketed as "Otto Silent Engine"—"the silence was comparative."[5] It was, however,

> ... the earliest recognizable ancestor of today's automobile engine, burned illuminating gas in a single horizontal cylinder, developed about three horsepower at 180 revolutions per minute, and weighed one or two thousand pounds per horsepower.[6]

There were a number of pioneers in the automotive field of various nationalities. Siegfried Marcus (1831–1898) was in the business of construction of vehicles powered by gas engines in 1873 and a year later George Bailey Brayton (1830–1892) got a patent on a two-cycle oil engine. Karl Friedrich Benz (1844–1929) produced the first reliable internal combustion engine in 1885 and made a number of improvements including electric ignition, a water cooling radiator, differential gears and the use of benzine for fuel.

> Enough has been said to show the impossibility of stating that any one person was the 'inventor' of the motor car. The idea germinated in many different minds and took many different forms, but if the meaning of 'inventor' is narrowed to signify the man who first designed and produced light self-propelled carriages for sale to the public, then Karl Benz of Mannheim (Germany) has the greatest claim to the honour.[7]

Essentially the automobile is the result of the combination of a horseless carriage, a bicycle, and the internal combustion engine—the new technology rising from those of the past. One explanation of the motor car mania has been give by Hiram Percy Maxim (1869–1936) who was himself fired up about making his bicycle self propelled in 1895.

> ... we all began to work on a gasoline-engine-propelled road vehicle at about the same time because it had become apparent that civilization was ready for the mechanical vehicle. It was natural that this idea should strike many of us at about the same time. It has been the habit to give the gasoline-engine all the credit for bringing the automobile, as we term the mechanical road vehicle today. ... this is a wrong explanation. We have had the steam engine for over a century. We could have built steam vehicles in 1880, or indeed, in 1870. But we did not. We waited until 1895.
>
> The reason why we did not build mechanical road vehicles before this was because the bicycle had not yet come in numbers and had not directed men's minds to the possibilities of independent long-distance travel over the ordinary highway. We thought the railroad was good enough. The bicycle created a new demand which it was beyond the ability of the railroad to supply. Then it came about that the bicycle could not satisfy the demand which it had created. A mechanically propelled vehicle was wanted instead of a foot-propelled one and we now know that the automobile was the answer.[8]

It might be added that the amount of leisure time available to the middle class and its size would be factors in stimulating the response to the bicycle since they were not in a position to own a horse.

In this country the first persons to design, build and produce a practical internal combustion automobile were the brothers Duryea. Charles Edgar (1862–1938) and L. C. J. Frank (1869–1967) were also in the bicycle business and as a rather quarrelsome couple between them had an automobile operating on the streets of Springfield, Massachusetts about September 20, 1893.

> In general appearance the 'carriage' was much like a horsedrawn vehicle—without, of course, the horses. The engine was electrically ignited and water cooled. It was located in the rear of the car, the two cylinders 'opposing' each other—that is, lying opposite to each other— and both parallel to the end of the carriage. The transmission, which took the power of the engine to the wheels, was by belt. Frank had a clutch, which enabled him to shift the motor to different gears, producing different degrees of speed as the gear was large (and turned slowly) or small (and turned rapidly). Having the clutch, Frank could shift to any one of three degrees of forward speed and power, a reverse, and a neutral position ... Daimler, at first, had only one foreward speed and a reverse, while Benz had foreward speed only. The cars of these men had been improved, but they did not have the wide choice that Duryea's new machinery offered.

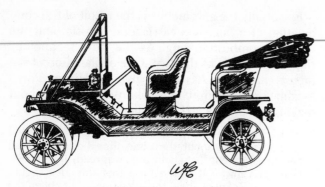

Figure XVI-1. Early Automobile.

The car would go 18 miles an hour. No machinery could be seen . . . To avoid noise in changing speeds, the gears in the clutch were lined with leather.[9]

One of the problems of the new cars was that of production. Parts were not interchangeable, although they were supposed to be, and the why and order of assembly caused more than one company to go out of business. Some of the companies were essentially carriage manufacturing companies and others were producers of bicycles. These facts influenced the appearance and construction of the early automobiles. Some used carriages for bodies while others had a tendency to use the tubular construction of a bicycle as well as the sprockets and chains. In some instances they made use of bicycle tires. Both the bicycle industry and the automobile industry were heavily dependant on the development of the pneumatic tire for their future.

Two giants of the automotive industry were to tackle the production problems—Henry Ford (1863–1947) and Ransom Eli Olds (1864–1950). Both had many of the same ideas to produce cheap cars and make a lot of them.

What Olds said he wanted was a car with production costs of about $300 that would retail at $650 or about half of what many cars then were selling for. He also had the idea that the automotive company should be the assembly plant—not the manufacturing company—and to this end they contracted a number of small machine shops around Detroit to make the parts.

Production began in the fall of 1901 and 425 cars were assembled in a few months. In the following year 2,500 cars were turned out. Such production was unheard of. The Oldsmobile one-cylinder had a 4-1/2 inch bore and the piston had a 6 inch stroke. The engine had two forward speeds and a reverse. It was easy to drive and repair (something not thought of today). Over long distances, it could average 14 miles an hour with the power supplied by a 5 horsepower engine.

Ford's work with motor cars was similar to that of many others in the field. His first satisfactory car was a belt driven vehicle with a buckboard type carriage which he made in 1896. He resigned his job as engineer of the Edison Company and took a position as superintendent of the Detroit Automobile Company that failed a short time later. To gain financial backing, Ford built a racer that won and founded the Ford Motor Company in November, 1901.

He, too, was convinced that the market lay with a cheap car and his Model A of 1902 was evidence of this. The machining was done by the Dodge Brothers and was a two-cylinder motor of 8 horsepower with two-speeds that sold for $800. There followed Models C, F, N, K, R, and S—most of these are relatively unknown today. Olds had lost the argument about a cheap car with his stockholders while Ford was able to win.

In the first few months of 1908, Ford was producing Models N, R, S, and T but when he was certain that he had found what he was searching for, he gave up all models but T and proposed that only that model would be produced and would rely on its cost to sell in large numbers. Large scale production began in 1909.

In order to meet the growing demand for his cheap cars, Ford innovated (he was not really an inventor) a change from what was called the "progressive assembly" method. In this older method:

A logical sequence was followed. Some workermen finished the car's chassis, or over-all frame. Others brought and attached the motor. Still others came with axles and wheels. The clutch and the transmission were installed. The steering wheel was connected with the proper elements. Finally, the windshield, seats, top, and side curtains, or, later, the enclosed body, were added.[10]

The new method involved taking the work to the men. The idea came from the approach used by Chicago meat packers to process a large number of carcasses. The cattle are attached by their hind feet to wheels and pulled upside down onto an overhead trolley and is moved past men, each of whom do only one job on the carcass. For automobile assembly, the process was more difficult.

The first thing that was done was to work out assemblies for seven separate units. These were magnetos, engines, transmissions, the chassis, the dash, the front axle, and body-and-top assemblies. The general plan was that the car moved down a line of workmen, and at various points the sub-assemblies fed in. Without moving a step, at the right time a worker performed an act—such as putting on a wheel, tightening a nut, or fitting a part. When all these aperations were perfectly timed, the car went forward without a stop. Starting as a frame, it was driven off at the end of the line as a complete automobile . . .

Before the main line could be put in motion, the special assemblies had to be worked out. They began with the magneto assembly. This job had previously been done by individual workmen, each averaging from 35 to 40 magnetos in a nine-hour day. The task of assembly was now broken down into 29 operations, each performed by one man.

The workers were spaced along a moving belt. The average previous time for assembling a magneto had been about 20 minutes. Under the new system it took 13 minutes, then 7, finally five. Three-quarters of the time formerly taken had been saved.[11]

In the new process the semi-skilled man replaced the skilled craftsman each working at a job from which time could be constantly pared by time and motion studies. This new method became known as "mass production" and its story is reflected by the numbers. In the year before Ford introduced this method to the automotive industry, his plant made 78,440 cars and in 1912 to 1913, with the introduction of the assembly line, it was 168,304 and by 1915–1916, with the process working smoothly, it had risen to 472,350.

Rising production meant that the price could be reduced and the 1908 cost of $825 had been cut to $260 by 1926. It was a car that most could afford and was affectionately named the "Tin Lizzie." In addition to cutting prices Ford, in 1914, reduced the work day to eight hours and set the minimum wage at his plant at $5 a day which was almost double what others were paying. This was not union negotiation, but what Ford regarded as good economic sense. It gave his workers a higher standard of living as an incentive to a dull job and made it possible for his own workers to become part of a market. The automobile had grown from a novelty and a toy for the rich to a creature of the common man.

Nearly all automobiles of today still make use of the four-cycle engine whose theory was described by de Rochas over a hundred years ago.

In the four-stroke cycle, it takes four strokes of the piston to give one power stroke. This is the reason that there are usually several pistons each operating in a cylinder. The piston slides up and down in the cylinder and is attached to the crankshaft by a connecting rod. The power is transmitted to the crankshaft which changes the up and down motion of the piston into rotating motion. The rotary motion has its direction changed in the differential gears and thus the power is supplied to the wheels. Provision is also made so that the wheels can turn separately since in going around corners the wheels do not travel at the same speed. On top of each cylinder are two valves and a spark plug. The valves let the gases in and out and the spark plug provides an electric spark that ignites the gas on the power stroke. Across the top of the cylinders is the camshaft which turns in connecting with the camshaft and operates rocker arms which open the valves. They are shut by springs. The gasoline or other fuel is vaporized and mixed in proper proportions with air in the carburetor.

The valve closes in the second or compression stroke and the piston rises compressing the fuel and air mixture to 8 times that of atmospheric pressure. Early engines had compression ratios as low as 4 to

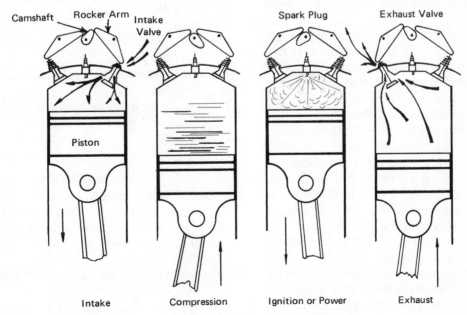

Figure XVI-2. Four-Cycle Engine.

1 which means that the piston in moving to the top of its stroke compressed the fuel-air mixture to 1/4 the volume that the total cylinder contained. Today's engines have compression ratios of 8 or higher.

With the valves still closed, the spark plug supplies the electric spark that ignites the fuel and the rapid burning give heat which causes the waste gases to expand. These expanding gases force the piston downward in the third or power stroke.

In the exhaust, or fourth stroke, with the exhaust valve open the piston again comes up forcing the waste gases out.

The heat and pressures developed in the cylinders of the internal combustion engines posed several problems. The burning gases produced temperatures on the order of 3,000°F or about 1,650°C. Since cast iron melts at 1150°C, it was necessary to reduce the heat and this was done by using cooling water which kept the heat of the cylinder walls at about 200°F but the water was constantly evaporating and if allowed to go dry would destroy the engine. The piston head was not water cooled and developed temperatures of 900°F. As you know, metals expand when heated and contract when cooled so this created a problem in the type of metals to be used for pistons and piston rings. Piston rings are separate rings of metal around the pistons that can expand or contract while keeping a tight fit. Metallurgy was constantly challenged to provide motors which would keep up with the fuels. Some hint as to the importance is indicated in a story of Ford that he had a piece of engine metal analyzed from a French racing car to find out its special qualities and then afterwards vanadium showed in the alloys made for Ford cars.

Engine design problems also included the arrangement and number of the cylinders. Early cars had cylinders which were placed horizontally and if more than one cylinder was used, they were placed oppositely. Then later, they were placed vertically and finally in a V formation so that the cylinders were inclined. In airplane engines the cylinders were arranged radially or around in a circle.

Mention must be made of another type of internal combustion engine which was developed by Dr. Rudolf Diesel (1858–1913) about 1892. His first successful engine was produced in 1897. His early engine replaced stationary steam engines but application to vehicles was not made until the 1930's because of the high weight and slow operation.

In the Diesel engine, there is no ignition system for with a higher compression ratio the air is heated by the great pressure and ignites the fuel when it is squirted into the cylinder. It is somewhat slower in starting than the gasoline engine but is more efficient. It replaced the steam locomotive and later was used for trucks and buses.

STUDY AND DISCUSSION QUESTIONS

1. In what ways were the development of the steam engine and the internal combustion engine related?
2. How did developments in the airplane and the automobile overlap?
3. How did developments in the illuminating gas industry, bicycle and carriage manufacture contribute to the introduction of the automobile?
4. Why would bicycle manufacturing tend to be related to both the persons developing the airplane and those developing the automobile?
5. What factors are necessary for the development of the concept of "mass production"?
6. What is the operation of the internal combustion engine?
7. What is the role of the capitalist in advancing an invention?
8. What technical advancements made possible the development of the internal combustion engine?

FOOTNOTES

1. Robert S. Woodbury, *History of the Lathe to 1850* (M.I.T. Press, 1961), p. 117.
2. See Robert S. Woodbury, *History of the Gear Cutting Machine* (M.I.T. Press, 1958).
3. Melvin Kranzberg and Carroll W. Pursell, Jr., (eds.), *Technology in Western Civilization* (Oxford University Press, 1967), p. 26.
4. Anthony Bird, *The Motor Car 1765–1914* (B. T. Batsford, 1960), p. 26.
5. *Ibid.*, p. 27.
6. Kranzberg and Pursell, *op. cit.*, p. 653.
7. Bird, *op. cit.*, p. 30.
8. Hiram P. Maxim, *Horseless Carriage Days* (Harper and Brothers, 1937).
9. Frank Ernest Hill, *The Automobile* (Dodd, Mead and Co., 1967), p. 16.
10. *Ibid.*, p. 68.
11. *Ibid.*, pp. 69-70.

Suggested Reading

Automobile Manufacturers Association, *Automobiles of America,* Wayne State University Press, 1968.

Frank Ernest Hill, *The Automobile: How It Came, Grew, and Has Changed Our Lives,* Dodd, Mead and Company, 1967.

Anthony Bird, *The Motor Car 1765–1914,* B. T. Batsford, Ltd., 1960.

Lynwood Bryant, "The Origin of the Automobile Engine" *Scientific American,* March, 1967, pp. 102-112.

XVII

The Jet Set and the Pink Car-Nation

Both the airplane and the automobile are parts of man's dreams of liberation. The airplane fulfilled man's long term dream to fly like a bird and be liberated from the earth while the development of the automobile liberated man from the city. In the achievement of his dreams man has had to let some of the bad side effects creep in.

The effects of aircraft, which began as playthings, has developed long range consequences. From the winter of 1783 to today is under 200 years which is a brief space in the 5,000 years of man's recorded history. Yet in that space of time, man has risen from the surface of the earth to explore the ocean of air above him and beyond—even to the moon.

Using a balloon, Dr. John Jeffries, in 1784, put the new device to the practical application of obtaining basic information about air. He recorded the temperature, pressure; measured its density and even took samples at different altitudes. The balloon is still used by weather services to obtain meteorlogical data as to air currents, temperatures, and pressures of the different air strata for today's forecasting.

It was an early recreational device for fun and frolic. It was, and still is, used at county fairs and other public gatherings to give the public a bird's eye view of the surrounding country. In the last decade ballooning has become such a popular sport that the Federal Aviation Agency has recently issued rules governing the licensing of balloon pilots.

The balloon provided a new challenge in a new dimension. It was a means of exploring the air above us. Joseph Louis Gay-Lussac reached an altitude of 23,000 ft. as early as 1804. The attempts for increasing altitude have continued and the present record is held by Commander Malcolm Ross and Lt. Commander Victor G. Prather who reached an altitude of 113,000 ft. (about 20 miles) in May, 1961. Besides the challenge and the setting of records, ballooning has provided man with information about the stratosphere. Operation Skyhook, which was begun in 1947, was used to study cosmic rays and to provide information about sun spots and emanations from space to the communications industry.

We usually do not think of the eighteenth century as the beginning of air warfare, yet the balloon was used by the French at the Battle of Fleurus in 1793–94. It not only provided intelligence information but also demoralized the enemy. They were also used at Antwerp in 1815 and at Solferino in 1859. Combined with the telegraph it was successfully used by the Union Army in the defense of Washington in 1861, at the Battle of Fair Oaks.

During World Wars I and II, balloons were used for observation, for propaganda (dropping leaflets over enemy territory), to protect cities from attacks of low flying aircraft (they were a factor in the Battle of Britain), and to disperse incendiary devices. The last was used by the Japanese against the United States. So successful were these in starting forest fires in our vast Western timber resources that the government imposed a strict censorship about the fires hoping that the enemy would discontinue the attacks as not being effective. The Japanese did not hear anything about the success of their attack and did discontinue it.

Gliders were used during World War II to transport troops and supplies in order to give an element of surprise to many attacks. They were also not only the prelude to powered aircraft, but have gained a following of their own. Meets are now held in many parts of the country and around the world for the glider enthusiast.

Man's first successful powered flight was just under 70 years ago which puts the air age well within the life span of one man. The first commercial airline began service in Florida in 1914 but most of the development of commercial aviation was during the 1920's and 30's. Charles Lindbergh (1902–   ) (he and many other aviation pioneers are still living) made the first solo non-stop flight from New York to Paris in 1927. The plane was named "The Spirit of St. Louis" for the backers of the flight were there and Lindbergh flew there from New York after his successful flight. The author's wife, who was raised near St. Louis, can remember when his plane flew over their farm.

The world has become much closer knit today. How fast did it happen? It took Ferdinand Magellan's ships (1520–22) two years to circumnavigate the globe. Jules Verne (1828–1905), whose first novelette in 1863 incidentally was titled *Five Weeks in a Balloon,* wrote *Around The World in 80 Days* in 1873. This work of fiction was beaten by fact when Nellie Bly in 1889 went around the world in 72 days, 6 hours, and 11 minutes. John Henry Mears cut this in half in 1913 with a trip of 35 days, 21 hours, and 36 minutes. Two U.S. Army airplanes made the trip in 1924 in about 14 days. The "Graf Zeppelin" in 1929 took 20 days. The time was reduced to 18 days by H. R. Ekins in 1936. While in 1947, a Pan American four-engine Lockheed Constellation made it in four and a half days. Ten years later, three U.S. Air Force B-52's made the trip in a little under two days while Henry G. Beaird in 1966 took about 3 days for the trip. Military planes have advantages of bases and subsidy that are not available to civil aircraft so that their time should not be considered in the same category.

In 1942, Wendell Lewis Willkie, an American industrialist and Republican presidential candidate (1892–1944) in 1940, made a journey around the world visiting the different battlefields of World War II. He travelled 31,000 miles at an average speed of 200 miles an hour.

> Later, in a book called *One World,* Willkie told how distances seemed so short to him that he had promised the President of a Siberian republic to fly back some week-end in 1945 for a day's hunting. 'And I expect to keep the engagement', he wrote.[1]

He realized how the earth had shrunk and waged battle within his party attacking isolationism.

It is now recognized that we cannot be isolated from the affairs of the world as we were between 1918 and 1941. From a military position we are but minutes from any place on the globe.

On the beneficial side, you can have almost any commodity from anyplace on earth on your table or in your home within hours. Fresh pineapple from our 50th state and other dishes give even the common man variety on his table that was impossible a few years ago except to the very rich who had to travel to get it. Today's rich, the jet set, may have breakfast in Rome and dinner in India with a midnight swim in the Pacific. Businessmen deal all over the world in increasing numbers and plan weekly trips around the continent. Many companies maintain their own aircraft for executive travel, to speed delivery of parts for production, and to give faster consumer service. Wealthy families have their own aircraft while large numbers of middle-income families belong to flying clubs. Flying enables far flung families to reunite more frequently so that the air age which scatters the family unit also provides the means for reunion. A weekend vacation at some remote place is possible and many fly just for the fun of flying—it is a recreation in itself.[2]

Isolation provided by remoteness, snow, floods, political, or military means can be broken by an airlift to provide vaccines, food, drugs, and fodder to save thousands or millions of lives. In the air age there is no physical reason why famine should stalk the land anywhere in the world. The Viet-Nam conflict has demonstrated that wounded can be moved swiftly to a hospital but as yet the helicopter has not been used regularly around urban areas as wreckers and ambulances although the physical means is available.

Aircraft have been used in law enforcement and fighting forest fires for the better protection of lives and property. Many underdeveloped countries are developing airports where there are no roads not only for use but also for status.

By suddenly moving their primitive society into the twentieth century, they are creating additional problems for their own countries. This is only one factor on the debit side of the ledger. The same speed that is beneficial in saving man from isolation may also introduce new diseases. A human (or any animal) disease carrier has spanned the bounds of space before the disease which he has contracted has shown symptoms. Some years back five million New Yorkers had to be vaccinated against smallpox because the disease was transported here. On slower means of transport the disease would have been detected in transit and quarantine could have been enforced.

The process of disturbing the balance of nature is accelerated by air travel. The rabbit was introduced to Austrialia by ship but many plant seeds and plants, both beneficial and dangerous, are transported by aircraft all over the world at a fast pace.

The speed of travel has also had its physiological effects on businessmen and diplomats. Traveling over wide distances rapidly seems to upset our inside clock and it takes our systems some hours or days to readjust to the new time relationship. Most businesses have instructed their representatives not to do anything the first day but to go to bed so that the benefit of getting to a place rapidly is lost by having to wait for readjustment.

The usefulness of speed could be challenged for most ordinary travel on the ground as well as in the air. The pleasure and enjoyment of travel is lost with modern travel either on jets or in automobiles. Traveling along freeways or at thirty thousand feet there is nothing to see and the travelers read or watch films —they are not "taking a trip." There is no contrast to be had. Automobiles, buses and airplanes have killed passenger rail service in this country and yet it is a method of mass transit that provides more of a sense of trip taking than does the car, bus, or plane. The pleasures of traveling have been subordinated to speed but man hasn't been taught how to use the time he has saved.

Both the automobile and the airplane in attempting to get places faster have created their own strangulation through congestion. Large numbers of people traveling by both these means may lose the time they save through traffic jams. Time lost by sitting in a lane of traffic or waiting in a pattern in the air may more than equal the amount of time saved by that convenience. Mass ground transit must be a strong supplement for more and wider highways and airports cannot keep up with the increased travel needs of an increasing population.

Pollution of the air and noise pollution are two side effects from the increased amount of air and automobile travel.

While on the ground, the modern jet requires 3 to 4 thousand pounds of fuel to taxi and in flight it uses 2,500 pounds per engine per hour. Plane passenger service increased by a third between 1966–1968 and in 1968 was about 114,000,000,000 passenger miles. Motor vehicles used 87 billion gallons of gasoline in 1968. This would be a little less than 17 billion pounds of gasoline or about 8 million tons. Each pound of fuel produced about 1-1/2 pounds of water vapor, and about 3 pounds of carbon dioxide and carbon monoxide using up 3-1/2 pounds of oxygen.

The internal combustion engine in the United States alone produces 24 million tons of carbon dioxide and carbon monoxide annually and uses 28 million tons of oxygen. This amount of carbon dioxide is either raising or lowering the temperature of the earth. Each theory has yet to be verified. According to one theory, the amount of heat absorbed increases with the amount of carbon dioxide and yet the amount of water vapor should increase clouds and hence keep the earth from the sun which would tend to cool the earth. It has been predicted that urban areas should have more rainfall than the rural areas and data seems to support that idea.

The automobile contributes to 60 to 85 per cent of the air pollution problem. Attempts to clean up the air must revolve around doing away with the pollution or doing away with the automobile. Improving the internal combustion engine by use of an afterburner on the exhaust or by burning natural gas has been one suggestion and some vehicles have been converted. The use of another form of power has been suggested and research is going on in the areas of producing automobiles powered by steam, electricity and gas turbines. Another solution is to encourage people to use other means of transportation around urban areas by making public transit so attractive that many people will prefer it to driving.

Not too much has been said about noise pollution that is primarily from trucks but in urban areas may be from planes and other sources. There are indications that noise pollution may be a large factor in circulatory diseases such as hardening of the arteries, and high blood pressure. Damage has been detected as low as 70 decibels—a heavy truck is rated at 90 decibels, a loud motorcycle at 110 decibels, an overhead jet plane at 110 decibels, and a rock and roll band at 120 decibels.

> Dr. Samuel Rosen, consulting physician at the New York Eye and Ear Infirmary and the Mount Sinai School of Medicine, and his colleagues have conducted comparative studies on the effects of noise on urban dwellers in the German city of Dortmund, on New Yorkers and on the primitive Mabaans . . . The urbanites came from an environment in which loud noise is commonplace. Their diets were rich in . . . animal fats, Coronary disease and hypertension are not uncommon among them. The Mabaans, on the other hand, live in virtual silence, are mainly vegetarians and rarely, if ever, have high blood pressure.

> When exposed to noise at 90 and 95 decibels . . . blood vessels constricted both in primitive tribesmen and individuals from industrial societies. Among tribesmen, however, constriction and relaxation of vessels were rapid, showing both quick response to and quick recovery from the stress of sound. Among Westerners, vessels remained constricted for longer periods, indicating a lesser degree of elasticity in their blood vessels and a diminished capacity to recover from the effects of noise.[3]

The automobile, while beautiful as a new car or antique, as junk and with the high price of labor and the low price of scrap constitutes an eyesore on every highway and byway of this country as well as adding to stream pollution. Automobiles are designed to last but a few years and suggestions either to do away with the automobile or to make it last longer are contrary to our economic good.

The auto industry supplies jobs for one out of every six workers—in 1969 nearly 14 million persons were manufacturing, distributing, and maintaining cars. Eight million cars and trucks were produced and one company, General Motors, sold $25 million dollars worth. The 1963 Census of Business showed that there were 211,473 filling stations that employed 732,542 persons. It also revealed 55,170 motels, tourist courts, and trailer parks. This does not include such motoring conveniences as drive-in theaters, eating places, banks, laundries, car washes, churches,—virtually any activity in which man wants to engage can be managed without leaving his car. Motor vehicle registration, in this country alone, totaled 101,-000,000 in 1968.

Changes have been wrought not only in the economic man, but also in the social man. He became more mobile, his dating habits and methods changed. The automobile has accelerated the growth of suburbs and the death of the cities. It has isolated man from his fellow man and made him a stranger in his own neighborhood.

> Let us look at the values clustered about the automobile. Most of them will seem good enough until I detail the values which the automobile has down-graded. The values up-graded by the automobile are: one's own pleasure; economic security; convenience in style of life; self-fulfillment; self-reliance; power; freedom from interference (liberty); privacy; novelty; human dignity. The values downgraded by the automobile are: love and affection; friendship; reasonableness and rationality; prudence; devotion to family; conscientiousness; law and order; freedom from interference (liberty); service to others; natural beauty; culture; reverence for life.[4]

Those who have a vested interest in the automobile will support the upgraded values as the most valuable and will ignore or minimize the downgrading, while those who favor mass-rapid transit or those who are antimechanistic will tend to support the downgraded values and try to reduce the emphasis of the upgrading.

> In a democratic world it is a proverb that liberty is not a license. . . . but the automobile has given its owner more liberty than his pedestrian fellow citizens. When man meets automobile, the machine prevails.

> Even among automobile owners the manner of use determines whose liberty prevails. Drive your car along any important street of any city. The street was paved and is maintained at great expense to you and to all citizens so that automobiles may pass that way. Yet any single person with his 125 square feet of closed space and his 3,000 pounds of weight (contemporary man's fortress) may immobilize that street in whole or in substantial part, and does it time and time again. If you ever walk, pass along one of those streets at a rush hour and observe the expression on the faces of the automobilists dismounting to pick up a packet of pins . . . immobilizing sometimes blocks of other motorists. The dismounted motorist's face will exhibit utter social unconcern. He has parked his castle, and other motorists cannot attack him without intolerable damage to their own castles.

> Presently, the American society has decided in favor of the values upgraded by the automobile unlimited, however, the automobile may be used and whatever values may be downgraded. The nonenforcement of parking regulations; the preemption of public streets for private, overnight parking; the failure to curb highway bill-boards, heavily used for automobile advertising, or roadside automobile junkyards; the assumption that 50,000 automobile deaths per year are all caused by bad driving, never by unsafe cars;[5] the fiction that because the states and federal government employ the taxing power to expropriate the gasoline tax to build highways, the users of automobiles are paying their own way (the gasoline tax does not pay for policing streets and highways, running traffic bureaus and traffic courts, providing morgues, hospital and ambulance services for the dead and injured)—the evidence is overwhelming that in the mid-twentieth century the American society is opiated by the automobile and that this folk addiction has been supported by whatever rationalization comes to hand. But the automobile unlimited is a recent aspect of American society.[6]

In spite of some grumbling and rumbling about automobiles, 80% of the families in the United States own one or more automobiles and drive 1 trillion miles a year. Not even fear of dying keeps drivers from their cars. No one has recorded the number conceived or born in automobiles but every year some 55,000 persons die in the automobile. Since 1900, 1,700,000 lives have been lost in the automobile—the count for those killed in action in wars since 1775 is 638,000. It should also be noted that about 2,000,000 are injured each year in motor vehicle accidents. These deaths and injuries affect each of us in increased insurance costs and in the pain we are caused by death or injury to those near to us—all of us know someone who was in an accident in the previous year and received an injury.

The influence of the internal combustion engine ranges far and wide. Although today the farmer constitutes a minority of the population, fewer than 4.8 million persons are in agricultural employment, he still provides the basis for the rest of us to live. The motor car and other mechanism has changed his way of life. Before the car he would often spend a day marketing his produce but with a truck he could do the same in an hour. His costs were reduced from thirty cents a mile with a horse to fifteen cents with a motor truck. He could provide better schooling for his children without it being necessary for them to walk or board in town. It meant more of a social life for his family and removed the isolation. It provided portable power for while it had been possible to use a steam engine for tasks there were disadvantages—the same for him as for small industries.

> . . . it was not fuel economy that sold the internal-combustion engine in the early years, but rather convenience of operation, and especially adaptability to intermittent operation. The steam engine had a long warm-up period and a dangerous boiler . . . Someone had to keep the fire going and keep an eye on the boiler whether the engine was working or not. The gas engine, on the other hand, could be started and stopped at will

(in theory), it did not have to be fed when it was not working, and required no fuel storage.[7]

This portable power was, of course, the tractor which appeared during World War I as the need for more food for a war spurred farmers to greater action. It was used for the full range from planting seed to harvesting. Today, mechanization has extended to the use of airplanes and helicopters for spraying insecticides and fertilizers. As mechanization has increased, the need for large numbers of agricultural workers has decreased. As the automobile use has increased, the number of acres of land lost forever to farming under asphalt or concrete highways has steadily increased.

It was not only the farmer who found his life changing. Doctors quickly took to the automobiles as they allowed him to complete his rounds faster to reach emergency cases—yes, there was a time when the doctor went to the patient. There were no more long buggy rides using an hour or more of time each way. Of course, the doctor could no longer sleep on the way back from an all night confinement case.

Others found that the new vehicle wouldn't allow as much freedom in petting but they soon invented the scheme of running out of gas. Drunks and milk delivery men have found motor vehicles less easy to train and have discovered that the horse was not entirely a disadvantage.

Motor vehicles were applied to many emergency situations such as ambulances, which at first were supplied to hospitals free by the companies, and police and fire vehicles. In the case of the fire vehicle the power not only moved the vehicle but increased the water pressure.

Trucks and buses ran on the streets of this country early in this century and were first important to factories, delivery services, large stores and the mail service. The trucks liberated factories from having to locate next to railroad track spur lines. During the thirties there was a gradual increase in the truck and bus traffic which began to eat into rail revenues.

Buses took the place of streetcars and inter-urbans in and between the cities until largely supplanted by the individual car. In 1907 the city of London had 977 buses and even snobbish Fifth Avenue in New York, with its millionaires and fashionable stores, installed a bus line to keep rails off their street in 1905.

The very building of the motor vehicle established a new pattern for American industry and the worker. The modern assembly line with standard interchangeable parts and specialized labor which began with the automobile industry spread to all types of manufacture. In World War I and World War II, industry demonstrated how quickly its plants could be converted to the job at hand. From producing almost no ships just prior to the World War II, American shipyards could turn out ships faster than they could be sunk and by the end of the war we were the leading shipbuilder of the world. Just as quickly after the war we shut down operations. Mass production means cheaper products and makes it possible for industry to pay a higher wage but the effects on the worker should also be noted. The loneliness and boredom caused by his work is difficult to measure in dollars. He has more leisure time since the beginning of the industrial revolution, but he has not learned how to manage his leisure time to the best advantage for himself and his family.

STUDY AND DISCUSSION QUESTIONS

1. List and discuss the uses of ballooning. The airplane.
2. Discuss the advantages and disadvantages of flight.
3. Compare horse travel to the automobile.
4. What has the automobile contributed to our lives? What problems have risen?
5. What are the pollution problems? How to solve them?
6. Is there more value in living at an accelerated pace?
7. Defend why you do or do not like to drive.

FOOTNOTES

1. Norman E. Lee, *Travel and Transport Through The Ages* (Cambridge, 1967), p. 180.
2. See Antoine de Saint Exupéry, *Wind, Sand and Stars* (Harbrace Paperback Library, 1967).
3. Barbara J. Culliton, "Noise Polluting the Environment" *Science News, 97* (Jan. 31, 1970), p. 132.
4. Leland Hazard "Challenges for Urban Policy" in Kurt Baier and Nicholas Rescher (eds.) *Values and The Future,* (The Free Press, 1969), p. 324.
5. See Ralph Nader's book, *Unsafe At Any Speed: The Designed-In Dangers Of The American Automobile* (Grossman, 1965).
6. Hazard, *op. cit.,* pp. 325-326.
7. Lynwood Bryant "The Beginnings Of The Internal-Combustion Engine" in Melvin Kransberg and Carroll W. Pursell, Jr., *Technology in Western Civilization,* Vol. I, (Oxford Univ. Press, 1967), p. 650.

XVIII

The Rise of Astronomy in the 19th Century

As with most events and movements in history, it is easier to explain the rise of astronomy in the 19th century as due to a plurality of causes and interwoven forces than by a single cause or force. This combination includes economic, social, religious, technological, and literary forces. In addition, there is a series of astronomical events rarely equalled in the history of man.

The economic picture in the America of the 19th century is one of rapid growth interrupted only by Civil War. It is a period of canal building, railroads, the Western expansion, and a growth of population to provide a market for the new industry. The good economic scene supplies the over-all optimism and money which will be needed for the observations to follow.[1]

Economic growth aided in forming an urban middle class which made up the core of the reform movement. The middle class reform movement was anti-slavery, and favored elevation of the lower classes; temperance activities and pushing free public education.[2]

> Adults were not neglected in this educational awakening. In all the cities and larger town's mechanics' institutes provided vocational courses and night schools. Free public libraries were very generally established. In towns, and even villages, the lyceums offered popular lectures, scientific demonstrations, debates and entertainments. The Lowell Institute for public lectures on literary and scientific subjects, was brilliantly inaugurated in Boston by Silliman in 1839, and led to similar foundations in other cities.[3]

The new astronomical events fit too with the religion of the middle class. They were a means of finding out more about the creator and the wonderful laws which He had set up to control His nature. Here could be seen God's handiwork on a grand scale.[4] As will be seen later, the evidences gathered by the astronomers lent itself to quite a traumatic change in the view of God's handiwork and the position of man in it.

Technological improvements and developments in the 18th and 19th centuries provided the tools for discovery and the means of disseminating informa-tion. Ever since the telescope was invented—Galileo's lens was smaller than those used in glasses—men's minds have been concentrated on producing better and larger instruments to fathom the universe to its innermost depths.

One of the problems of the telescopes of Galileo's time was chromatic aberration caused by the prism-like shape of the lens and giving any object viewed fringes of color. Following Galileo, this difficulty was overcome to an extent by making the telescope longer. Some attained the length of two hundred feet which must have made them a bit cumbersome. In 1758, John Dollond made the achromatic lens by using lenses of different kinds of glass together, which removed the color effects.

Galileo's telescope, and those of the following fifty years was a refracting telescope. That is, the light traveled through a series of lenses to the eye. By the time of Newton, 1670, and as a matter of fact partly developed by Newton, astronomers were using reflecting telescopes. In these instruments, light from the stars was reflected by a curved mirror made of speculum metal (composed of tin and copper) to a small flat mirror set at a 45° angle and hence to a lens for viewing. During the 18th century, James Short constructed excellent reflecting telescopes which collected light much better than the refractors, although the reflectors were much more easily affected by temperature changes and weather than were the refractors. By the beginning of the 19th century, the reflecting telescopes were of large dimensions. Herschel, for example, used a telescope with a mirror four feet in diameter, in 1801. Metal mirrors were, however, rather heavy, a six foot mirror weighed four tons. In 1857, Leon Faucault devised mirrors of glass with a thin coat of silver deposited chemically on its surface. The silver was also a better reflecting surface than was the speculum metal.

In 1830, Niepce and Daguerre founded photography and it was suggested almost at once that this be used to record the positions of stars. Sir John Herschel carried out a set of experiments in 1839 for making wet plates suitable for use in astronomy, but it was J. W. Draper, a year later, who succeeded in the first photograph of a celestial object—the moon.[5]

Astronomers were quick to apply the techniques of photography with the earlier techniques of spectrum analysis. Joseph von Fraunhofer (1787–1826), a man of no formal education, had observed in 1815 a series of dark lines in the solar spectrum (when light passes through a prism and is broken into its separate colors) and had also observed these in other stars. It was not until 1859, when Gustav Kirchhoff and Robert Bunsen proved that when the light from any heated body is passed through a prism, the spectrum is an indication of the chemical composition of the light source. Kirchhoff was later (1861) able to indicate that the sun was no longer a habitable globe, but one with its visible boundary at a fierce heat—hot enough for metals to be gases. Thus, the telescope, photography, and spectrum analysis were powerful tools for the astronomer of the 19th century and gave him knowledge hitherto impossible in the history of man.

Three inventions improved greatly the means of disseminating information. The first of these was the invention of a practical telegraph by Samuel Finley Breese Morse in the 1840's which was the first really rapid means of communication over long distances. The next two inventions are related to printing. In 1886, Ottmar Margenthaler invented and patented the linotype which did away with hand setting type. The other invention was the rotary press invented by Friedrich Koenig in 1811 and introduced to this country in 1845. It could turn out 8,000 newspapers in an hour and made possible the penny newspapers of the 1840's, (a reduction to one-sixth the previous price). This reduction was also reflected in the price of books.[6]

All of these various factors which have been discussed above played a great role in popularizing the sequence of astronomical events which began in the mid 18th century and following. These events included: the rare, not to be repeated for two-hundred years, conjunction of a number of comets; meteorite showers, transits of the sun by various planets, solar eclipses, and the discovery of planets and their moons.

Here a listing of these events may prove useful. Halley's Comet returned, as predicted by him, in 1759. There were two transits of Venus—one in 1761 and the other in 1769. The discovery of Uranus, March 13, 1781 and discovery of two of its moons (Oberon and Titania) on January 11, 1787. Both the discoveries were by Herschel. Schroter observed a transit of Mercury on May 7, 1799. The perihelion passage of Pon's Comet, September 15, 1812. In 1815, Fraunhofer mapped 324 dark lines in the solar spectrum. The earth passed through the tail of a comet June 26, 1819. On May 24, 1822, the first calculated return of the Encko's Comet occurred. Biela discovered the comet bearing his name February 27, 1826. On November 12 and 13, 1833, there was a great star-shower visible in North America. Halley's Comet made a "canals." passage November 16, 1835. A great comet was visible to the naked eye at noon February 28, 1843. Galle discovered Neptune, September 23, 1846. On September 30, 1858, Donati's Comet made its perihelion passage. Kirchhoff describes the chemical make-up of the sun December 15, 1859. June 30, 1861, the earth is involved in the tail of a great comet. There is a transit of Venus, December 8, 1874. On August 16 and 17, 1877, Hall discovers the two satellites of Mars. Also in 1877, Schiaparelli describes lines on Mars which he calls "canals." A year later, 1878, Jupiter's Great Red Spot is observed. A transit of Mercury is observed May 6, 1878. Tebutt's Comet makes a perihelion passage June 16, 1881. In the same year on August 22, there is a perihelion passage of Schaeberle's Comet. In 1882, there was a perihelion passage of two comets: Wells on June 10 and a great comet September 17. Later the same year, December 6, a transit of Venus occurred. Pon's Comet makes a perihelion passage January 25, 1884. On May 9, 1891, Mercury makes a transit. Two comets are discovered in 1892: Swift's on March 6, and Holme's on November 6. The meteor shower November 23, 1892.

These natural events, visible both to literates and illiterates, helped popularize astronomy and gained for astronomy a number of amateurs who were and are of great assistance in the discovery of comets and asteroids. Formation of astronomical societies, such as the Astronomical Society of London (1820) and a similar German institution (1863), and astronomical periodicals, the *Monatliche Corresponding,* (1800) and *Astronomische Nachrichten,* (1822) as well as the transmission of news of astronomical events by the telegraph kept the professional informed and disseminated information to interested amateurs.

One of the first events, for which there is considerable documentation, is the great meteoric display of November, 1833. "It was visible . . . throughout almost the whole of the United States, from Maine to Louisiana, and from Lake Huron to the Gulf of Mexico, they were also seen in Mexico, in the islands of Cuba and Jamaica, off the Bermudas, and on board the brig Francis, in the Atlantic Ocean, at a distance of three-hundred miles northeast of the Bermudas, and five-hundred miles from the American coast."[7] An observer in Missouri had this to say of the display,

"Above, and all around the firmament, thicker than the stars themselves, which were uncommonly bright, large, and beautiful, we beheld innumerable fire-balls of a pallid color, rushing down, and to appearance, across the sky, drawing after them long luminous traces, which clothed the whole heaven in majesty and gave the air and earth a pale and death-like appearance."[8] Another comment was that the "stars seemed to fall like snowflakes." The excitement generated by the event was given additional force two years later with the appearance of Halley's Comet.

Halley had first observed the comet which bears his name in 1682, and noticing that the comet had a similar orbit to one's observed in 1531 and 1607, he predicted that it would re-appear at intervals of about 75 years. It appeared in 1835, the year of Mark Twain's birth, and again in 1910, the year of Twain's death. The 1910 appearance is the last impressive naked-eye comet this century.[9] Comets have, throughout the history of man, excited extreme superstitious fear and as Dr. M. Wilhelm Meyer, the director of the Urania Observatory commented in 1879, "The comets have remained mysteries as they were thousands of years ago."[10]

In the 19th century there is no dread at the appearance of a comet. We are not taken by surprise; we are not alarmed at the novel spectacle; we were fully prepared for it, were expecting it almost impatiently; we had indicated the time when, and the spot where it was to make its first appearance; had traced among the stars, the path it was to describe; foretold the rate of its progress from day to day, and the general increase of its magnitude and brightness; when it was to be seen with the telescope, and when with the naked eye.[11]

Even the common man can participate with pride in the achievements of Man. "The most illiterate are capable of appreciating this evidence, the undeniable and irrefragable proof of the high advancement of what science, which has thus enabled us to penetrate the future, and to forewarn mankind of events that are to come."[12]

However, lest we think that this is the 20th century, we have only to quote the *Knickerbocker* commenting on Halley's Comet. "It is both interesting and gratifying to observe the universal care and foresight which pervade every object in nature. The celestial world bears, in its order and harmony, the signs of the wisdom and providence, as well as the sublime magnificence of its Maker."[13] The article continues pointing out superstitions of the past in regard to comets and eclipses.

In the same magazine, which is largely literary, there are several witnesses to the effect of comets on literature. A poem entitled, "The Comet's Address To The Earth," which reads in part:

I passed, and saw the holy hill
Where the Redeemer did fulfill
    Jehovah's pledge to man:
The sun refused to lend his light,
And, hast'ning from the fearful sight,
    Again my course I ran.[14]

There is also a sort of prose poem or short story titled, "A Voice From The Comet," which ranges through history and comments, if a comet had a voice, on what it sees.[15]

The comet having the most immediate and traceable influence was the Great Comet of 1843 that was visible to the naked eye at noon.

The appearance of the comet coupled with the lectures by Ormsby MacKnight Mitchel on astronomy

... stirred an impressionable audience to the pitch of providing him with the means of erecting at Cincinnati the first astronomical establishment worthy the name ... To the excitement caused by it (the comet of 1843) the Harvard College Observatory (erected in 1847 with a 15" refractor) directly owed its origin ... Corporations, universities, municipalities, vied with each other in the creation of such institutions; private subscriptions poured in; emissaries were sent to Europe to purchase instruments and to procure instruction in their use. In a few years the young Republic (the United States) was, in point of astronomical efficiency, at least on a level with countries where the science had been fostered since the dawn of civilization.[16]

In point of fact by January 1, 1882, there were 144 observatories in the United States.

Again, a comet generates a poem, "Outlaws" (part only)

A prowling comet steamed along with outer seas of Night,
An ancient pilot grasped the wheel, and guided its frantic flight:
A grim, gigantic engineer stood by the furnace door,
And red fire shone through many gates, those vast black empires O'er[17]

As a further example of the influence of astronomy on literature of the 19th century, one of the best examples is that of Jules Verne (1828–1905), sometimes called the father of science fiction. Most of you are probably familiar with his *Twenty Thousand Leagues Under The Sea* at least. He also wrote three interplanetary novels. Two of these, *De la Terra a la Lune,* (1865) and its sequel, *Autour de la Lune,* (1866) were published as one book in English under the title *From The Earth To The Moon and Round The Moon* in 1877. In this book, a group of disgruntled cannoneers (there is no war) fire a projectile to the moon with three persons aboard. The site of the cannon is in Florida some 60 miles from the present Cape Kennedy (Cape Canaveral). His third novel, *Hector Servadac,* written in 1877 and translated as *Off On A Comet,* Ace Books, New York, 1957, deals with a group of men who are carried off when a comet grazes the earth and are later returned on the next passage of the comet.

The two moons of Mars have their beginning and end in literature. Johannes Kepler in his *Narratio de Jovis Satellitibus,* (1610) guesses that Mars had two moons. Jonathan Swift, (1667–1745) apparently read Kepler's book and uses the idea of the two moons in Gulliver's Travels, "A Voyage to Laputa."

> They (the astronomers of Laputa) have likewise discovered two lesser stars, or satellites, which revolve about Mars; whereof the innermost is distant from the center of the primary planet exactly three of his diameters, and the outermost five; the former revolves in the space of ten hours, and the latter in twenty-one and a half; so that the squares of their periodical times, are very near in the same proportion with the cubes of their distance from the center of Mars ...

Francois Marie Arouet, better known as Voltaire, (1694–1778), borrowed from Swift when he wrote *Micromegas,* which is a story of a giant from Sirius visiting our solar system and who notes that Mars has two moons. In 1887, Asaph Hall discovered the two moons of Mars and found that they had periods around Mars closely corresponding to those which had been given by Swift. At the suggestion of a Mr. Madan, of Eton, England, the two moons were named from the 15th book of the Iliad:

> He (Ares, the Greek name for Mars) spake and summoned Fear (Phobos) and Flight (Deimos) to yake his steeds, And put his glorious armour on ...

And so is completed the circuit—from literature to literature.

With the improved telescopes provided by a stimulated public and with a large number of amateur assistants, more than 100 planetoids are discovered and their orbits mapped between 1801 and 1868. Neptune is discovered in 1843, and gives rise directly to the observatory at the University of Michigan in 1852.[18] The transits of Venus provide a tool for gauging distance of the sun and since they occur only two to a century the importance of timing and technology cannot be overlooked.

What is the importance of all this for the big picture? By the 1860's the spectrograph and spectro-analysis supported a materialist view of the universe. The universe is uniform—the same elements are to be found everywhere. The stars are not different,

essentially, from the earth. The relation of the temperature of the stars could be related to spectral studies to permit the classification of stars on the basis of type. The stars were typed to age and the idea of stellar evolution (remember that Darwin's *Origin of The Species,* appears about 1860) is begun with the stars arranged in an evolutionary scheme with the whole cosmos passing through stages of development.

With the advent of larger telescopes and photography, the heavens are enlarged from a few thousand stars to stellar astronomy and galactic systems containing 100,000 to 200,000 stars each. As the heavens enlarge, man must shrink. He is a product of a process and will die inevitably in unyielding despair and doubt. Man is a microcosm in a macrocosm—there is space, infinitely large, but without room for God.

STUDY AND DISCUSSION QUESTIONS

1. Discuss those economic, social, religious, technological, and literary forces as well as the astronomical events contributing to a rise of astronomy in the nineteenth century.
2. Investigate the nearest observatory. When was it founded and by whom? Does it fit the pattern?
3. What was the final result of nineteenth century astronomy as far as we of today are concerned?
4. What was the role of Darwin's theories in astronomy?
5. What general types of astronomical activity were occurring?

FOOTNOTES

1. For a more detailed account of the American economy in this period see Ross M. Robertson's, *History Of The American Economy,* (Harcourt, Brace and Company, 1955), pp. 78-215.
2. Further discussion of the age of reform may be found in Alica Felt Tyler's, *Freedom's Ferment,* (1944), and in Carl Russell Fish, *The Rise Of The Common Man,* 1830–1850, (1927).
3. Samuel Eliot Morison and Henry Steele Commager, *The Growth Of The American Republic,* (Oxford University Press, 1962), p. 518.
4. See Richard H. Niebuhr's, *The Social Sources of Denominationalism,* 1929. Especially the chapter "The Churches Of The Middle Class," and also Gillespie's work, *Geology and Genesis.*
5. Helmut and Alison Gersheim, *History Of Photography,* Oxford University Press, 1955), may be examined by those interested in further information.

6. For more details on these and other inventions see Jerome S. Mayer's, *Great Inventions,* (The World Publishing Co., 1966).

7. Sears C. Walker, "Researches Concerning The Periodical Meteors Of August and November," *North American Review,* (April, 1833). Vol. 56, pp. 409-435.

8. *Ibid.,* p. 410.

9. Willy Ley, *Watcher Of The Sky,* (Viking Press, New York, 1963).

10. *Ibid.,* p. 154.

11. M. Arago, "Arago On Comets," *North American Review,* (January, 1836). Vol. 42, pp. 196-216.

12. *Ibid.,* p. 197.

13. "Comets and Eclipses," *Knickerbocker Magazine,* (March, 1836). Vol. 7, pp. 262-265.

14. "The Comet's Address To The Earth," *Knickerbocker Magazine,* (October, 1836), Vol. 8, pp. 446-447.

15. "A Voice From the Comet," *Knickerbocker Magazine,* (December, 1835), Vol. 6, pp. 491-494.

16. Agnes M. Clarke, *A Popular History Of Astronomy During The Nineteenth Century,* London, 1893. p. 8.

17. "Outlaws," *Knickerbocker Magazine,* (March, 1850), Vol. 35. pp. 224-225.

18. Lectures of Professor Losh at the University of Michigan.

XIX

"In the Beginning ..."

In the beginning God created heaven and earth. The earth was without form and there existed a void which was deep and dark. The spirit of God moved over the face of the waters ... God said, let the waters under the heaven collect in one place and allow dry land to appear. God called the dry land Earth and the collected waters He called Seas. In essence this is what scripture tells the Hebraic-Christian man of the formation of the earth.

The revival of Greek antiquity in the Renaissance did nothing to change this view. The Greeks had believed that the universe was ordered from chaos which fit the overall description from Genesis. The spark from which the flame of geology was to grow was created in a physical, technical process.

Before the 16th century it was generally believed that fossils were formed at the time of the deluge. Some, it is true, did hold that fossils were due to a formative quality or a stone making force.[1] But in 1517, excavations were made to repair the water systems of the city of Verona and a number of fossils were found which caused considerable speculation as to their nature and causes. Among those engaged in the speculation was Girolamo Fracastoro (1483–1553) an Italian physician who wrote that the fossils were once living organisms and that the flood was too transient to have caused the material which covered the fossils. As the geologist Lyell sums it up, "... the talent and argumentative powers of the learned were doomed for three centuries to be wasted in the discussion of these two simple and preliminary questions ... were fossils ever alive and if they could have been caused by the deluge?"[2]

In the meantime, many theories old and new were advanced to counter his suggestion. Andrea Mattiolo (1501–1577) a botanist, embraced the ideas of Agricola, that fossils were produced by fermentation of "fatty matter," and yet, from his own observations he concluded that porous matter, such as bone, might petrify by a "lapidifying juice." Fallopius (1523–62), professor of anatomy at Padua and the discoverer of the fallopian tubes, thought that fossils were earthy concretions. Along similar lines, Mercati (1574) thought that fossils were stones which were in their peculiar shape due to the influences of the heavenly bodies. While Oliva of Cremona considered them a mere "sport of nature."[3]

The quarrel spread to Cambridge in England. Thomas Burnet (1635–1715) starting from an idea of Antonio de Torquemada that the earth, at the time of the flood, was level proposed that when the flood occurred the earth cracked open releasing the waters and creating the mountains, the rough earth surface and the seas. He published his ideas in Latin in 1681 and in English in 1684. Some theologians objected that not enough time was available under Burnet's method but most persons agreed with him that the Flood had remade the earth for sea shells had long been found buried in the stone of mountains.

John Ray (1628–1705), another clergyman from Cambridge, was a skilled naturalist and he thus was capable of noting that the fossil shells of deep-lying mountain strata matched beach shells exactly. He calculated that the Flood lasted ten months, thirteen days and that animal bodies should have been strewn in a thin layer. Instead they were in great clumps.

His explanation, published in 1693, was that the great internal fires in the hard core of the earth were responsible for thrusting up blocks of stone and that waters contained and connected in underground networks, like the human circulatory system, burst during the Flood.

Others, such as John Woodward (1665–1728), a physicist at Gresham College, had differing theories although nearly all were concerned to relate the physical evidence with scripture.

With the publication of Sir Isaac Newton's *Principia* in 1687, new hope arose of solving questions such as how the earth came into being using mathematics and equations.

One of the earliest rational explanations was formulated by Georges Louis Leclerc comte de Buffon (1701–1788), a French naturalist, who proposed that the earth was formed by a comet's glancing collision with our sun. In the eighteenth century Immanuel Kant (1724–1804) about 1750 and Pierre Simon Laplace (1749–1827) in 1796 proposed that in the beginning the sun was surrounded by a thin gaseous envelope—shaped like two saucers—one inverted on the other—which rotated and from which the planets were condensed. This later theory, somewhat modified, is the one that is believed today.

Using Buffon's idea, the Copernican theory and Newtonian astronomy, William Whiston (1667–1752), a professor of mathematics at Cambridge, "proved" that the deluge was caused by the passage of a comet.

During the late 17th and early 18th centuries and even well into the 19th century, the English constantly attempted to reconcile the deluge with geological discoveries. This is true of Hooke (1705), Woodward (1695), and Whiston (1696), as well as many 19th century figures who introduce religion in science to show that the great over-all plan of the creator can be found by the study of nature.

On the other hand, the state of Italian geology by the mid-18th century can be summed up by a selection from the work of Circilo Generelli (1749), who notes that the bowels of the earth have made a record of past events; fossils were once living creatures that the universal deluge does not explain; earthquakes make the major changes of the surface of the earth, such as mountain formation; fossils are not in a random formation but in orderly formations. By 1750, Vitaliano Donati and Baldassari had shown a progression from fossils to living animals of the same species.[4]

The importance to geology from these early attempts to theorize about the formation of the earth and its relation to the Flood was that these men studied the fossil record and the strata of the earth as they attempted to fit fact to fancy.

Although it may be said that sciences like geology spring from the basic sciences, it is interesting to note that they are at least spurred by their branches. "One can build a hierarchy of sciences starting with the most fundamental ones (physics and chemistry) on the bottom leading up to the biological and medical sciences, sociology, etc."[5] Such was the case as materials from geology did challenge chemistry to make identifications.

First, however, was the collecting stage. An example might be Jean Etienne Guettard (1715–1786), who was a physician naturalist and had a patron, the Duke of Orleans, who was an enthusiastic naturalist. He went over the French countryside collecting minerals and mapping the distributions of rock formations. He noted among other things that the distribution of plants were dependant on the occurrence

of rocks and minerals—a device later used by prospectors. He was practical as well for he sought out minerals which enabled France to move to the forefront in the manufacture of porcelain.

Other great collectors and observers turned their attention to the question of how these rocks were formed.

In 1749, Buffon had argued that the mountains were made up in the ocean out of sediments washed from the land and was forced ro recant by the Theological Faculty of Paris, the Sorbonne. Nicolas Desmarest (1725–1815), on the other hand, from his studies found that many rocks, especially basalt, seemed to arise around and from volcanic origin.

Within a few years, in 1787, Abraham Gottlob Werner (1749–1817), concluded after numerous years of study that no basalt is volcanic, but that they are of aqueous origin. This divided the geologists into two camps—the Vulcanists that argued for the volcanic origin of rocks and the Neptunists who argued that rocks were formed by the action of water.

The Neptunists considered that all rocks and minerals were deposited in water action except the most recent which was due to volcanoes caused by ignited coal beds. It was this group which tended to explain the whole geological development in terms of the providence of God or God's plan as revealed in scriptures. Contrariwise, the Vulcanists argued that the lower rock layers were formed by heat and that the upper layers were deposited by the action of eroding forces still working.

There had been a few Englishmen who had amassed geological data in the 18th century but the first real comprehensive synthesis of geology, based on facts and not flights of the imagination, was presented by James Hutton (1726–1797) to the Royal Society of Edinburgh in 1785, and published in 1795 as his *Theory of the Earth.* In his book, Hutton would only credit those causes which could be seen operating as changing the surface of the earth. He also stated that the process of change took place over a very long period of time although, since he had no means of determining or estimating this, he did not give a speculative figure. He concluded that he could find no evidence of an origin for the earth. A statement which his contemporaries interpreted to be a denial of God and Genesis.

The Vulcanists were to emerge supreme about the 1820's but the Flood story died slowly for the English speaking world. It was not until 1855 that Charles Lyell (1797–1875) wrote that the idea that Noah's flood was responsible for fossils had long ago been rejected by all who had carefully investigated.[6] Even in 1863, James D. Dana (1813–1895) who wrote the standard work on mineralogy, although he does not mention the Flood, gives a chapter in the back of his book in an attempt to reconcile geology and the Bible. The same chapter appears in the 1869, 1875, and 1880 editions. The chapter is finally cut to a closing paragraph in the 1895 edition.[7]

As we turn toward today's view of the earth, we find that gradually more rational ideas supplant the attempt to force facts to conform to a theological picture. Theories must stand up to the test of physical sensory evidence and those that do not must be rejected.

Most of the ideas following Kant and Laplace about the formation of the earth used the sun as a source of materials and it was not until spectroscopy had shown the composition of the sun that these views were brought into question. Such an origin for the earth would have meant that temperature of the forming earth would be in the tens of thousands of degrees. A return to the idea of cosmic dust being the source of materials from which the earth was formed in the 1940's permitted another type of explanation based on a cold earth.

Such a proposal was made by Harold Clayton Urey (1893–   ).[8] He was a chemist who pointed out that if the earth's temperature was as high as tens of thousands of degrees that it could not have retained its water not its atmosphere. Admitting the creation of a cold earth gave other problems such as the source of molten rock. What was proposed was that as the earth collected materials, the heavier materials began to shift towards the center but due to the motion of the earth and to the convection currents all of the heavy material did not settle. In a hotter earth it would be difficult to explain why heavy materials should not all accumulate in the center. The surface would cool much faster since the earth was not as hot as previously believed.

It is believed that the heat from radioactive materials under the crust of the earth keeps the interior hot. But with the cooling of the surface liquid water could form and the atmosphere was possibly one of carbon dioxide, methane, and ammonia—quite different from our present one. The reason it is believed that this was the atmosphere of the earth in the early stage is because this is the atmosphere of the other planets.

The action of lightning in producing organic compounds from the type of primitive atmosphere that has been described could have given rise to the substances necessary to form life. This was shown in an experiment under Urey's direction:

In an air-tight vessel a mixture of ammonia, methane, and carbon dioxide was circulated over boiling water to simulate the early atmosphere of Earth. Through this mixture an electric spark was discharged for a week. The vaporous mixture first turned pink, then a turbid earthy red. Upon analysis it was shown that a whole range of organic compounds had been created in the vessel, entirely at random by chemical and physical means, including measureable amounts of such complex amino acids (the building blocks of proteins) as glycine and alanine.[9]

With the presence of living materials on earth all of the factors necessary to its exterior change were present and functioning. The earth began to be changed as soon as it was formed.

Those factors that effectively change the earth may be divided into external and internal ones. The external forces can be divided into those activities that break up and remove materials (denudation) and those which are responsible for piling up material (deposition).

The breaking up and removal process include those that are biological, chemical, and physical and each of these major divisions has its own agents that are constantly at work.

Probably the most important physical agent is water. It is the great remover. The water cycle in operation puts huge volumes of water over the land masses and the action of the tides rolls the oceans against the shores which constantly wears at the land and land masses. In the process of becoming a solid, ice expands and in so doing may split apart the rock. In large masses glaciers move over the face of the earth and break up and move large masses of earth and rock. Water not only breaks up and removes rock and earth material, but also removes the products of other agents and exposes a new surface to their action. Sometimes it carries the agent to a new area for action.

Related somewhat to water is the physical agent of temperature. The fall in temperature causes the water to become ice and to split masses but if the change in temperature is drastic enough it may split the rock material by itself. Of course, if the temperature is hot enough and cold water runs or drops on the hot rocks they may split or break apart. Or if water is trapped in the rock mass and the rock mass is heated then the material may explode as water is turned into steam.

Another physical agent, gravity, is also related to water. It causes the water to flow toward the sea. In addition to its effect on water, it will move unbalanced masses of rock downward which may cause other rock formations to be broken up. Earth masses that become soaked with water may also be acted on by gravity and slide down revealing another rock surface to be acted upon. As you can see, many of the physical agents work with each other. In rare instances meteors are pulled to the earth's surface, break rock material or explode and throw earth and rock material about.

Wind, a fluid in motion, like water removes earth material and using sand as a cutting agent also erodes rock masses. In arid regions wind may be a more important factor than water and although it may be somewhat slower than the water action in time will still cut up and remove the rocks.

While relatively minor as far as the range is concerned, lightning did and does strike the earth and its action would tend both to fuse the earth and to cause parts of it to explode due to overheating.

There are several chemical agents which bring about the break up of rocks. Chemical changes may affect the physical nature of the rock material. It may make parts of the rock softer than other parts so that the water may wear it away or parts of the rock may be made soluble in water and so the water dissolves and removes the material. The oxygen in the air reacts with some materials forming oxides and these oxides have their own chemical and physical properties, they may be softer or more soluble, e.g., plants as they decompose produce acids which can react with some rock materials to make a soluble product. Lightning produces nitric and nitrous acids that may act in the same way as the plant acids. Carbon dioxide dissolves in water to form a weak acid which may react with some of the rocks. The carbon dioxide may be formed from burning materials set aflame by lightning. After a forest fire the wood ashes which contain substances like lye may be dissolved by water and moved to rock material containing minerals which may react with the lye to form gels or other soluble products.

Besides producing acids and bases when decayed or burned plants, a biological agent, may produce acids while living and eat into the rock surface to gain a root hold. Through growth roots and other parts of the plant may split rocks or push them aside. In a sense the force of the roots is water power because of the osmotic pressure built up in the roots.

Other biological agents burrow to permit the entry of water and air to rock surfaces. The chief such agent is the earthworm but it may also include burrowing mammals who undermined rock structures so that they fall over or down inclines and are further broken up.

Possibly the greatest biological factor today is man who strips and burns forests from the land; removes the soil cover as he plows, builds roads, strip mines; restructures or removes hills and mountains; tunnels in mines; dams and dikes rivers; and performs a multitude of activities speeding the break up and removal of rock and soil.

The other external factor is one of piling up or deposition and again makes use of the chemical, biological, and physical agents.

As the physical agents of water and wind lose their force or are retarded by an obstacle the material which they were carrying is deposited and builds up. This gives rise to sand bars in the oceans or rivers and to soil and sand deposits on the open stretches of land where stumps or other obstructions have lessened wind power. It also gives rise to the deposits of the glaciers as they have retreated. The evaporation of water may be considered another agent for it is by this means that anything carried in solution by water is left behind. The change in the temperature of the air or water may cause materials to be deposited. As water warms some substances dissolved in it become less soluble and precipitate out of solution (carbonates and bicarbonates e.g.).

Water may carry substances dissolved in solution or as solids in suspension or as solids in a colloidal form. When in colloidal form the substances do not settle as the water slows down. However, when the water carrying the colloidal materials enters the ocean the salt water discharges the charge on the colloidal particles, which kept them separated and from settling, and they clump and settle out. In this way deltas are formed at the mouths of rivers. The precipitating and colloidal effects may be considered chemical agents. And again may be noted the overlapping effects of several agents.

The gradual addition of cosmic dust and meteorites over the millenia have added to the mass of the earth and constitutes a relatively minor physical agent to the pile up process.

Biologically the life and death of living matter causes vast accumulations and pile ups which will be dealt with in greater detail elsewhere. Many of the Pacific islands are topped with coral which is billions of dead bodies. Even the modern cities of the Near East are built on the tombs and rubbish of ancient cities.

These external forces we can see at work or at least their workings can be more easily imagined than those internal factors which must be figured out by indirect methods. The existence of internal factors is revealed through earthquakes and volcanoes. While many theories have evolved the amount of direct evidence about the interior is scant.

Highway cuts and the works of man have but scratched the surface. The Grand Canyon shows a mile or so but this is only the reworked surface materials. The deep mines of man to depths of several miles and even oil well drillings go down but five miles or so. What we do know is that in descending the temperature increases. In diamond mines of South Africa it was found that the temperature increased one degree centigrade for every 70 feet of depth in the first half mile (in sedimentary rocks) and in the next half mile one degree for every 137 feet (in igneous rocks). In a 9,000 foot oil well the temperature at 7,200 feet was 100°C. and at the bottom, 120°C. Molten lava is as high as 1,200°C. But if the same proportionate increase is used, the rather absurd figure of 1,456,000°C. is obtained for the center of the earth. This is higher than the surface of the sun and hence does not seem reasonable.

Most of the ideas about the structure and changes due to internal factors are based on the belief that the interior of the earth is a fluid or semi-fluid material. Original beliefs centered on the idea that as the earth cooled, a crust like material became the continental land masses. While we retain the term "crust" the molten surface is no longer a central belief. It was early observed that not all of the earth's crust has the same density—continents have an overall density of 2.67, mountains are less dense than sediment deposits at the mouths of rivers. On these differences rest one of the theoretical processes.

Clarence Edward Dutton (1841–1912) read a story in the rocks of the Grand Canyon that told of the repetitious cycle of uplift, erosion, deposit, sinking, and uplift. This must have taken millions of years and the uplift must have been considerable. In 1871, he proposed that as the lower layers of rock sank and came under the heat and pressure of the depths that it was changed or metamorphized into dense and hard rocks. This increase in density would cause further sinking. He later proposed a theory which he called Isostasy. This included the idea that if deposits were made in one place while being removed from another the change in overall weight would cause the heavier place to sink and the lighter place to rise. Rock layers are rather rigid in structure and do not bend easily. The shift of large masses of materials by erosion sets up stresses in the crust of the earth and these are relieved from time to time by the rock layer snapping. This is believed to be one cause of earthquakes.

Methods for studying earthquakes have been devised starting with the invention of the seismograph by John Milne (1850–1913). The instrument measured the earth movements and recorded them on a revolving paper drum. By setting up a number of stations it was possible to determine not only the quake but also the position of its center. The quakes produced waves which traveled through and around the earth—the same techniques are used for finding out where an A-bomb was detonated—in addition the waves tell some things about the interior of the earth.

The speed with which the waves travel depends on the type of material through which they are traveling and also the shock produces different types of waves whose speed varies. A good deal of work on these waves and their velocities was done by Sir Harold Jeffreys (1891–),[10] Beno Gutenberg (1889–1960) and others. From wave patterns they have deduced that the earth is composed of a solid core having a radius of 800 miles on which is a concentric layer the 1,780 mile mantle of solid, rocky material on top of which is the 5 to 30 mile crust.

The lower boundary of the crust was discovered about 1909 by Andrja Mohorovicic (1857–1936) for whom it is named but sometimes shortened to Moho. The rocks making up the continents are sedimentary (erosion products converted into stone), granites, and metamorphic (changed igneous and sedimentary rocks). These rocks are largely made up of minerals consisting of silicon and aluminum compounds and abbreviated as Sial (Si for silicon and Al for aluminum). The ocean floor which makes up the last layer of continental masses is basaltic which is made of rock high in silicon and magnesium compounds and is abbreviated sima (Si for silicon and Ma for magnesium).

The denser mantle is believed to be made largely of a mineral called dunite which is rarely found embedded in lava of some Pacific islands. Even though the mantle is believed to be a solid, from the velocity of the shock waves through it, it is believed to flow. This flow is believed induced by the heat coming from below and the relative slow conduction of heat in rock material. By applying the principles that heat travels from a hotter to a cooler place and knowing the amount of heat conduction in rock some idea can be obtained about the shifting mantle. Studies show that the temperature of rock under oceans exceeds that under continents by 100° down to 300 miles and produces a density difference which accounts in part for earthquake centers along the borders of oceans and continents. These also match a mathematical model set up of the earth's heat history on a computer.

> The high rate of heat production in the earliest stage of the earth's history led to a rapid thermal expansion of the planet. The formation of the earliest continents is associated with this period of rapid expansion and extension of the crust . . .

> The deep structure of continents place heavy restrictions on any theory of continental drift. A relative motion of the continents must involve the mantle to depths of several hundred kilometers; (c. 150 miles) it is no longer possible to imagine thin continental blocks sailing over a fluid mantle.[11]

Look at a globe and it can be seen that if the North and South American continents were tilted slightly and moved eastward the extension of Brazil would fit very nicely into the indentation on the continent of Africa. This was noticed more than a hundred years ago[12] and is part of a continuing controversy about how and why continents drift. There is evidence other than just the shape of the continents. The only way that land animals and certain plants could have been on the different land masses at one time was if at that time they were connected or very close. So there is the fossil evidence in each time period. There is also the geological evidence of matching rock strata or layers and the magnetic evidences of shifting magnetic patterns. At one time the continent of Antarctica was in a lush tropical zone as shown by the coal deposits that are there.[13]

Gradual mapping of the earth's surface including the ocean areas have shown that there are zones of mountain building including some islands and breaks or fractures in the crust. Looking again at the globe the mountain zone can be traced starting in North Africa with the Atlas, across Sicily and up Italy, down along the Adriatic and across Turkey, Iran, Afghanis-

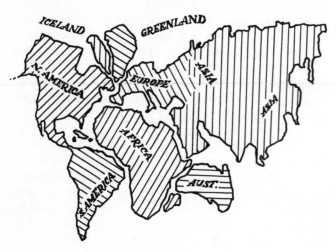

Figure XIX-1. Continental Drift.

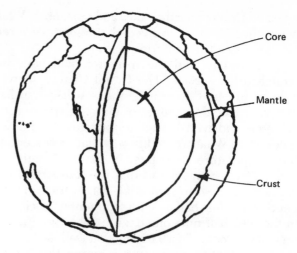

Figure XIX-2. Cross Section of Earth.

tan, Pakistan, Tibet, Burma, then south to Indonesia, North along the Philippines, Japan, then across the Aleutian string and south along Alaska, Western Canada, the United States, Mexico and along the West coast of South America. There is also a Middle Atlantic ridge and several lesser known ones in the Pacific area. In these zones occur most of the volcanoes, earthquakes and hot springs of the world. Also in the ocean near each of these zones are found some of the deepest trenches and fractures. With aerial photography and shots from our rocket programs some of the larger features of stress are becoming known.

It is possible that someday through these stress studies to predict when and where volcanic eruptions and earth quakes will occur. New studies in Japan and Hawaii are leading in this direction. Those living in California on or near the San Andreas Fault would appreciate such information.

Another problem in geology which for long has been a problem for man in general, is the age of the earth. In the Eastern World the Hindu's sacred book, The *Manusmiti*, shows that the age of the earth is about two billion years old since it has achieved the seventh cycle of the fourteen cycles of Brahma's "day" which was reckoned at lasting 4.32 billion years. On the other hand, the Western world had gone on the chronology of Archbishop Ussher who, early in the seventeenth century, calculated the age of the earth from the successive lives of persons recorded in the Old Testament. He fixed the date of the earth's creation at 4004 B.C. so that the earth is now about 6,000 years old.

In the nineteenth century scientists tried various methods for calculating the age of the earth. Based on the amount of salt washed into the sea, one person came up with 80–90 million years; while another, using the separation of the earth and the moon, arrived at 57 million years. Still others had used the thickness of erosion sediments to arrive at ages near 100 million years.

All these were pushed aside by the calculations of Lord Kelvin who used the original temperature of the earth based on the sun's heat and figured how long it would take the earth to reach its present temperature. He came out with a range of 20 to 40 million years. This was far lower than the others and was a blow to geology. This could not be overcome until the realization that there were radioactive materials and the amount of heat that they would produce on decaying.

Ernest Rutherford changed all by his announcement to a geologist colleague that a piece of pitchblende, the ore of uranium, was 700 million years old. This led to the methods used today.

Most of the modern methods of dating materials are based on the "half-life" of various radioactive materials. The half-life of a radioactive material is the time required for one-half of the substance to decay into another substance. This time seems to be fixed for each radioactive substance and its decay cannot be speeded up or slowed down by any chemical or physical means—for example: increasing or decreasing temperature or oxidation. We will be interested here in a limited number of radioactive materials. They and their half-lives are:

Uranium—238	4.51×10^9 yrs.
Uranium—235	7.13×10^8 yrs.
Carbon—14	5.7×10^3 yrs.
Potassium—40	1.3×10^9 yrs.
Rubidium—87	4.7×10^{10} yrs.

As an example of half-life dating, suppose we have a one pound block of Uranium—238. In a period of 4.51×10^9 yrs. we would have one half pound of the Uranium left and one half pound of decay product. In the next 4.51×10^9 year period we would have left one quarter pound of Uranium and three quarters of a pound of decay product. In each half-life time we would lose one half of the remaining Uranium. In the half-life method of dating if we know the half life of the radioactive material and the ratio between the radioactive material and its decay product in a rock then we can date the time that the rock was formed.

The usefulness of a radioactive material as a dater depends in part on the approximate age of the substance to be dated, that is, a substance of great age would need a radioactive material having a long half life. Also the material to be dated must contain the radioactive material to be used. Living material or material that was once alive would not normally contain radioactive Uranium or Rubidium so that these substances could not be used for dating once living things. On the other hand, radioactive carbon—14 has such a short half life that it is only useful for dating substances formed 50,000 to 100,000 years ago.

Using the general method described above, investigators have examined the oldest rock formations on the earth and have concluded that the age of the earth is about 4.7×10^9 years old.[14] If the assumptions are made that meteorites are the remains of an exploded planet formed at about the same time as the earth then we may date the formation of the solar system. Meteorites are made primarily of iron-nickel but contain other material as well. Studies of Strontium and Rubidium and their decay products reveal an age of 4.7×10^9 years for meteorites.[15]

Carbon—14, which is found in all living matter and matter which has been alive, is formed when Carbon—12 as a part of the carbon dioxide molecule is bombarded with cosmic rays high in the atmosphere. Plants use the carbon dioxide, some of which is radioactive, and when it dies the ratio of the carbon—12 to carbon—14 is fixed.[16] This method assumes that the rate of cosmic radiation is a constant and hence that the rate of production and ratio of Carbon—12 to Carbon—14 is a constant.

Recent evidence indicates that the rate of cosmic radiation may be influenced by explosions of supernovae. If true, this would weaken the dating assumptions for Carbon—14 dating methods. This same type of explosions has been used by some as a partial explanation for the periodic disappearance of various species of animal life especially if the exploding supernovae were within 100 light years distance from the earth. Such an occurrence is estimated to have occurred about every 50,000 years.[17]

You will note that the modern dating procedures are based on a number of assumptions: (1) that the decay rate of radioactive materials has always been a constant (2) that over a relatively short period of time it is possible to determine these long half-life periods (3) that the rocks contained only the radioactive materials such an uranium and thorium and lead not formed from radioactive decay at the time they were formed, (4) that, except for radioactive decay, there has been no change in the ratios of the materials, and (5) that there has been no contamination of lead from any other source.

Eras and Periods	X 10⁶ Years Ago	Length of Period X 10⁶ Years	Predominant Life Form or Significance
Cenozoic			
Quaternary	0-1	1	
Plicocene	1-12	11	Stone Age.
Miocene	12-28	16	Mammals climax (Elephants in U.S.).
Oligocene	28-40	12	Great apes in Eurasia.
Eocene and Paleocene	40-60	20	Modern orders of mammals appear. Older forms of mammals.
Mesozoic			
Cretaceous	60-130	70	Small mammals. Donosaurs climax.
Jurassic	130-155	25	Dinosaurs dominate.
Triassic	155-185	30	Dinosaurs appear.
Paleozoic			
Permian	185-210	25	
Pennsylvanian carboniferous	210-235	25	Reptiles, insects, coal making.
Mississippian	235-265	30	Shell crushing sharks.
Devonian	265-320	55	Lung fishes, amphibians.
Silurian	320-360	40	Coral reefs, first land life.
Ordovician	360-440	80	Shallow seas, invertebrates.
Cambrian	440-520	80	Marine life, trilobites, and brachiopods.
Precambrian or Cryptozoic	520-2,100 plus	1,600 plus	
Keweenawan			
Huronian			Lake Superior iron ore deposited.
Timiskamian	1,050		
Keewatin			Oldest known sedimentry rocks.

STUDY AND DISCUSSION QUESTIONS

1. Since scientists have never been to the center of the earth, how can they suggest that it is composed of nickle and iron? Why those particularly?
2. What was the basis for the argument between the Vulcanists and the Neptunists?
3. What was believed to be the origin of fossils? What is believed today?
4. What are the external factors that change the surface of the earth? Internal factors?
5. What was the origin of the atmosphere? Chemicals for life?
6. What is the evidence for continental drift?
7. What was the early Eastern and Western views for the age of the earth?
8. How are rocks and the earth dated?
9. What assumptions does the scientist make in the dating game? Can you think of others not listed in the text?
10. Why are two of the periods referred to as carboniferous?

FOOTNOTES

1. A. D. White, *New Chapters In The Warfare of Science*, (D. Appleton and Company, 1888), p. 13.
2. Charles Lyell, *Principles of Geology*, (J. I. Kay and Co., Philadelphia, 1837).
3. Lyell, *op. cit.,* pp. 38-39.
4. Lyell, *op. cit.,* pp. 52-57.
5. Uno Krask, *Chemistry,* (Barnes and Noble, 1969), p. 8.
6. Charles Lyell, *A Manual Of Elementary Geology*, (John Murray, London, 1855), p. 4.
7. James D. Dana, *Manual of Geology,* (Theodore Bliss Company, Philadelphia, 1863), p. 746.
8. H. C. Urey, *The Planets: Their Origin and Development* (Yale University Press, 1952).
9. Laurence P. Lessing, *Understanding Chemistry,* (Interscience Publishers, 1959), p. 23.

10. See H. Jeffreys, *The Earth,* 2nd edition, (Cambridge University Press, 1959).

11. Gorden J. F. MacDonald, "The Deep Structure of Continents" *Science,* 143 (Feb. 28, 1964), p. 928.

12. See A. Snider, *La Creation et ses mysteres devoiles,* (Franck, Paris, 1858).

13. For further information on continental drift see:
    G. D. Garland (ed.), *Continental Drift,* (University of Toronto Press, 1966).
    Articles in *Science:*
    Vol. 163, pp. 528-532, Feb. 7, 1969.
    Vol. 163, pp. 176-179, Jan. 10, 1969.
    Vol. 166, pp. 609-611, Oct. 31, 1969.
    Vol. 164, pp. 1229-1242, June 13, 1969.

14. *Science,* (Dec., 1965), pp. 1805-1807.

15. *Science,* (Dec., 1965), pp. 1814-1817.

16. "How Old Is It," *National Geographic* (Aug., 1958), describes the method use.

17. *Science,* (Jan., 1968), pp. 421-423.

Suggested Additional Readings

Ruth Moore, *The Earth We Live On,* (Alfred A. Knopf, 1963).

William C. Putnam, *Geology.* (Oxford University Press, 1964).

William J. Miller, *An Introduction To Historical Geology.* (D. Van Nostrand Company, 1952).

A popular book with good illustrations is Kirtley F. Mather, *The Earth Beneath Us* (Random House, 1964).

James Gilluly, A. C. Waters and A. O. Woodford, *Principles of Geology,* 2nd ed., (W. H. Freeman Company, 1959).

Bernhard Kummel, *History Of The Earth,* (W. H. Freeman, 1961).

John A. Skimer, *This Changing Earth: An Introduction to Geology,* (Harper & Row, 1968).

John H. Hodgson, *Earthquakes and Earth Structure,* (Prentice Hall, 1964).

For rockhounds and other interested persons the following are suggested:

Richard M. Pearl, *American Gem Trails,* (McGraw-Hill Publishing Co., 1964).

Herbert S. Zim and Paul R. Shaffer, *Rocks and Minerals,* (Golden Press, 1963).

More technical books on rocks and minerals for the serious amateur include:

Carroll Lane Fenton and Mildred Adams Fenton, *The Rock Book,* (Doubleday and Company, 1940).

John Sinkankas, *Mineralogy For Amateurs,* (D. Van Nostrand & Company, 1964).

James Furman Kemp, *A Handbook Of Rocks,* (D. Van Nostrand and Company, 1940).

L. E. Speck, *Guide To The Study Of Rocks,* (Harper and Row, 1953).

XX

The Three in One Problem

There are three general areas in which physical, sensory evidences from geology are considered by many to conflict with man's faith as presented by scripture. These are the age of the earth, what fossils prove, and evolution.

Edward: Sea-shells, did you say, mother, in the heart of solid rocks, and far inland? There must surely be some mistake in this; at least it appears to me incredible.

Mrs. R: The history of the shells, my dear, and many other things no less wonderful, is contained in the science called Geology, which treats of the first appearance of rocks, mountains, valleys, lakes, and rivers; and the changes they have undergone, from the Creation and the Deluge, till the present time.

Christina: I always thought that the lakes, mountains, and valleys, had been created from the first by God, and that no further history could be given them.

Mrs. R: True, my dear, but yet we may without presumption, inquire into what actually took place at the Creation; and, by examining stones and rocks, as we now find them, endeavor to trace what changes they have undergone, in the course of ages.

Edward: This will indeed be romantic and interesting.[1]

As we have seen from the Renaissance on, there is an effort to find God in nature and to find His laws. The beauties and wonders of nature were used to show God and the order that was sought for and found was further evidence for the existence and might of the Creator. Much of the early science work was done by the clergy and monks with the idea that they were appreciating the wonder and glory of God. The dilemma that modern man finds himself in is that while he would like to have the material benefits and knowledge of the universe from science, he would also like to retain the concept of a personalized God. These arguments climax in the nineteenth century and occur chiefly where Protestant groups insist on a literal interpretation of the Bible. We can see some of the difficulties from the following arguments.

DIALOGUE AT DAYTON[2]

(Dayton, Tennessee was the scene of the Scopes trial to determine the legality of the law against teaching evolution in the public schools.)

Modernist (hereafter M): I don't see what difference it makes to faith and hope whether the world was created as set forth in Genesis or not. I could join you in insisting upon the moral truth of the Sermon on the Mount, but why do you insist upon the scientific truth of Genesis?

Fundamentalist (hereafter F): If I deny the teaching of Genesis, what assurance have I about the teaching in Luke, Matthew or Mark?

M: The teachings of the Gospels are verified by the religious and moral experience of men. That is your assurance. But the teaching of Genesis runs counter to the knowledge of men.

F: This comes down to saying that I am to pick and choose what parts of the Bible I shall believe.

M: Are you not a Protestant? Do you not believe in the right of private judgment?

F: Had you studied history a little more carefully you would not ask this question. You might examine Luther's views on Zwingli and on the Anabaptists.

M: You mean that the right of private judgment was not a principle in Luther's teaching, but a method, provisionally adopted, for dealing with the authority of the Pope?

F: I do. For me the hope of salvation depends upon the authority of the Scriptures.

M: The account of creation in Genesis has nothing to do with the promise of salvation.

F: You are quite wrong. It would make no difference if the Bible had said that the world was created in seven million years rather than in seven days or that man was descended from an ape. I would believe that as readily as I believe what I now believe. The important question is not what the Bible says about creation, but that the Bible says it. If the Bible is wrong about creation, how am I to know that it is not also wrong about salvation? I say to my children: You must not steal. . . . You must not lie. . . . You must keep the Ten Commandments. . . . You must follow the teachings of the Sermon On The Mount . . . They say to me: Why should we do that? . . . I tell them it is God's will. They say to me: How do you know it is God's will? I reply: Because Scripture is the word of God divinely inspired. And then they say: Yes, but teacher says the Bible is wrong when it tells that woman was made from Adam's rib.

M: Must your children then believe what is untrue in order that they may believe what is true?

F: I don't know that Genesis is untrue. Neither do you. You weren't there at the creation and neither was Darwin. Your scientists are all at sixes and sevens, and none of them even pretends to know how the world came to be created.

M: They know it could not have been created as set forth in Genesis.

F: They do, do they? Well, what if they do?

M: Don't you want your children to respect the truth?

F: Indeed I do. That is just where I quarrel with modernism. It undermines the respect of my children for the truth. They learn a lot of half-baked theories about evolution in school, and then they come home disbelieving the whole religion and morality of their fathers, and recognizing no standards of conduct except their own wilfulness.

M: You are going to ask me to believe that the whole of religion and morality rests upon an old Jewish legend about the creation?

F: I am going to ask you to tell me what guarantee there is for religion and morality if you first reject the authority of the Catholic Church, as Protestants do, and then proceed to reject the authority of the Bible, as you modernists do?

M: We do not reject the authority of the Bible. We hold that it is profoundly inspired.

F: Is the whole Bible inspired?

M: Not all of it. The Bible is a vast literature not all written at one time and by no means all of the same quality. The Bible is not a single book.

F: Then how do we know which parts of the Bible are, as you say, profoundly inspired?

M: Those parts are inspired which are verified by the religious experience of mankind.

F: And how do you know that the religious experience of mankind is reliable? Do all modernists agree on what they will regard as the profoundly inspired parts of the Bible?

M: There is some disagreement.

F: So that one modernist might call one part of the Bible inspired and another modernist might not.

M: That is so.

F: Then perhaps you will tell me how I am to convince my children that any part of the Bible is inspired.

M: You must enlarge their experience and train their judgment. Then they can decide for themselves.

F: They are to decide for themselves what is moral and immoral?

M: They must learn to be guided in their decisions by the accumulated wisdom of the race.

F: Guided by it. But not bound by it?

M: It is impossible to lay down absolute rules which will be valid in all cases.

F: So we have come to the conclusion that it is impossible to know for certain what God's will is. It is equally impossible to know for certain what the so-called wisdom of mankind is. Each youngster, therefore, is, under your system, to face the temptations and the perplexities of the world with nothing more than a tentative moral code which he is at liberty to revise as he sees fit. How do you distinguish this beautiful theory from sheer moral anarchy?

M: You have got to have some faith in the common sense and decent instincts of your fellow men.

F: Have you such faith?

M: Yes. Haven't you?

F: No. I haven't. Such a faith is contradicted by that experience of mankind to which you are so fond of appealing. Your natural man is a natural barbarian, grasping, selfish, lustful, murderous. Your psychoanalysts will tell you that. The religious teachers knew it long before the psychoanalysts rediscovered it. They called it original sin. They knew man was unworthy until he had been redeemed, or as you would say, educated.

M: And what does that prove?

F: That even if the modernists could agree upon a moral code, they could not inculcate morality. For the moral life is due not to the acceptance of a set of rules but to a transformation of the will.

M: And what have you to offer that will transform the will except kindly advice and the good example of the wise?

F: I have the knowledge that I am part of a universe governed by a divine plan to which, if I wish to be everlastingly happy, I must make my will conform.

M: In order to be good is it necessary to believe all that? Have there not been good men who disbelieved it?

F: No doubt there have been. Yet you will not deny that the great mass of mankind has always insisted upon believing that it was in communication with, and subject to, a will greater than its own. Do you know any popular morality in the history of the world which has not had some sort of supernatural sanction?

M: I do not know of any popular morality. But I know that there have been men who lived nobly without supernatural sanction.

F: Is it your hope, then, that what a few men have done all men might ultimately do?

M: That is, I suppose, the faith of the modernist.

F: Are you not rather an optimist?

M: In what respect?

F: You admit that all history shows how few men have been able to live a moral life without the conviction that they were obeying a divine will. You then point out a few unusual men, a few stoics perhaps, a few Epicureans, a few followers of Spinoza, a few pure and disinterested spirits among the scientists, and you ask me to believe that what this trifling minority has achieved through innate moral genius, the great humdrum mass of mankind is to achieve by what you optimistically describe as education. I do not believe it.

M: If I accepted your argument I should be forced to the conclusion that the mass of mankind are incapable of receiving the truth.

F: That horrifies you?

M: It does indeed.

F: I thought you had solemnly sworn to follow the truth wherever it might lead. But when you are confronted with the possibility that mankind cannot receive truth, you shrink from that truth. Does that not rather tend to confirm my suggestion that the appetite for truth is not so strong after all?

M: That is an ingenious paradox and I haven't time to unravel it. Come back to the main point. You were saying that only a small minority has ever found the moral life possible without the absolute assurance that it was obeying a divine call.

F: And you were admitting that the task of modernism is to teach the vast majority to live as only this small minority has previously lived. I was expressing my doubts as to whether the modernists would succeed.

M: Are they not compelled to try?

F: I'll let you answer that.

M: Then I should say that nothing can now arrest the penetration of the scientific spirit into every field of opinion. You do not seriously think your laws against Darwin and the rest are effective?

F: Apparently they are not.

M: Then what you call unbelief—that is, disbelief that there exists an authoritative account of destiny —is bound to infect the whole population. That be-

ing the fact, is it not our task to prepare men to lead independent, rational lives?

F: I might grant you that unbelief is bound for the time to spread. But I cannot believe you will succeed in teaching the great mass of men to lead independent, rational lives.

M: You believe that all men are the children of God?

F: I do. And that most of them are children.

M: That they are endowed by their Creator with reason and conscience?

F: Yes.

M: Why then deny to the humblest that salvation which some of the greatest have found?

F: You are argumentative, and you are very naive.

M: If there is one thing we modernists are not, it is naive. We are highly sophisticated people.

F: I withdraw the word *naive*. You are merely ignorant. You have never studied the history of religions. Naturally you are ignorant.

M: I have studied the history of religions.

F: Then how could you have failed to observe that the greatest teachers, Jesus, St. Paul, Buddha, Plato, Spinoza, all of them, taught that the perfect life was too difficult for the mass of men, that it required discipline and renunciation which most men could not endure, that it was in fact beyond their moral capacities?

M: You mean that "narrow is the gate"?

F: I do. Spinoza said that the way of salvation must be hard or it would not be by almost all men neglected.

M: What you are saying now deems to destroy what you were saying before. You began by asking me how you could convince your children that it is God's will. Now you are saying that it is extremely difficult for most men to obey God's will.

F: The great religious teachers, my friend, reached up to heaven, but their feet were firmly planted upon the earth. They had no illusions, as you have, about the capacities of men. That is why every important religion contains every grade of teaching from grossest superstition to the purest expression of the spirit. Nobody who examines Buddhism or Christianity can have the slightest doubt that they have been adapted in the course of time to all the different intellectual and moral levels of mankind.

M: And what do you deduce from these rather broad and doubtful generalizations of yours?

F: I deduce the extremely important conclusion that modernists are trying to grow grapes from thistles. And that is a rather naive thing to do.

M: I told you once that I objected to the word *naive*.

F: Well, I don't know what to call it. But are you not in fact blithely proposing to teach the scientific method to a mass of people who haven't the remotest chance of understanding it?

M: I said we were compelled to try because the supernatural systems of authority were ceasing to be credible in the modern world.

F: That you are helping to break down the supernatural systems of authority I won't deny. But that you are teaching men to do without them I am not so sure.

M: Why should you be so doubtful? The western world was once pagan. Then it became Catholic. Then a part of it became Protestant. Why should it not now become scientific?

F: Because this last change calls for the profoundest change in the habits of the human mind which has ever taken place. It means that men must learn to act with certainty upon premises which are uncertain. Your scientific method cannot guarantee its conclusions beyond all shadow of doubt.

M: Of course it cannot.

F: But the mass of men won't tolerate this much uncertainty. They are not sufficiently disinterested. They will make incontrovertible dogmas out of scientific hypothesis. You are not teaching science when you teach a child that the earth moves around the sun rather than, as the Bible teaches that the sun moves around the earth.

M: I know that.

F: So what your scientific education comes down to is this. A minority, perhaps a slowly growing minority, acquire the scientific habit of mind and learn to live the disinterested kind of life which science demands. But the rest merely acquires odds and ends of more or less obsolete information which, while it destroys the authority and the majesty of their inherited religion, is in itself morally worthless. It fits no plan, it supports no ideal of life. It provides no background for the human spirit.

M: I find all this rather depressing.

F: I remind you again that as a modernist you are a devotee of the truth.

M: What truth are you talking about?

F: A truth which shatters many pretences and illuminates some of the modern mind.

M: I am all agog to hear this truth.

F: Then you shall hear it. In our public controversies you are fond of arguing that you are open-minded, tolerant and neutral in the face of conflicting opinions. That is not so. You are fond of arguing that the conclusions of science can be reconciled with the conclusions of theology. That is evasive. These claims deceive nobody. They are merely adopted as convenient pretences by the politically minded, the timid, and the superficial. You know and I know that the issue is not whether Adam was created at nine o'clock in the morning or whether he descended from an ape. The issue is whether there exists a Book which, because it is divinely inspired, can be regarded by men as the "infallible rule of faith and practice," or whether men must rely upon human reason alone, and henceforth do without an infallible rule of faith and practice. What you challenge is not Genesis but revelation, and what I am defending ultimately is the faith of men that they know the Word of God.

I would ask you therefore not to be confused by the incidental ignorance of your partisans and of mine. It is of no consequence in itself whether the earth is flat or round. But it is of transcendent importance whether man can commune with God and obey His directions, or whether he must trust his own conscience and reason to find his way through the jungle of life. On that question it is impossible for him to be open-minded.

M: I don't see that.

F: You can say: Maybe Darwin is right . . . maybe Lamarck is right . . . perhaps Einstein is right . . . perhaps he isn't. . . . That is your scientific spirit. But you can't say: Perhaps the word of God is right and perhaps it isn't.

M: Why can't you?

F: Because the authority of the Bible rests upon revelation, and if you are open-minded about revelation you simply do not believe it. Doubt is an essential part of the method of science, but it is the negation of faith. To say that you are open-minded about the inspiration of the Bible is nonsense. Open-mindedness in this connection is a perfectly definite rejection of the Bible's inspiration.

M: I don't understand.

F: What do you mean by open-mindedness?

M: A refusal to reach a conclusion on the ground that the evidence is not conclusive.

F: Exactly. But those who believe in the divine authority of the Bible believe it on grounds which are beyond the reach of human inquiry and evidence. Once you subject that authority to the test of human reason you have denied the essence of its authority. You have made finite understanding the judge of the infinite. You are then, in the historic meaning of the word, an unbeliever.

M: Is there no way out of this conflict?

F: That is not for me to say. I speak for the ancient and established order of mankind. You are the newcomer. You are the rebel. I do not know what you are going to do about it. You can disguise this issue but you cannot obliterate it.

M: We can at least discuss it like gentlemen, without heat, without rancor.

F: Has it ever occurred to you that this advice is easier for you to follow than for me?

M: How so?

F: Because for me an eternal plan of salvation is at stake. For you there is nothing at stake but a few tentative opinions none of which means anything to your happiness. Your request that I should be tolerant and amiable is, therefore, a suggestion that I submit the foundation of my life to the destructive effects of your skepticism, your indifference, and your good nature. You ask me to smile and to commit suicide. . . .

The argument does not hold so much emphasis within the Roman Catholic Church and battle is not given after 1862.

[Where did you come from?] Or did you slip into existence by chance? . . . There is no such thing as chance. Its existence is contrary to reason. It is an empty word, often used, indeed, by those who deny the existence of God, in order to get clear of Him and His divine providence. Chance is the god of fools. . . . Were I to tell you that my watch just came by chance, just happened to grow in my pocket, you would laugh at me. And rightly so. It was made—by a watchmaker.

"But," some people say, "that is not what I mean by chance. I don't mean that everything just happened to be all at once. I mean that the whole world, and man along with it, just gradually developed during the course of thousands and millions of year." Without admitting that the statement is true, let us take it for gradual development, or evolution, do away with the necessity of a Creator? By no means! On the contrary, the wisdom and might of Him Who could place in the original elements such germs of wonderous power become all the more apparent. But who created the original elements? Who first set them in motion? Who gave them natural law and perfect order? It is useless to say that the world itself is eternal and has developed itself out of itself. Science tells us of heavenly bodies that are masses of liquid fire, of others that are bleak and cold in the grasp of icy death; and science has even reckoned the time when this earth of ours, now gradually cooling off, will be a ball of frigid lifelessness. Had the world evolved from eternity, it would long since have been doomed to destruction. No, that which is subject to change and development cannot be eternal. We can indeed admit a certain evolution within the same species. But no *creative* evolution, whereby new species are produced, has thus far been proven. Now, if not even a blade of grass ever formed itself, much less a meadow; if not even a single tree, then much less a forest; if not even one single man, then neither the race of men. From nothing, nothing is made.

Who made you? The laws of nature? Without a lawgiver there can be no law; so likewise no natural law, from which all other laws depend.[3]

In the preceding argument evolution within species is allowed if only for the sake of argument, as is an earth of an extensive lifetime. The Protestant view is, if anything, more generous to the strength of the sensory evidence of science.

The Scripture tells us that God, in the Beginning, created the heavens and the earth. . . . But we are living in our world of today, a world that has been carefully researched by many generations of scientists. Gradually man has developed a quite complete record of observations of nature. He has found in the rocks signs of ancient life, obviously much older than the Bible indicates. There are regular clocks built into nature, counting time by the inner changes that took place and by the marks left by the past.

The story of the creation was written long before men thought as scientists, and therefore one cannot judge these statements by scientific standards. Ways of expressions change. But the message carried by the ancient creation stories is as true today as it was at the time when these prophetic glances back into history were first recorded. The scientist has to work all over this wide world, to gather as many observations and facts as he can, and then attempt one great view of the universe. The Bible works exactly the other way around. It begins with the jubilant proclamation that "in the beginning was God" and that the vastness of the universe came out of him. Naturally the ways in which this is told are the words and thoughts of that time. You are as much tied to the present as those men were tied to their present.

What, then, is the common message the two reports proclaim? First of all, that there is nothing outside and before God, that God is the beginning of all. Secondly, that we are created by His commanding word. God is not a kind of craftsman who puts materials together. He simply thinks and speaks His divine thoughts, and this universe comes into existence. Thirdly, we learn that the creation is one, step by step, in an orderly proclamation. It does not appear suddenly from nowhere. God has made a world where everything depends on everything else. Sometimes Bible-believing Christians get upset about what we call evolution. We are afraid that evolution becomes a kind of automatic mechanism, and that therefore we do not need God any more. But must we think of evolution as something without God? Could this not be God's way of continuously creating? The message of the Bible is not that God stopped creating after "the first weekend." His work is continuing creation. How it was done is barely told. Evolution is not contradiction of report—as long as we adhere to the clear Biblical statement that the one who continuously makes all this is the One God.[4]

There have been numerous attempts to reconcile the evidences of scripture to fit the geological evidences but nearly all of these and in some cases all have been rejected by various groups.

In an attempt to solve the problems both of time and of the fossils, it has been suggested that there is a gap between Genesis 1:1 and Genesis 1:2. In the first verse is a statement of the creation for angels and when they fell into sin that world was destroyed. The first world contained many of the plants and animals that are preserved as fossils.

> . . . there is no evidence of a double creation in Scripture. . . . (and) Finally, the last verse of Genesis 1 reports that God saw everything that He had made, and beheld it was very good. This could hardly have been said if the angels had already fallen into sin and if their world had already been destroyed.[5]

One of the most popular proposals has been that the six days of Genesis 1 were long periods of time—millions or billions of years—and that evolution might be God's device for creation.

> Yet there is no reason for taking "day" in any sense other than that of an ordinary day. The repeated emphasis on "evening and morning" indicates that ordinary days are meant. God compares the creative week with the Jewish week in Ex. 20:11, thus indicating once more that these were ordinary days.[6]

Still others have maintained that the first part of Genesis is a sort of allegory and presents in poetic form the drama of creation as performed by God.

> Yet if the creation account and the account of the fall are allegorical, is the promise of the Savior also allegorical?

> If evolution is accepted, then Adam did not exist as an individual. Rather he must represent an evolutionary population . . . Yet if this is the case, Paul's whole argument in Romans 5 falls. For he speaks of the offense of one and the righteousness of one, and of the disobedience of one and the obedience of one (Rom. 5:18f). He refers to a single Adam just as he refers to a single Christ.[7]

A summation of the two positions on these questions might be seen in the following imaginary dialogue.

THE FUNDAMENTALIST AND THE POPULARIZER

F: The earth was created in 4,004 B.C.

P: How do you know?

F: This is the revealed word of God, as interpreted.

P: How do you know you have interpreted what God said?

F: By the light provided by Grace.

P: Well, I think the earth is many millions of years old.

F: How do you know?

P: From the salinity of the ocean and the thickness of the soil layers.

F: And how do you know you have read what is true?

P: From my senses, or measurement which is an extension of the senses, and from reason.

F: But as everyone knows the senses can deceive, so how can you be certain?

P: How do you know you have Grace?

F: I know from an inner feeling and because I have Faith.

P: Well, I feel that the senses can be used because I *believe* in the methods of science which are reliable if used with care.

P: How do you account for fossils?

F: I don't. They may be part of the original creation or they may have been deposited in the Great Flood.

P: I will fight this argument on your grounds. If fossils were a part of the original creation, then why were they created?

F: They were created for man's use.

P: But deposits of the same kind could have been made without putting them in the form of fossils. Isn't it true that God is Truth.

F: Yes.

P: But if God put these deposits in the form of fossils then God must be a deceiver.

F: Not at all. God could have put the fossils here to test the Faith of men.

These arguments of straw should illustrate that the essential basis of both religion and science is faith. The differences are based on their concepts of reality and on what they will accept as truth.

P: Could they have been placed on earth by a miracle?

F: Yes, indeed. God established the laws that govern the universe and when He does suspend them, we say that a miracle has happened. Since God is omnipotent He cannot be bound by His own laws nor does chaos result when He suspends one in order to work His will.

P: But one of the basic assumptions of science is that in order to know there must be uniformity in nature and that nature is immutable.

Yet these differences are quite strong and real. We, as individuals, in order to live in a world of science and materialism have suspended the argument by using one of the preceding devices so that the question has little importance to us today.

The question had considerable magnitude because the 19th century is one of popularization of science due, in part, to middle class reform ideas. Geology, like astronomy, is a science easily accessible to the amateur and could thus generate considerable interest.

It is a question that also generates considerable emotion. To deny the flood is to make God a clockmaker who wound things up at the time of creation and has since been sitting on His haunches—just watching. On the other hand, one of the basic assumptions of the scientist is that there are natural laws that man can know—that there is uniformity in nature. But if God can change the laws then there can be no uniformity.

Newton was willing to have God correct things periodically, but this is three centuries earlier. An examination of geology books in the late 18th and early 19th centuries show that they wanted to keep God—many of the geologists were clerics themselves. Churchmen during the same period exult science as a means of seeing God's great plan.

Aside from the religious implications of geology there are a number of ways that this field influences our lives either directly or indirectly.

Geology, as with astronomy, is popular because its subject matter is easily accessible to all. Almost every community have people banded together for the purpose of collecting rocks and providing themselves with an interesting and inexpensive means of utilizing leisure time. It also provides a means of demonstrating craftsmanship in an age when more and more we need to express our individuality.

For many it is a vocation instead of an avocation. Oil companies and mining companies hire geologists to seek out new deposits and to advise on the extraction from existing deposits. They are hired by civil engineers to determine the type of strata below in order to know what kind of structure can be placed and how.

Our minerals are a one time resource that is not renewed and hence needs to be conserved or mapped out as to present and future use. The rapidity of use of fossil fuels (coal and oil) point to the fact that use is increasing while the supply is fixed.

A growing field for geology is in the promise of prediction for earthquakes and volcanic eruption. The displacement and death of men from time immemorial due to these catastrophies can perhaps be brought to a halt or reduced. Since 1900 about 680,-000 people have died in earthquakes. Of these, about 900 were in the United States, 250,000 in China, 150,000 in Japan, 104,000 in Italy, and 61,000 in India.

STUDY AND DISCUSSION QUESTIONS

1. What sensory evidence can be used to demonstrate the truth of geology as to the age of the earth, fossils formed by former living plants and animals?

2. What Scriptural evidence contradicts the physical evidences?

3. What is the position of your church on these questions?

4. How can science and miracles be reconciled?

5. On what are both science and religion founded?

6. In what ways does geology play a role in our lives?

7. What flaws do you see in the arguments of the person arguing for a literal interpretation of the Bible? What flaws do you see in the arguments constructed by those who want a liberal interpretation of the Bible?

8. What can science say of the origins of the materials from which the earth or universe was created?

FOOTNOTES

1. Granville Penn, *Conversations on Geology,* (London, 1828), pp. 1-3 as cited in Charles Coulsten Gillespie's *Genesis and Geology,* (Harper and Row, 1959). The same book is excellent for a full discussion of the question in England from 1790–1850.

2. This is from Walter Lippmann, *American Inquisitors, a Commentary on Dayton and Chicago,* (New York, 1928), pp. 37-66, and is quoted by permission. Also see *Evolution and Religion: The Conflict Between Science and Religion in Modern America,* (D. C. Heath and Co., 1957).

3. Winfrid Herbet, *Answers: A Book of Catholic Information on Religious Topics,* (The Society of the Divine Savior, 1946), pp. 104-105.

4. Hagen Staack, *Genesis.* A non-copyrighted booklet produced by the National Council of the Churches of Christ in the U.S.A. and based on a series of eight television programs on the book of Genesis presented on "Frontiers of Faith" over the NBC Television Network, 1966.

5. John W. Klotz, *Modern Science in the Christian Life,* (Concordia Publishing House, 1961), p. 111.

6. *Ibid.,* p. 111.

7. *Ibid.,* p. 113.

XXI

Gathering the Elements

How does it happen that some elements and compounds are found concentrated in deposits? We would normally expect that no matter how the earth was formed or what its state in its early condition that all of the compounds and elements would be completely mixed and that on picking up a handful of rock or dirt that it should yield all 88 of the naturally occurring elements. Such, however, is not the case and the task of the chemist is thereby made easier than it might be otherwise.

One of the most potent forces in concentrating compounds is water. During the "water cycle,"[1] salts are dissolved by the water, and are carried and deposited. The great Salt Lake of Utah and the Dead Sea are examples today of the never-ending process. Such lakes, those without an outlet, when no longer fed by streams are covered and provide us with salt deposits. The salt bed under Detroit is one such deposit.

The oceans were made salty in the same way. Even though streams and rivers do not taste salty they contain small amounts of various salts which concentrate over millions of years in the oceans. Sea water is by weight about 34.40 parts, per thousand, solids. The solid part has the following composition in per cent:[2]

Sodium chloride (table salt)	77.758
Magnesium chloride	10.878
Magnesium sulfate	4.737
Calcium sulfate	3.600
Potassium sulfate	2.465
Magnesium bromide	0.217
Calcium carbonate	0.345

Our presentation, so far, accounts for the concentration of these compounds which are water soluble. How can water concentrate insoluble compounds? Carnotite, an ore of uranium, is frequently found in ancient river beds, and is not very soluble in water. So that its concentration must be explained in terms other than solubility. The mineral is soft so that the water by physical action could loosen and carry it in suspension. At curves in the river where tree trunks would hang up, the current would be slowed. Since the carnotite is fairly heavy it would sink to the bottom of the stream. Over millions of years the stream could change its course or disappear, leaving the ore deposits in pockets.[3]

Not only do streams disappear in geologic time, this is also true for seas. The region denoted as the Middle West, our own, was at one time a shallow sea —a warm shallow sea. This condition and the fact that carbonates are not as soluble in warm water as they are in cold accounts for the thick beds of limestone (calcium carbonate) that underlies the midwest. You have no doubt noticed the deposits of lime built up inside a tea kettle caused by the same sort of thing.

Another great force in the concentration of compounds and elements is living organisms.[4] The smaller size of the organism does not restrict the size of the deposit. Take the coral polyp as an example. By itself the single organism is quite small yet it can build islands of large size. The great barrier reef off the coast of Australia is an example of its handiwork, as are most of the islands of the South Pacific. Each organism concentrates calcium compounds from the dilute concentrations in sea water to build a protective coat itself. When the organism dies, the insoluble calcium shell is left for other organisms to grow upon, and over a period of millions of years great structures rise. The chalk cliffs of Dover and the large deposits of diatomaceous earth are the result of small organisms.

Nearly all lower marine life, such as the oyster, concentrate a considerable amount of copper which acts in the same manner as the iron in our blood. The lobster concentrates both copper and iodine, up to 200,000 times as much as is in sea water. While the sea cucumber may have large amounts of vanadium in its structure. These minerals, elements, and compounds which are concentrated by the organisms frequently become a part of the sea bottom when the organism dies and if the sea bottom is elevated by some geological disturbance, such as an earthquake, provides a strata which thus contains a concentration of the substance larger than is found elsewhere.

Plants as well as animals act as chemical concentrators. In the past, the chief source of iodine was from kelp washed up on the shores during storms. But perhaps the plants that are the greatest concentrators are the bacteria. The great deposits of coal, iron and sulfur are due to organisms too small to be seen with the naked eye.

Coal is formed from decaying plants which are made up largely of celluose and lignin having the generalized formula $C_6H_{10}O_5$. It takes about 15 feet of rotten trunks, branches, and decayed leaves to produce one foot of coal. As the plants die and sink in the surrounding bog, the bog is necessary for the formation of coal, the large amount of organic material (decay) which makes up the plant prevents ordinary oxidation by removing the gaseous oxygen in the water. Bacteria that can operate without oxygen, anaerobic, break apart the cellulose producing carbon dioxide, carbon monoxide, methane and water. The coal that is formed is classified depending on the amount of hydrogen and oxygen removed. (i.e. peat, lignite, sub-bituminous, bituminous, semi-bituminous, semi-anthracite, and anthracite.) The anthracite has had the greater amount removed and the peat the least. Peat is formed in bogs today but most of the grades of coal have been formed in the long distant past with the aid of pressure from overlaying strata of soil and rock. We will discuss the formation of petroleum, the other fossil fuel, in a later chapter.

Sulphur deposits are formed primarily by three bacterial genii: *(Thiothrix, Beggiatos,* and *Thioploca)*. Again, these reactions take place in swampy areas where the lack of free oxygen prevent usual decay patterns. The decomposing vegetable matter produced quantities of hydrogen sulfide gas or gas may come from so called "sulfur" springs. The bacteria used the hydrogen sulfide as a source of energy and produce free sulfur as a by-product. The large sulfur deposits of Louisiana are presumed to have been formed in this manner.[5]

The large bog iron ore deposits of Birmingham, Alabama, Chattanooga, Tennessee, and central New York are believed to have been formed by various bacteria (such as genii: *Spherotilus, Clonothrix, Leptothrix,* and *Crenothrix*). Springs or streams with some concentration of iron compounds are the home of the bacteria which build a sheath of iron hydroxides forming a brown gelatinous mass on the bottom of the spring or stream. When the bacteria dies the iron concentration remains. There is, however, some argument as to the extent that bacteria are responsible for iron deposits.[6]

Our next substance is described in one of Poe's stories:

> The vaults are insufferably damp. They are encrusted with nitre ... but observe the white web-work which gleams from these cavern walls ... The nitre ... see, it increases. It hangs like moss upon the vaults. We are below the river's bed. The drops of moisture trickle among the bones ...[7]

The substance is potassium nitrate. In Poe's day the same name, nitre, was used for the chemical. Again it is the work of bacteria which change dead animals or animal wastes. (The genii: *Nitricystis, Nitrosomonas,* and *Nitrobacter* oxidizes nitrite to nitrate.) In arid regions, bird and bat wastes (guano) may provide a good source of nitrates. Birds, such as the masked booby, are machines for converting fish, such as anchovies, into guano. Between 65 and 75 anchovies have been found inside one bird. Each day the bird eats about a pound of fish and deposits about an ounce of guano. This is deposited on islands off the west coast of Peru and Africa. Bird deposits seem to be made on land everywhere but only grow into huge deposits where there is no or little rainfall since the nitrates, in the form of ammoniam oxalate and urate, and the phosphorous, in the form of phosphates, are soluble in water and would wash away. Why sea birds tend to have the urge over the land nobody seems to know or if they have the urge elsewhere it doesn't build up.

During the Revolutionary War powder was scarce and nitrates were needed to make gunpowder so the Continental Congress had a committee write a booklet to tell people how to extract nitrates from the soil.

This booklet told those interested in extracting nitrates to collect barnyard dirt in containers with holes in the bottoms and pour water through. The water was then evaporated and the crude soluble nitrates remained.

Most of the existing ore bodies were, of course, formed when the earth was young. Gravitational pressure and the heat formed from the decaying radioactive materials caused the sub-crust materials to melt. Heavy metals would, by and large, sink toward the center of the earth. Silicates, oxides and sulfides of the metals would rise or stay towards the surface. In observing this process at a smelter it can been seen that the molten metal sinks while the lighter silicate, a sort of scum or slag, floats on the top of the melt. The term "magma" will be used to designate molten rock material within the earth that produces igneous rock when it surfaces and cools. From time to time after the earth was formed this magma would be forced to the surface by unequal pressures through faults or cracks in the crust.

It has been observed, that as any crystalline material cools, faster cooling produces small crystals while slow cooling produces large crystals. Further crystallization occurs at different temperatures for different substances.

Victor Maritz Goldschmidt (1888–1947) showed that this crystallization occurs depending on the arrangement of the atoms of the particular material into a pattern. Other atoms, if they are of the right size and geometry, can slip in between the atoms of the crystallizing atom-pattern. This approach of sorting by the temperature and crystal structure seems to check out in a number of deposits. A large deposit of iron ore at Kiruna, in northern Sweden, one of the most important iron mines in Europe, appears to have been formed from the magma in this manner.

In some deposits the original rock has been changed by magma forced into it. Frequently, it is indicated, hot gaseous fluids coming from the magma, rather than the magma itself, seems to have caused the change. This change seems to be a replacement in which the mineral grains in the rock are replaced by ore.

A word here about terms. Rock, like weed, usually means that the material is not usable by man while ore, like flower, means that the surface is useful to man. Whether something is an ore depends not only on the amount of concentration of the wanted material but also on the ease of extracting what is wanted and hence the cost of extraction. For example, while aluminum ore is 20% aluminum and iron ore is 25% to 70% iron, gold ore may be only 0.001% gold.

Another method by which materials are concentrated is from hot water and gases originating from the magma that percolates upward and carry materials in solution or act on them, in the case of gases. Deposits of copper, as those of Butte, Montana, are of this nature. Large proportions of the world's gold deposits are apparently formed in this manner. The gold rush to California was spurred by the discovery of this kind of deposit as well as Kirkland Lake in Ontario and Kalgoorlie in Australia.

Besides the evidence of the ores themselves, geologists have tried to duplicate, in the laboratory, some of the above conditions to see if this is actually what happened. The close correspondence between what happened in the laboratory tests and the appearance of the rocks examined in the field, have led geologists to conclude that ores were concentrated in these manners. Thus, the main forces for sorting out and concentrating elements and compounds are those of water, biological organisms, the slow cooling rate of the original molten material, and intrusion of fluids from magmas.

STUDY AND DISCUSSION QUESTIONS

1. What factors are responsible for the concentration of elements and compounds?
2. How are salt deposits formed?
3. What is the difference between rock and ore? What is a gem?
4. How is coal formed? Sulfur? Iron?
5. How are elements concentrated in the ocean?
6. What caused the original concentration of some metals?
7. What reasoning would have been given for the concentration of elements and compounds by the fundamentalists?
8. What is a mineral?

FOOTNOTES

1. In the "water cycle" rain falls, runs off the land into rivers, is carried by the rivers to the ocean, is evaporated by the sun, condenses into clouds which move over the land and again to rain. This is, of course, an over-simplification and a text on meteorology should be consulted for a complete understanding.
2. Thomas C. Chamberlin and Rollin D. Salisbury, *A College Textbook of Geology,* (Henry Holt, 1909), p. 289.
3. This idea is presented in "The Petrified River-The Story of Uranium," a film available from Motion Pictures, Bureau of Mines, 4800 Forbes Avenue, Pittsburgh, Pa.
4. For a more complete discussion of the action of living organisms, see C. C. Furnas, "The Formation of Mineral Deposits" in Samuel Rapport and Helen Wright (eds.), *The Crust Of The Earth,* (Signet Key Book, 1955), pp. 146-154.
5. See Marjory Stephenson, Bacterial Metabolism, (Longman, Green and Company, 1950), pp. 255-266, for a more complete treatment of the reaction.
6. For further discussion see John Roger Porter, *Bacterial Chemistry and Physiology,* (John Wiley and Sons, 1946), pp. 646-657.
7. Edgar Allen Poe, "The Cask of Amontillado."

Suggested Readings:

F. H. Day, *The Chemical Elements In Nature.* Reinhold Publishing Corp. 1963.

The role of sea water in chemical deposition is discussed in Peter J. Brancazio, and A. G. W. Cameron, *The Origin and Evolution of Atmosphere and Oceans,* (John Wiley and Sons, 1964), pp. 1-55.

A rather technical explanation of the deposition of minerals is to be found in Paul Niggli, *Rocks and Mineral Deposits,* (W. H. Freeman and Company, 1954).

XXII

Alchemy—Prologue to Chemistry

Alchemy had in many respects the same relation to chemistry as astrology had to the development of astronomy. Both are based in the stuff that dreams are made of. Astrology was one method of attack to seek to know the future. On the other hand, alchemy sought the "philosopher's stone" which could change base metals, such as lead, into gold and which would produce an "elixir of life" to prolong youth or extend life. Alchemists also used astrological signs to represent the materials with which they were working. Seven bodies in the heavens and seven materials: the sun—gold ☉; the moon—silver ☾; Mars—iron ♂; Mercury—quicksilver ☿; Saturn—lead ♄; Jupiter—tin ♃; Venus—copper ♀. Astrologists and alchemists were men of wealth, members of the clergy, or supported by noble patrons and later by wealthy commercial men. Astrology had its practical side—the determination of position and time—so too did alchemy grow in part from the practical recipes which went back to prehistory.

The practical arts were well developed in the cradles of civilization—Mesopotamia, Egypt, India, and China—at an early date. But just as the Egyptians did their calculations by rote and recipe so they did their practical arts. The Egyptians could handle practical problems in land measurement by a crude geometry but they had no system. It remained for Euclid to gather the ideas into a system of geometry. So too with the practical arts. It was the Greeks who began to make the theoretical ideas which underlaid alchemy.

The Greeks looked for an underlying constant of matter and believed that it was possible to make changes in matter from one form to another. They had finally evolved the idea of four basic elements—air, earth, fire, and water—which we have seen that Aristotle was able to build into his total system.

During most of the history of man there has been an intense separation of the man of ideas from the artisan. Neither could recognize the other as an equal and neither could learn from the other. This was true in Greece, except during the Golden Age of Pericles, and Egypt, where alchemy had begun in the first century A.D. in Alexandria, fell before Rome and the Romans being of a practical bent took the arts minus the ideas. The rise of Christianity and the fall of Rome meant the further fall of ideas from Greece either through neglect or the active attacks of Christians on anything of a pagan source. The ideas survived in the eastern empire until the closing of the academy by Justinian. The men of ideas were driven to Syria and the beginning Islamic Empire.

In Islamic writings is the statement that the ink of the scholar is more revered than the blood of the martyrs. This injunction plus the injunction that the *Koran* could only be read in Arabic not only preserved the Greek learning but meant the absorption and spread of Hindu and Egyptian learning. The word Alchemy itself is Arabic in origin and is apparently a blend of the practical arts of the Egyptians (chem referring to Egypt) and the theories of the Greeks.

The Arabs made many contributions to Alchemy including the development of much of the apparatus, the distillation of alcohol (another Arabic word) and a number of acids. The sum total was gradually transmitted to Europe during the period of the Crusades—scholars, then as now, were international.

One reason that Europe was ready to accept the ideas of Alchemy was a religious one. A basic tenent of Christianity was the general acceptance of the doctrine of trans-substantiation—the change of wine and bread into the blood and body of Christ.

Another was the practical arts contained in alchemical writing. Recipes used in brewing, glass making, metallurgy dying, paint pigments, mordants for textiles, etc.

Most of the alchemists claiming to be able to make gold were charlatans and quacks. Many were persecuted as warlocks and others, when they failed to convince their financial supporters that they could make gold, were put to death. For this reason, as well as the fact that if one could make gold or thought he could, he would work and write in secret. Most of the recipes were written in a secret code or used allusions to hide the real processes. In addition, there were mystical and magical incantations and prescriptions that had to be followed for success.

Literary works are filled with references to the alchemists ranging from Geoffrey Chaucer (c. 1340–1400)[1] to Johann August Strindberg (1849–1912)[2] so that certainly the influence of alchemy spreads beyond chemistry and for a considerable period of time.

> Also there was a disciple of Plato
> That on a tyme seyde his maister to,
> As his book *Senior* wol bere witnesse,
> And this was his demande, in soothfastnesse,
> "Telle me the name of the privy stoon."
> And Plato answerde unto hym stoon.
> "Take the stoon that Titanos men name"
> "Which is that?" quod he. "Magnesia is the same."
> Seyde Plato. "Ye, sir, and is it thus?
> This is ignotum per ignotius.
> What is Magnesia, good sir, I yow preye?"
> "It is a water that is maad, I seye,
> Of elementes foure," quod Plato.
> "Telle me the rote, good sir" quod he tho,
> "Of that water, if it be your wille."
> "Nay, nay" quod Plato, "certain, that I nille";
> The philosophres sworn were everichoon
> That they sholden discovere it unto noon,
> Ne in no book it wryte in no mannere;
> For unto Crist it is so leef and dere,
> That he wol nat that it discovered be,
> But where it lyketh to his deitee
> Many for tenspyre, and eek for the defende
> Whom that hym liketh: lo, this is the ende.[3]

In this quotation the pupil is asking the name of the "privy stoon" or first stone (the Philosopher's stone). The answer given is rather ambiguous. Titanos is the name for lime, which could be calcium carbonate or calcium oxide. Magnesia was used as a name for various substances—today we would say magnesium oxide but it is doubtful if that is what was meant. From the fact that magnesia is described as a water that is mad it could mean mercury. The reference to four elements could mean: earth, air, fire, and water; or it could refer to the four spirits: quicksilver, arsenic, sal ammoniac, and brimstone. If it meant the latter then these are principles rather than substances, which described the state and relation of an activity or ingredient rather than speaking of an actual unique substance. The alchemist by his mumbo-jumbo hid his secrets well.

Even though they are somewhat related there is a sharp distinction between alchemy and chemistry, just as there is between astrology and astronomy.

> Alchemy is distinguished from chemistry first by its purpose and secondly by its method. The purpose of alchemy is the perfection of all things in their kind and most especially of metals; that of chemistry is the gaining of knowledge concerning different kinds of matter and the use of this knowledge for all manner of ends.[4]

> The method of alchemy is to find and study old manuscripts and books on alchemy and to use the recipes; while that of chemistry "... is the accurate description of changes of all kinds of matter and the classification of such changes in order to discover general laws."[5]

The beginnings of the change from alchemy to chemistry had their origins in the Renaissance in the person of Philippus Theophrastus Bombastus von Hohenheim (1493–1541). He assumed the name of Paracelsus from "para"—above and celsus from the name of a famous Roman physician; thus he meant that he was greater than Celsus. The beginning of chemistry was in the field of medicinal chemistry (also called iatrochemistry).

It is interesting to note that he, like others of the Renaissance, respected and learned from the craftsman. In the case of Paracelsus some of his instruction came from miners.

Basically he claimed that each disease had a cure in medicines already supplied by God but that the cure was not in a prepared form so that the natural preparation had to have its "quintessences" extracted to be of the most benefit to the patient.

In place of the Aristotelian four elements he introduced their three principles—mercury, sulfur and salt. The concept of salt was new but the other two were already present in alchemy. Mercury, sulfur, and salt were not the same as the elemental concepts we have today. They did not stand for actual substances and still concealed what was being spoken of.

Progress within chemistry remained very slow and was, in fact, sidetracked by the development of the "Phlogiston" theory. This theory is dead end and only served to confuse the major issues. By and large the four elements with the seven principles and the phlogiston theory held the field until the late eighteenth century.

The man who has sometimes been referred to as the father of chemistry was Antoine Laurent Lavoisier (1743–1794). He used quantitative methods in his experiments—i.e., he weighed and measured materials with which he was working very carefully. In his experiments he showed the nature of combustion (burning) and destroyed the phlogiston theory. Perhaps more importantly he laid the basis for chemical nomenclature (naming) and formulated the basis for distinguishing between elements and compounds.

Only a few years previously Karl von Linne, usually referred to as Linnaeus (1707–1778), had formulated the binomial (two name) method of designating plants and animals (1758). It was, perhaps, with this system in mind that Lavoisier formulated his method of naming compounds such that the elements composing them were indicated. All of his findings were published in 1789 in his *Elements of Chemistry.*[6] With modifications the basic approach of both Linnaeus and Lavoisier has held to the present day.

It is with the publication of Lavoisier's book that the field of chemistry separates from that of alchemy, although alchemy does not disappear and may still be practiced by some. But the main stream of investigation is stripped of the mystical and the last vestiges of the Aristotelian four elements together with dealing the death blow to the phlogiston theory.[7] He also listed some twenty-three authentic elements, i.e., those substances which could be broken down no further by chemical means. It might be noted, however, that the alchemist's symbols for the elements were retained.

SEEKING ORDER

With the stimulus of Newton's *Principia* about a century back and the organization of known chemical knowledge and an organized system of naming coupled to the ideas that chemistry could be quantified led to the development of several empirical laws in the field of chemistry.

The law of equivalent proportions was formulated in 1791 by Jeremiah Richer (1762–1807), who was a student of Kant and believed as did his master that the physical sciences were all branches of applied mathematics. He discovered that there seemed to be a definite relationship in the weights of substances that combined with one another. A table of equivalent weights was drawn up which showed the relative quantities of the chemical elements that would combine with each other.

Eight years later, Joseph Louis Proust (1754–1826) discovered that no matter how a compound was formed that weight proportion in which the elements combined was always the same and hence formulated the law of constant proportion.

Democritus' atomic theory had been revived during the Renaissance and had been incorporated into the Newtonian-mechanists view of the world since it was assumed that the microcosm was a reflection of the macrocosm. Newton had tried to explain Boyle's law (volume and pressure of gases) by an application of his inverse square law by assuming that the gas atoms were nearly stationary and repelled each other with a force that varied inversely with the distance.

John Dalton (1766–1844) in 1803, made modifications of the atomic theory to meet the needs of chem-

istry. He kept Newton's idea but introduced the idea that the atoms of different elements differed. He held that they differed in size, weight, and the number in each unit volume. He had also discovered that when it was possible for two elements to form more than one compound that there was always a whole number relationship between the two compounds. This led him to formulate the law of multiple proportions.

He felt that one of the most important property of the atoms was their relative weights so he made a table of weights using hydrogen as value one. It was this approach which led others to invent order. Before we can explore order in chemistry further we might note that the definition of element by Lavoisier in 1789 was one of the factors leading to the discovery of further elements as can be noted by the peak about 1790 on the graph "Decade of Discovery." Let us investigate the effects of the development of various techniques on the number of elements discovered. Before we can order chemistry we must have a few elements.

DISCOVERING THE ELEMENTS

The first elements extracted were metals which were oxides that broke down easily on heating or were non-metals which already existed in the free state in nature. Many of these had been known as substances from antiquity but had not been classed as elements until after Boyle and Lavoisier.

In 1799 Volta produced a battery (called a voltaic cell) and a year later two Englishmen, William Nicholson (1753–1815) and Sir Anthony Carlisle (1768–1840) found that passing electricity through water produced hydrogen and oxygen. Humphry Davy (1778–1829) applying this technique about 1807 to melted salts isolated in quick succession the elements sodium, potassium, strontium, calcium and magnesium.

Geology and chemistry become tied together in the so-called "Heroic Age" of geology between 1790–1830 when geologists were discovering many new minerals for the chemists to analyze. Just as one example, between 1810–1820, Baron Jons Jakob Berzelius (1779–1848) described the preparation, purification, and analysis of over two thousand compounds. In addition, he developed our modern chemical symbols and discovered the elements selenium, thorium, and cerium.

The development of the spectroscope by Bunsen and Kirchoff in 1822 gave a new technique for the identification of elements and their discovery. How much the techniques of electrolysis and spectroscopy aided in the discovery of new elements may be ob-

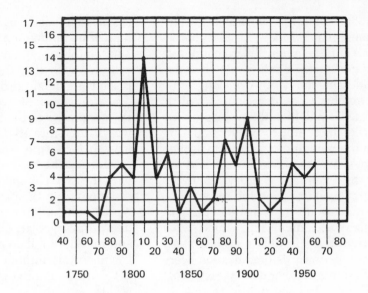

Figure XXII-1. Decade of Discovery.

served by and examination of the peaks on Figure XXII-1.

While the oxides of the rare earths, those elements between 57 to 71 on the periodic table, had been found and identified by the spectroscope it was not until the development of the technique of ion exchange about 1935, that it became possible to separate them as elements. They are so similar in all their properties that it is difficult to find a separation method.

In 1895 William Hampson (1854–1926) in England, and Carl von Linde (1842–1934) in Germany, were able to liquify air by cooling and using high compression. Baron John William Strutt Rayleigh III (1842–1919) noticed that nitrogen which had been prepared chemically differed from the nitrogen obtained when oxygen was removed from atmospheric air, in 1892. This led him to the discovery of Argon in 1898 and opened the way for the discovery of Krypton, Neon and Xenon by Sir William Ramsay (1852–1916) in 1898, after the technique of liquifying air had been developed.

Antoine Becquerel's (1852–1909) discovery of the phenomena of radioactivity in 1896 led to Marie Sklodowska Curie's (1867–1934) discovery of Polonium and Radium in 1898.

The theory of radioactivity in terms of atomic structure was established by Ernest Rutherford (1871–1937). He found that the radiation from uranium consisted of two types: alpha (which he later identified as the nucleus of the helium atom) and beta. He realized one of the dreams of the alchemists when in 1919 using radium as a source of alpha parti-

cles (they are given off by radium during the natural decay of the element) to bombard nitrogen and changed it into oxygen and hydrogen. Thus, the first artificial transmutation of an element and the beginning of the age of nuclear physics.

Particle accelerators of various types were developed during the 1930's to speed up particles to bombard various elements and to produce artificial ones. Although the major breakthrough in the synthesis of new elements had to wait for World War II and the development of atomic piles and bombs. This is shown by peaks in the late 40's and 60's on the graph in Figure XIII-1.

TOWARDS A PERIODIC LAW

We can return to seeking order in chemistry now that we have some elements. It had been observed, even in antiquity, that certain substances could be grouped together because they were somewhat similar in some of their properties. As an example one might cite the coinage metals—Au, gold; Ag, silver; and Cu, copper. These metals were all used in making coins, had similar properties, were found free in nature or when they were not they were easily extracted from their ores.

With the relative profusion of elements that were being discovered plus the need of man to find order and to group things, ideas were not lacking. As a number of elements were discovered, it was found that if one used hydrogen as one and then compared the other relative weights of the elements with hydrogen that the weights of the other elements were either whole numbers or nearly whole numbers. This fact caused William Prout (1785–1850) to hypothesize in 1815 that all of the elements were condensations of the hydrogen atom. For a while this idea stimulated a considerable flurry of work along these lines but as the weights were determined a little more exactly it was found that the values of the atomic weights were more apt to be fractional instead of whole number multiples of the hydrogen atom. So the "Prout's Hypothesis" as it was called was abandoned until the twentieth century when the work of Harold Urey again revived the idea.

There have been many attempts made in the past 150 years to organize the elements into a table. What follows is not an exhaustive study but merely to show that the faltering steps towards organization were not the product of a single mind or nationality or that all of the steps even went in the right direction.

One of the important steps leading to the great generalization—the Periodic Law—was the discovery made in 1829 by Johann Wolfgang Dobereiner (1780–1849), a correspondent of Goethe (German Poet), that there were little groups of elements, which he called triads, in which the properties of the middle element was the mean of the two extremes. These triads included the following: calcium, strontium, and barium; lithium, sodium and potassium; and sulfur, selenium and tellerium. Even the atomic weight of the middle element seemed to be the average of the two extremes, e.g. Calcium is 40 and barium is 137 so that strontium should be about 86 —it is 87. Lithium is 7 while potassium is 39, therefore, sodium should be 23—it is. Sulfur is 32 and tellerium is 127 thus selenium would be 78—it is.

Some twenty years later, in 1853, John Hall Gladstone (1827–1902) arranged the elements in order of their increasing atomic weights but so many of the weights at that time were faulty that no broad generalization was possible. A year later, Josiah Parson Cooke (1827–1894) proposed a classification system in which the elements were divided into series. The basis for the series was the general chemical similarities of the elements, the types of elements in relation to the compounds which they formed, the crystal relationships as well as the general chemical and physical properties of the elements. This was the first attempt at classification based on a study of all the available chemical facts.

The main stickler was the lack of valid atomic weights. This lack was provided by the work of Stanislao Cannizzaro (1826–1910) in 1858 and that of Jean Baptiste Andre Dumas (1800–1884) and Jean Sorvais Stas (1813–1891) about 1860.

Based on the revised and more accurate atomic weights, Alexandre Emile Beguyer de Chancourtois (1819–1886) made an attempt to include all the then known elements in a single classification system in 1862. In his system, he constructed a spiral line on a cylinder such that the symbol of the element was placed at a height proportional to its atomic weight. The atomic weights were plotted as the ordinates on the generatrix of the cylinder whose circumference was divided into 16 parts since the atomic weight of oxygen is 16 and was being used as the standard reference. The helix, or spiral, made an angle of 45° with the axis. The spiral crossed a given generatrix at distances from the base which were multiples of 16. As a result, lithium, sodium, and potassium fell on one perpendicular, while oxygen, sulfur, selenium, and tellurium fell on another perpendicular. For de Chancourtois number is the controlling factor and in this sense he is a Pythagorean—number (or figure) is reality.

A year later, John Alexander Reina Newlands (1838–1898) published a table in Chemical News which listed sixty-two elements in order of their atomic weights. They were divided into eight vertical columns and seven horizontal rows. The rows related the families. The relationship as stated by him was that similar elements are separated from each other by seven and that they have the same relationship as the octaves in music—hence the "Octaves of Newlands." This again is mystical and bears a strong resemblance of the ideas of Kepler to be found in his harmony of the spheres.

The philosophical implications of both de Chancourtois' and Newland's systems caused them to be rejected by their contemporaries although it was from the idea in Newland's system that the periodic law was to come.

Lothar Meyer (1830–1895) made a graph which showed the relation between the atomic weights and the atomic volumes of the elements. He also pre-

Figure XXII-2. Mendelejeff.

H	1	F	8	Cl	15	Co + Ni	22	Br	29	Pd	36	I	42	Pt + Ir	50
Li	2	Na	9	K	16	Cu	23	Rb	30	Ag	37	Cs	44	Tl	51
G	3	Mg	10	Ca	17	Zn	24	Sr	31	Bd	38	Ba + V	45	Pb	52
Bo	4	Al	11	Cr	18	Y	25	Ce + La	32	U	39	Ta	46	Th	53
C	5	Si	12	Ti	19	In	26	Zr	33	Sn	40	W	47	Hg	54
N	6	P	13	Mn	20	As	27	Di + Mo	34	Sb	41	Nb	48	Bi	55
O	7	S	14	Fe	21	Se	28	Ro + Ru	35	Te	43	Au	49	Os	56

Figure XXII-3. Law of Octaves.

(1869)

		Ti	Zr	?	
		V	Nb	Ta	
		Cr	Ms	W	
		Mn	Rh	Pt	
		Fe	Ru	Ir	
	Ni =	Co	Pd	Os	
		Cu	Ag	Hg	
H		Cu	As	Hg	
	Be	Mg	Zn	Cd	
	B	Al	?	Ur	Au
	C	Si	?	Sm	
	N	P	As	Sb	Bi
	O	S	Se	Te	
	F	Cl	Br	I	
Li	Na	K	Rb	Cs	Tl
		Ca	Sr	Ba	Pb
			Ce		
		Er	La		
		Yt	Di		
		In	Th		

Figure XXII-4. Mendelejeff's Chart.

pared other curves which showed the relation of certain physical properties—such as fusibility, volatility, malleability, and brittleness—of the elements to their atomic weights. On the basis of these curves, in 1896, he arranged fifty-six elements in a table consisting of groups and sub-groups which left spaces for undiscovered elements.

At about the same time and working independently from Meyer, Dmitri Ivanovich Mendelejeff (1834–1907) also prepared a periodic table which was more complete than any others and was more thoroughly founded on experiment.

In 1887, Mendelejeff made a statement concerning the importance of the periodic law.

> The law of periodicity first enabled us to perceive undiscovered elements at a distance which formerly was inaccessible to chemical vision; and long ere they were discovered, new elements appeared before our very eyes possessed of a number of well-defined properties.

Not only then does the periodic relationship allow the organization and relationship of the known elements to be shown but also enables the properties of undiscovered elements to be determined. The two key words are organization and predictability.

Let us compare the properties of one of the elements as they were predicted with those that were

H	(1870)									
Li	Be	B	C	N	O	F			
Na	Mg	Al	Si	P	S	Cl			
K	Ca	—	Ti	V	Cr	Mn	Fe	Co	Ni
Cu	Zn	—	—	As	Se	Br			
Rb	Sr	—	Zr	Nb	Mo	—	Rh	Ru	Pd
Ag	Cd	—	Sn	Sb	Te	I			
Cs	Ba	—	—	Ta	W	—	Pt	Ir	Os

Figure XXII-5. Early Periodic Table.

Properties Predicted by Mendeljeff	Properties Found for Gallium
Atomic weight c. 68.	Atomic weight 69.9
Metal. Sp. Gr. 5.9	Metal. Sp. Gr. 5.94
Low melting point.	m.p. 30.15 (below body temperature).
Non-volatile.	Non-volatile.
Should dissolve slowly in acids.	Dissolves slowly in acids and alkalies.
Oxide properites: M_2O_3	*Oxide* properties: Ga_2O_3
Sp. Gr. 5.5	Sp. Gr. unknown.
Should react with binary acids to form salts of type, MX_3.	Reacts with binary acids to form salts of type, GaX_3.
The hydroxide should dissolve in acids and alkalies.	The hydroxide dissolves in acids and alkalies.
Salts: Should form alums.	*Salts:* Alums are known.
Sulfide should ppt. by H_2S.	Sulfide does ppt. by H_2S.
The element will probably be discovered by spectroscopic analysis.	Gallium was discovered with the aid of the spectroscope.

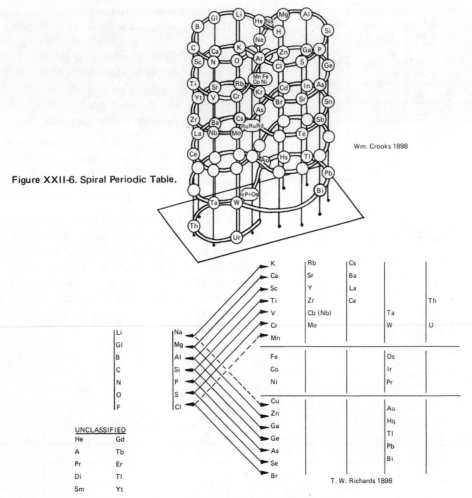

Wm. Crooks 1898

Figure XXII-6. Spiral Periodic Table.

T. W. Richards 1898

Figure XXII-7. Early Periodic Table.

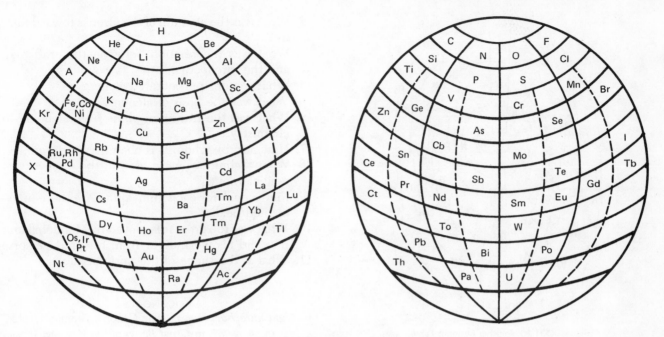

Figure XXII-8. Spherical Table.

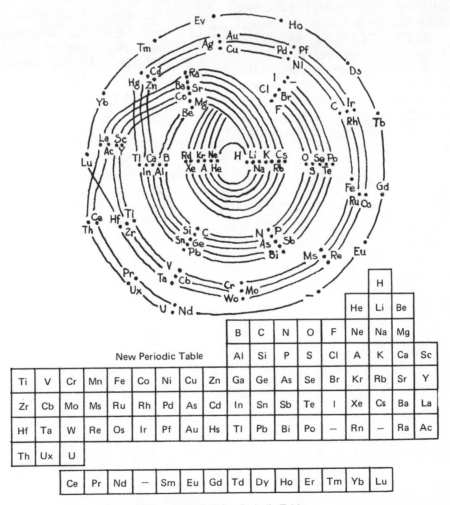

New Periodic Table

Figure XXII-9. Other Periodic Tables.

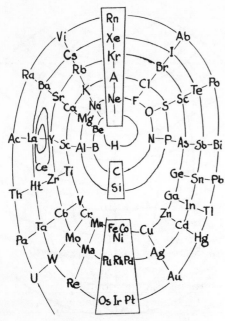

Figure XXII-10. Another Periodic Table.

found when the element was isolated. We can also note, in the following figures, that, like "the" scientific method, there is no "the" periodic table since it has taken many forms and shapes over the years.

As can be seen the predictions closely parallel the properties that were actually found for the element. This predictability gave a strong push towards discovering elements to fill in the blanks of the periodic table. This can be readily seen from Figure XII-1 which shows a peak about 1900.

STUDY AND DISCUSSION QUESTIONS

1. What was the relations between alchemists and astrologers?

2. What was the attitude of the Greeks toward matter?

3. What was the relation between the craftsman and the intellectual?

4. What reasons perpetuated alchemy?

5. What was the basis of alchemy?

6. What was the transition phase from alchemy to chemistry.

7. How and by whom was chemistry ordered?

8. What techniques were used in the discovery of new elements?

9. Find a peak for each innovation. Are there unexplained peaks?

10. What was the great role played by Mendeleff and how did his work differ from his predecessors?

11. What is the shape of the periodic table?

FOOTNOTES

1. *The Canterbury Tales,* "The Canon's Yeoman's Tale," pp. 474-487, in *Great Books of the Western World,* Vol. 22.

2. Leta Jane Lewis, "Alchemy and the Orient in Strindberg's Dream Play," *Scandinavian Studies,* Vol. 35, No. 3, (Aug., 1963), pp. 208-222.

3. F. Sherwood Taylor, *The Alchemists,* Collier Books, 1962, p. 151. In general, a most readable book on alchemy which the lay reader should find interesting.

4. *Ibid.*

5. An English translation is available in *Great Books of the Western World,* Vol. 45.

6. More details about the phlogiston theory and the work of Lavoisier may be found in Ch. 26, of Stephen F. Mason's, *A History of the Sciences,* Collier Books, 1962.

XXIII

Play Twenty Questions—Elements, Mixtures, or Compounds?

REACTIONS AND ORGANIZATION IN CHEMISTRY

Between 1800 and the middle of the twentieth century chemists of a number of nationalities labored to bring order to the chaos of matter. They organized millions of substances, studied their make up, and how they interacted with each other. The following is merely a condensed summary of their work.

All substances are classed as elements, compounds, or mixtures. Elements are subclassified as metals or non-metals. In general, metals are shiny, malleable, ductile (they can be pounded flat or drawn out without breaking), and are good conductors of electricity. On the other hand, non-metals are brittle and are not good conductors of electricity. Compounds are subclassified as organic or inorganic. In general, organic compounds have a backbone of carbon or silicon atoms which are linked to each other in chains or rings. All living or once living substances are organic but the converse is no longer true. Organic compounds do not generally act as electrical conductors. The primary sources for organic compounds are petroleum, coal, and cellulose from plants. Living or once living organic compounds may be roughly subclassified as lipids (fats), carbohydrates (sugars), or proteins. Organic compounds differ not only in the arrangement of the atoms within the molecule but also the order in which the atoms occur and the arrangement of parts of the molecule in space. This makes the difference among cabbages and kings and ships and sealing wax. The elements making them up may be exactly the same but the order of the arrangement and the way the arrangement is made in space are the key differences. Organic chemistry defies the rule that the whole is equal to the sum of its parts.

Inorganic compounds are generally classified as acids, bases, or salts. Acids are formed by the reaction of the oxide of a non-metal with water. Bases are formed by the reaction of the oxide of a metal with water. (Both of the preceding statements are true only if the oxide will dissolve in water.) An acid reacted with a base will form two compounds—one known as a salt and one known as water. Acids, bases and salts are called electrolytes because in a water solution they will conduct electricity.

Two other types of reactions should be mentioned. In general, metals will react with non-metals to form a salt, and a more-active element will replace a less-active element in a compound.

For the time being, let us say, that when one atom "sees" another atom it is repelled because they have the same outer charge so that in order to force them together into a molecule it is necessary to get them close together. This may be done by heat (they bounce together faster), by pressure (this pushes them together more often), or by a catylst (this gives them a close meeting spot). When they get together they may find that they do not need as much energy to stay together and so energy is given off (exothermic reactions). This energy given off is usually in the form of heat which helps other atoms get together and react. If, on the other hand, they need more energy to stay together than they produce by joining then they must take energy in (endothermic). This last type of reaction requires that heat constantly be applied in order that other atoms will join to form molecules. No matter which type of reaction is occurring there is usually a small amount of heat necessary to start the reaction. The speed with which a reaction takes place may vary considerably. Some reactions, like the rusting of iron, takes place slowly while others, such as burning leaves, may take place rapidly; and still others may take place almost immediately, such as the burning of gunpowder. These are all the same kind of reaction but they take place at different speeds.

ELEMENT, MIXTURE, OR COMPOUND

Reactions are necessary to form compounds and all chemical changes are reactions. In general, an equation can be written to describe what happens in a reaction. A compound is different than the elements that went into its making. It not only looks different, as a rule, but also has completely different properties.

It cannot be separated easily or without having another reaction. A compound has a distinct make-up that is always the same for a particular compound.

Sodium chloride is a common substance (table salt) that we use every day. The parts that make up the compound are quite different. Sodium is a soft, silvery metal that could be cut with your finger nail except that it would burn your hand. It reacts with the air so easily that great heat is produced and has to be stored under oil to keep the oxide from forming. It reacts with water to form lye (sodium hydroxide) and if taken internally would be a quick way to commit suicide. The other constituent is chlorine which is a yellowish gas at room temperature. It is extremely poisonous and was the first poison gas used during World War I. It is dissolved in water to kill bacteria in our drinking water and in a moist surrounding it acts as a bleach. Chlorine is very reactive and forms compounds with most metals. Two nasty elements go together to form a compound which we need to continue life. Salt is so precious that it has been used, like money, as a medium of exchange and yet that which makes it up is poisonous. Certainly then compounds differ in properties from the elements that make them up—they are completely different.

Mixtures, on the other hand, can be made in any proportions without a reaction and can usually be separated by some simple physical means such as sifting, magnetic attraction, or selective solubility (a solvent is chosen that dissolves one of the ingredients without dissolving the other).

Definitions of mixtures may vary but they run in a similar direction. A mixture consists of two or more elements and/or compounds with each constituent keeping its own characteristics. On the other hand, a compound is usually defined as a substance which can be decomposed into two or more simpler substances, either elements or another compound and elements, by ordinary chemical means. A compound is differentiated from an element by stating that the element cannot be broken down by ordinary chemical methods. These definitions are, however, not operational.

Another way in which elements, compounds, and mixtures are classified is by appearance. If something looks the same throughout then it is called homogeneous and includes substance and solution. Solutions are always mixtures because they are made up of the solvent and the solute. Substances include elements and compounds and both are homogeneous. Contrariwise, if something does not look the same throughout, that is if there are visible particle differences, then it is heterogeneous and a mixture.

Appearances can, however, be deceiving. While it can be stated, with few exceptions, that something which is heterogeneous in appearance is a mixture, it does not follow that a solid in which no visible particle differences can be seen is a compound or element. There are some 50,000 inorganic salts, most of which are white crystals. Two or more of these compounds placed together would make a mixture, the ease of error has been noted by most of you if you have ever mistaken salt for sugar and vice-versa, or if someone has mixed the two in the same container. The sugar that you know is just one kind, (as the salt you know is just one kind in thousands), in hundreds and is organic which makes it one compound in millions—most of which are white crystals. Some of these, like moth balls, have a distinctive odor while most inorganic compounds have no odor. Organic compounds if strongly heated will turn black like overheated fudge or bread or they may simply evaporate and disappear. Most inorganic compounds if heated will do nothing unless they get hot enough to become liquid or decompose as a few do.

Most (about 70%) of the naturally occurring 88 elements are metals. In some books this figure is mistakenly given as 92 but in the first 92 elements four are man made. These are Technetium 43 (1939); Astatine 85 (1940); Francium 87 (1939); and Promethium 61 (1947). All except the last were made by the bombardment of another element with subatomic particles and it came from the pile of the first atomic reactor.

Pure metals are shiny but if finely divided or in their oxide form they may be a variety of colors. Metals and their oxides can be dissolved by a reaction with *aqua regia* (a combination of hydrochloric and nitric acids). Some of the pure metals can be dissolved by mercury the only metal a liquid at room temperature and, indeed, this was the method used in antiquity for the extraction of gold from its ore.

Of those elements that are clearly not metals 12 are gases and one is a liquid at room temperature. This leaves only a few solid non-metals. The common ones are boron, carbon, silicon, phosphorus, sulfur, and iodine.

Most of the earth, from the crust up at least, is a mixture. Air is a mixture of gases while dirt and rocks are mixtures of minerals. Water is, for the most part, not water but a solution of water and something or somethings else. The purest form of natural water is rain water and even it has dirt in it since the rain washes out the sky. Ocean water, as we have seen, is a solution containing many compounds. Enough water will dissolve anything in nature in time. Water contains dissolved gases as well as dissolved solids.

A list of the commonly known elements as determined by a survey.

Aluminum	Hydrogen	Platinum
Arsenic	Iodine	Potassium
Calcium	Iron	Radium
Carbon	Lead	Silicon
Chlorine	Magnesium	Silver
Chromium	Mercury	Sodium
Cobalt	Neon	Sulfur
Copper	Nickel	Tin
Fluorine	Nitrogen	Tungsten
Gold	Oxygen	Uranium
Helium	Phosphorus	Zinc

The 88 elements and 15 man made ones with their atomic numbers.

Actinium	89	Gadolinium	64	Strontium	38
Aluminum	13	Gallium	31	Platinum	78
Americium	95	Germanium	32	Technetium	43
Antimony	51	Gold	79	Polonium	84
Argon	18	Hafnium	72	Terbium	65
Arsenic	33	Helium	2	Promethium	61
Astatine	85	Holmium	67	Thulium	69
Barium	56	Hydrogen	1	Radon	86
Berkelium	97	Indium	49	Tungsten	74
Beryllium	4	Silver	47	(now called Wolfram)	
Bismuth	83	Palladium	46	Rubidium	37
Boron	5	Sulfur	16	Xenon	54
Bromine	35	Plutonium	94	Scandium	21
Cadmium	48	Iodine	53	Zinc	30
Calcium	20	Iridium	77	Tellurium	52
Californium	98	Iron	26	Praseodymium	59
Carbon	6	Krypton	36	Thorium	90
Cerium	58	Lanthanum	57	Radium	88
Cesium	55	Lawrencium	103	Titanium	22
Chlorine	17	Lead	82	Rhodium	45
Osmium	76	Lithium	3	Vanadium	23
Sodium	11	Lutetium	71	Samarium	62
Phosphorus	15	Magnesium	12	Yttrium	39
Tantalum	73	Manganese	25	Silicon	14
Chromium	24	Mendelevium	101	Potassium	19
Cobalt	27	Mercury	80	Thallium	81
Copper	29	Molybdenum	42	Protactinium	91
Curium	96	Neodynium	60	Tin	50
Dysprosium	66	Neon	10	Rhenium	75
Einsteinium	99	Neptunium	93	Uranium	92
Erbium	68	Nickel	28	Ruthenium	44
Europium	63	Niobium	41	Ytterbium	70
Fermium	100	Nitrogen	7	Selenium	34
Fluorine	9	Nobelium	102	Zirconium	40
Francium	87	Oxygen	8		

Most river water contains suspended and colloidal solids. Even the pure driven snow has a dust particle at its center around which it was formed.

Things which are the most common to us are mixtures. Of what we take internally all are mixtures save salt (sodium chloride) and sugar (sucrose). The gasoline that drives our cars, the clothing we wear, the materials we build with—all are mixtures. As we have already seen in thermodynamics and the principle of entropy, that the tendency of nature is in the direction of mixtures and not purity that would be found in the selection of compounds or elements.

Of the four elements of the ancients, air, earth, water, and fire, two are always mixtures (air and earth) while one is nearly always a mixture (water) and the last is a process. Possibly the greatest factor in the slowness of the development of chemistry was related to the fact that while the organization of chemistry was ultimately to rest on the elements as the building blocks most of what the chemist knew was about mixtures and compounds.

Even with the definitions of elements and compounds as proposed by Lavoisier there remained the question of how to tell when something had reached the point where it could no longer be broken down. If something could be separated the chemist knew that what he had started with was a compound or a mixture but if it could not be separated he didn't know into what category to place what he had. Only by the laborious gathering of information on the properties of substances—both compounds and elements—was the chemist to begin to get ideas of how to classify and separate. It is one thing to define a process and another to perform it.

STUDY AND DISCUSSION QUESTIONS

1. How are elements classified? Subclassified?
2. How are compounds classified? Subclassified?
3. What types of reactions are there?
4. How do compounds, elements, and mixtures differ?
5. How do reactions occur?
6. How many naturally occurring elements are there? How many are common? How many man-made?

XXIV

The Fourth State of Matter

If one uses the criteria of temperature then there are three states of matter as is commonly taught. Classically there are solids, liquids and gases. Solids have a definite shape and volume which they retain but the same substance at a higher temperature becomes a liquid which have a definite volume but whose shape is that of the container. At a still higher temperature the same substance is a gas which has neither a definite shape nor volume.

On the other hand, if the ability to flow, particulate size, or change in properties is used as criteria then there are more states of matter. Today, the scientist recognizes two additional states—colloids and plasma. The plasma state is used in experimental physics as a possible technique of harnessing hydrogen fusion which we will be discussing briefly toward the end of the course.

The reason that colloids are considered to be a separate state is that they have peculiar properties not shared by the other states of matter. The colloidal state consists of particles which are less than 10^{-5} cubic centimeters in size and larger than molecules. Molecules are made up of atoms and are the smallest part of matter that retains the identity of a substance. Both molecules and colloidal particles are invisible—even to the microscope.

One of the effects of reducing the size of a particle is that the surface of the total mass is increased. If a cube of matter measuring one centimeter on an edge and hence one cubic centimeter in volume is repeatedly reduced in size ten fold each time the total volume remains the same but the surface area increases ten fold with each reduction in size.

If we continued to reduce the particle size until they were one atom thick the 1 cm.^{3} would cover an area of 5×10^7 cm.^{2} or over one acre.

As the colloidal size is attained or as the colloid is dispersed in a medium the particle acquires a charge. Since the charge is the same for each particle they repel each other and do not settle out. The factor of size and that of similar charge accounts for those properties which are observed.

If a solid is reduced to the colloidal state it conforms to the shape of the container, like a liquid, and can be poured. No solid or liquid will burn but if placed in a colloidal state and mixed with air it will burn explosively. The rate of reaction is dependant on size and one of the first steps in any chemical industry is the reduction in size of the materials that will be worked with.

Frequently the colloidal system is of more interest than the colloid itself. e.g.

Individual Size	Surface Area
1 cm.^{3}	6 cm.^{2}
0.1 cm.^{3}	60 cm.^{2}
0.01 cm.^{3}	600 cm.^{2}
0.001 cm.^{3}	6,000 cm.^{2}
0.0001 cm.^{3}	60,000 cm.^{2}
*0.00001 cm.^{3}	600,000 cm.^{2}
0.000001 cm.^{3}	6,000,000 cm.^{2}

*Limit of visibility for light microscope.

Dispersion Medium	Dispersed Substance	Examples
Gas	Liquid	Clouds, mist, fog
Gas	Solid	Dust, smoke
Liquid	Gas	Whipped cream, foam
Liquid	Liquid	Milk, butter, mayonnaise
Liquid	Solid	Milk of magnesia, paint, starch (sols)
Solid	Solid	Some alloys, gems having color
Solid	Liquid	Gels, jelly, pearl (calcium carbonate and water)
Solid	Gas	Floating soap, pumice stone

Thomas Graham (1805–1869) was the first to use the term "colloid" in 1861 to describe substances which exist in a non-crystalline or gelatinous condition such as gelatine or glue. While the term is over a hundred years old, most of the direct influence on changing the actions of man may be restricted to the last 30 years.

When one substance is dispersed in colloidal state within another, although it is not possible to see the individual particles, they may be observed collectively in the beam of a strong light. Similar to the way that dust may be seen in a darkened room when a light ray passes through the air and is diffused. This is called the Tyndall effect and named after John Tyndall (1820–1893).

There are a number of ways to make colloids. Among these methods is the homogenization of liquids and an electric arc; which is used to disperse metals in a liquid. Gold, for example, was first dispersed by Faraday in glass to produce a red colored glass. Mechanical grinding mills called colloid mills are used for some materials like paint pigments.

It is not only the ability of the particle to absorb charge on its surface from charged atoms or molecules that keeps the colloid from settling but also the energy that is given it by the molecules from the surrounding medium. This motion, which is rather random, is called Brownian movement from the observations of the Scotch botanist Robert Brown (1773–1858) in 1827. Sometimes the colloidal particle has a coating which helps keep them apart and hence from losing their charge.

Faraday, in 1857, made some colloidal suspensions that have remained clear and in the colloidal state for over one hundred years. The colloidal state can, however, be destroyed in a number of ways such as heat, electric discharge, salt, alcohol. These act by removing the charge.

There are times when colloids are undesirable. F. G. Cottrell in 1911, found that smoke could be removed by passing it between highly charged electrical plates and thus invented the Cottrell precipitator which is used by many industries to eliminate smoke and hence a part of air pollution. The cost of such a precipitator is high and in many cases prohibitive. Explosions in flour and other mills, and coal mines are sometimes due to the colloidal size of the particles of flour or other material in the air which are ignited by a spark from the machinery.

Protoplasm, which is the basis of all animal life, is colloidal in nature as are all plant and animal tissue. Besides this most bacteria and viruses are within the colloidal size range and some of the methods of dealing with them are based on removal of their colloidal nature. Many processes of life result from or are related to the colloidal state.

Dirt, sometimes described as the housewife's friend because you can't clean house without it, is often coated with grease and hence resists cleaning. Sometimes ammonia or another base is used for cleaning which changes the grease to a soap and thus one end of a soap can be wetted and dissolved. Part of the cleansing action of soap is related to the fact that it can change from a solid gel to a colloidal suspension easily. The fat end of the soap goes into the grease while the wettable end goes into the water and causes the dirt to become a colloidal suspension or emulsion. Detergents act in a similar way except they do not form insoluble compounds in hard water which makes a scum. Once the way in which soap cleaned was understood, large numbers of compounds were created to do a better job than soap and since World War II most soaps have been replaced by other detergents of a synthetic nature.

The foam, which the housewife associates with cleaning, is no longer needed and fouls up sewage disposal plants because they are hard to get rid of. In addition, some of the detergents have phosphates which increase the amount of algae in lakes and increase pollution. Other detergents cannot be broken down by any bacteria and remain in the lakes constantly increasing the pollution problem. Yet this same idea of foam is used to concentrate low grade ores by a process called the floatation method. Each technical idea carries with it the good and the bad.

In addition to the undesirable features of some colloids, there are areas in which colloidal states are useful.

The blending of several technical advances during World War II made possible the new aerosols. Disease has always killed more in warfare than bullets and there were several that were carried by insects. D.D.T. had been developed so that there was a problem of dispersing it easily. By 1944, most of the bugs had been worked out of the system. An inert gas that was non-toxic, freon, which had been developed in the 1930's as a refrigerant, was used as a propellent. The gas developed pressures inside the container of 90 pounds/square inch and required a strong cylinder. Hence, the device was called a bug-bomb. The other problem was to devise a means of making a small enough hole to produce a spray of colloidal size —the D.D.T. would stay in the air longer and would be more effective. Another war-time problem was the lack of blood plasma to treat wound cases. A new chemical was made by forcing acetylene together under pressure to form a large molecule. It was called polyvinylpyrrolidone or PVP for short. This could be used with blood plasma to reduce the amount of plasma needed.

After the war it was found that PVP when sprayed on hair not only forms a light transparent film on each exposed hair but also swells the surface of each hair and makes it more easily managed and adds to the luster. The aerosol technique quickly spread through the cosmetic industry and other areas. Today they are used for a wide variety of products—paints, lacquers, whipped cream, shaving cream—regular, menthol, lime, heated, etc., insecticides, deodorants, powders.

STUDY AND DISCUSSION QUESTIONS

1. What is the basis for calling colloids the fourth state of matter?
2. How does matter in the colloidal state differ from matter in the traditional states?
3. How can matter be made colloidal?
4. How can one determine if a solution is a colloidal?
5. What are some undesirable features of colloids?
6. Why have colloids become more important to the average person since World War II?
7. Why are colloids lively topics?
8. How does soap work?
9. Explain the formation of deltas.

Suggested Reading:

"Symposium on the Teaching of Colloid Chemistry" *J. Chem. Ed.* 26 (1) 18-31 (1949).

C. P. Neidig and A. B. Hersberger "Organic Synthetic Detergents" *Chem. Eng. News* 30 (35) 3610 (1952).

E. A. Hauser "The History of Colloid Science" *J. Chem. Ed.* 32 (1) 2 (1955).

F. L. Schaffer "Poliomyelitis Virus" *J. Chem. Educ.* 36 (9) 469 (1959).

Peter Debye "How Giant Molecules Are Measured" *Scientific American,* (Sept., 1957), p. 90.

"The Teaching of Colloid and Surface Chemistry" *J. Chem. Ed.* 39 (4) 166-195 (1962).

XXV

The Genie's Magic Lamp

While not as influential on the thought of man as some aspects of physics, astronomy, and geology that we have looked at nevertheless chemistry has deeply influenced the actions of man as it has made possible for man to remake his world. In this section we shall see how the chemist organizes the riot of matter for the use and abuse of mankind.

From the beginning it should be apparent that few things could be changed for man's use without involving chemical as well as physical changes. Chemistry had a strong relation in all the practical arts that came with the origins of civilizations. These included the home arts of baking and brewing as well as tanning, weaving and dying of textiles, metallurgy, glassmaking, and the formulation of drugs. So that the uses of chemistry occur long before chemistry or even alchemy has been developed and science follows art.

BLACK IS BEAUTIFUL

Coal was the first fossil fuel used by man and by the eighteenth century had replaced charcoal (growing scarce as forests disappeared) when it was heated without air to make coke. One ton of coal made 1,400 lbs. of coke and the following other products:[1]

Coke-oven gas (11,200 ft^3)
    Hydrogen (52%)
    Methane (32%)
    Carbon monoxide (4-9%)
    Nitrogen (4-5%)
    Ethylene (3-4%)
Ammonia or its compounds (5.5%)
Light Oil (2.85 gal.)
    Benzene (65%)
    Tolune (15%)
    Xylenes (7%)
    Light solvent naphtha (10%)
Coal tar (150 lb.)
    Naphthalene (11%)
    Pyridine
    Heavy solvent naphtha
    Creosote oil

Benzene could be changed to nitrobenzene which in turn could be changed to aniline. A derivative of aniline has the formula $C_{10}H_{13}N$. William Henry Perkin (1838–1907), an eighteen year old chemist, was working under the direction of another man who was studying the composition of quinine. Quinine was used in the treatment of malaria and was important to England if she were to subdue the tropics. The formula that had been worked out for quinine had the formula $C_{20}H_{24}O_2N_2$—look back to the formula for aniline. Yes, this young man thought that if two anilines could be put together using an oxygen reaction and taking out water (which would remove the two hydrogens) he could make quinine instead of getting it from the bark of trees in South America. His reaction did not produce quinine, which was finally synthesized in 1944, but it did produce a beautiful purple dye. Like the story of *The Ugly Duckling,* a substance smelly and black became beautiful. The accidental invention of aniline dyes began a whole new industry and also stimulated an invention in the area of medical arts.

Petroleum, sometimes called black gold, has gradually replaced coal as the starting material for putting together new compounds. Therefore, we will reserve the discussion of further important useful items to man until we have looked at the formation, rise in use, discovery, refining methods, and composition of petroleum.

The story of the formation of petroleum is not as clear as that of coal. Apparently some of the story is similar or at least laboratory work to reproduce conditions and the general observations of oil deposits would suggest this. Small plants, such as diatoms, each contain a small drop of oil, or oil like substance, which would be carried to the bottom of the sea when the plant died. It seems that oil was laid down under salt water, as opposed to fresh for coal. There is always salt water or a salt dome associated with oil deposits. This minute drop of oil multiplied by the large number of organism was further acted upon by some kind of salt water anaerobic bacteria to remove oxygen. Analysis of oils shows about 85% carbon,

14% hydrogen and 1% oxygen. Oil deposits are found only where a porous layer of rock is trapped between two impervious layers of rock and further that the porous layer must be tilted.

The most frequent type of structure is called an anticline where all three layers have been folded so as to produce a dome like structure. Sometimes the structure trapping the oil is a fault through which oil can seep to the surface or a tilted fault which while not allowing oil to the surface has still trapped it. The oil, gas, and salt water migrated through the porous rock towards the top of the dome or fault making the oil deposit. When a drill goes through the impervious layer, the gas comes up first since it is usually under pressure. After the gas is the oil which lays over the salt water and the salt water acts to force the oil upward as well as the trapped gas. Such a deposit should be thought of as a sponge rather than an underground pool.

Surface seepage of oil was known from antiquity but had little use except the bitumin deposits which were used for waterproofing or that which was burned but the light was dim and had a foul odor. Animal fats were used in some places for light but the burning fat smarted as you know if you have been near a frying pan that is on fire.

In the early days of this country people went to bed at dark or used the fireplace for light. Candles were known but they were expensive and seldom used except by the middle class and wealthy. As the middle class grew, the need for candles grew as well.

Candles, lubricating oils, and lamp oil were supplied largely by the whale. The growth of the market and faster ships threatened the whale with extinction. An improved lamp coupled with the refining of crude oil for kerosene produced a shift by the mid-nineteenth century. The gasoline and heavy portions were thrown away.

Just before the Civil War, a surface pool of oil was discovered near Titusville, Pennsylvania. While crude "Rock oil," as it was called, had been bottled and sold as medicine and a few had used it as a lubricant it was not used as an illuminant until George H. Bissell (1821–1884), an American promoter, sent samples to Professor Silliman at Yale College. Silliman wrote him that a lighting material could be obtained at low cost.

Bissell hired a retired streetcar conductor, Col. Edwin Lourentine Drake (1819–1890), who drilled the first oil well and brought in oil on August 27, 1859.

It became obvious about 1912 with the increased use of petroleum products by the automobile that the surface indicators and divining rods could not find enough new oil wells. By the 1920's the search for oil had become the job of professional geologists using a number of new techniques. They used drill core samples to reveal the type of strata below the surface and which way the strata sloped. They employed techniques of seismology in which explosive charges were detonated and the shock waves recorded on sensitive instruments also told them of the underlying strata. Two other techniques used were gravimetry and magnetometry. In the first one by noting on the instrument the changes in the force of gravity it told them how far below the heavy rocks were located or if the rocks were slightly magnetic the other technique could be used. In any of these techniques the most that they knew was where oil was likely to be found and not where it could be found— that rested on actually drilling the hole.

At the beginning of the century crude oil was still processed chiefly for kerosene. About half of the output was kerosene, 10% each for lubricating oils and gasoline, 12% fuel oil and the rest was lost.

The refinery batch heated the crude oil and cooled and collected the parts as the temperature was raised. The process was one of separating a mixture into other mixtures. It was not long before the losses from heating and cooling in the batch process was realized and a change made to continuous stills piped together in a series with each one maintained at a given temperature to get a particular part of the mixture.

In 1913, the process of "thermal cracking" was devised. In this process some of the parts of the mixtures made of long chain carbon compounds were heated under pressure to crack them into shorter chains and increase the amount of gasoline and decrease the heavier less valuable compounds. In this way refiners increased gasoline production to 25% of the crude oil by 1920.

It was soon realized that savings in the process could be effected in equipment cost and heat if the stills were made into one vertical one to produce what is today called a fractionating tower. Different parts, or fractions, of the mixture are drawn off at each tower level ranging from the heavy material at the bottom to vapors at the top.

Further changes in the yields of gasoline were effected in 1937, when the idea of catalytic cracking was introduced. A catalyst (usually nickel or platinum) is used which does not take part in the reaction but which speeds it up at a lower temperature and pressure. The reaction was again to break up the longer chain molecules of the heavy fraction to produce lighter fractions of gasoline. Another method that was introduced was alkylation in which the short chain molecules of waste gases could be changed into the longer chains of gasoline. Just before World War II better airplane motors demanded higher octane fuels than were available in quantity. A process which was just the reverse of cracking was developed in which the lighter fractions are combined by means of a catalyst (sulfuric or hydrofluoric acids) into a gasoline component called alkylate. This is mixed with other gasoline to produce aviation or other high test gasolines.

By the 1960's the chief products from petroleum were gasoline 44%, kerosene 5%, fuel oils 36%, and petrochemicals 3%. The major fraction had shifted due to the declining use of kerosene as it was replaced by gas light and electricity and the increasing use of gasoline with the rise of the motor car and the use of fuel oil to replace coal in heating.

Crude oil is a mixture of hydrocarbons; i.e., compounds that are composed of hydrogen and carbon. The basic list of hydrocarbons is as follows:

CH_4	Methane
C_2H_6	Ethane
C_3H_8	Propane
C_4H_{10}	Butane
C_5H_{12}	Pentane
C_6H_{14}	Hexane
C_7H_{16}	Heptane
C_8H_{18}	Octane
C_9H_{20}	Nonane

There are others in the series increasing by one carbon atom and two hydrogens for each going up into the hundreds of carbon atoms to a compound.

In the naming of organic compounds the basic name is for the longest chain of carbon atoms. Then any side chains drop the -ane suffix of their name and -yl is used. Thus, 2 methyl pentane would be a compound five carbons long with a methyl group hooked to the second carbon. If you look in your medicine cabinet or elsewhere if a list of ingredients is on a label and you will see that some of the long names are merely combinations of these short prefixes and suffixes.

The length of the chain also determines if the substance will be a gas, liquid or solid at room temperature. This can be noted in the following chart of fractions.

Petroleum Fractions

Fraction Title	Number of Carbon Atoms	Uses
Gases	1-4	Bottle gas, synthesis
Petroleum Ether	5-6	Solvents, synthesis
Naphtha	6-7	Solvents, fuels
Gasoline	6-12	Engine fuels
Kerosene	11-18	Jet and Diesel fuels, cracking
Gas oil	16-up	Heating fuels, cracking
Lubricating oil	17-up	Lubricants
Residuals		Asphalt, road oils, paraffine wax, Coke, vaseline

A rating scale was developed, octane rating, for motor fuel based on how badly the gasoline knocked in the engine. N-heptane, $CH_3(CH_2)_5CH_3$, knocked very badly and was designated as zero octane while Isooctane, $CH_3C(CH_3)CH_2CH(CH_3)_2$, hardly knocked at all and was designated as 100 octane. Mixtures of n-heptane and isooctane were used by comparison with other gasolines to give an octane rating. About 1922, several substances were found by Charles Kettering (1876–1958), *et al* which would reduce the knocking of gasolines, e.g., lead tetraethyl.

A field which may be more important to us in the future than gasoline is petrochemicals. These are elements or compounds derived directly or indirectly from the petroleum industry and used for the synthesis of organic compounds. The first use of petrochemicals was for the formation of isopropyl alcohol in 1920. (If you have rubbing alcohol in the house it is probably this compound.)

Year	Petrochemicals in Millions of Pounds
1945	3,300
1950	10,000
1955	20,000
1960	60,000
1965	100,000
1970 (est.)	130,000

Some of these petrochemicals are ethylene, propylene, butenes, pentenes, benzene, toluene, and xylenes. From these basic compounds a number of products are made including: plastics, drugs, synthetic fibers, enamel, detergents, weed-killers, and fertilizers, synthetic rubber, photographic film, waxed paper, car polish, ointment and cream bases, and explosives.

The chemist can frequently design new products to replace items in short supply. Until the late 1950's Japan could supply her own lumber needs—then the need outstripped the supply and she was forced to begin importing large amounts of wood. Recently chemists in Japan used oil to produce acrylonitril butadiene styrene (ABS). They then blew air into it and hardened it to produce a synthetic lumber to which grain can be added if desired. Predictions have been made that by the end of the '70's about 20% of Japanese lumber may be produced in this way. Perhaps "only God can make a tree" but lumber can be made from oil.

IT'S TAKEN A SHINE TO US

About 70% of the elements are metals, yet most of them are not found in the free state, i.e., not combined with other elements. The following elements do occur in the free state: (Those with an asterisk were not known to the ancients) Antimony, Sb (stibium, mark); Arsenic, As; *Bismuth, Bi; Copper, Cu (Cuprum, from the island of Cyprus); Gold, Au (Aurum, shining dawn); Iron, Fe (ferrum); Mercury, Hg (hydrargyrum, liquid silver); *Platinum, Pt (platina, little silver); *Rhodium, Rh (Rhodon, rose); Silver, Ag (Argentum); and Tin, Sn (stannum). You will note that only three of the elements found free in nature were unknown to the ancients. The only other metal known to the ancients was lead which is not found free.

Even these metals that are found free, and those which do not occur in their free state, are more often found as ores. The metal ores are usually composed of the metal combined with sulfur or oxygen, as the sulfide or oxide of the metal, and a certain amount of undesirable material called rock. The natural state of the metal is the oxide as the body of an old automobile demonstrates. The whole story of corrosion is one of the metal returning to its natural state. Free iron rusts or oxidizes to form iron oxide and silver forms a tarnish from substances containing sulfur to form silver sulfide.

The whole process of refining consists of separating the useful metal from the useless rock and to chemically remove the sulfur or oxygen combined with it thus to reverse the trend of nature. The first step is to physically crush the ore to reduce the size of the particles so that the reactions will occur easier, faster, and cheaper. The step of separating the useless rock from the metal compound varies depending on the particular ore being processed but wherever the step does occur it is a physical one based on the difference in density between the rock and the metallic compound. The metallic compound is usually heavier than is the rocky material or the magnetic property of the metal. The chemistry can be represented by two types of chemical equations. Let M stand for the metal, C for carbon, S for sulfur, O for oxygen. The carbon is usually supplied in the form of coke, i.e., coal which has been heated with little air and which leaves mostly carbon. The oxygen is supplied with a blast of air.

Metal sulfide plus oxygen yields the metal or metal oxide plus sulfur dioxide.

$$MS + O_2 = M + SO_2 \qquad \text{Heat must be supplied.}$$
or

Metal oxide plus carbon yields the metal plus carbon dioxide.

$$2\,MO + C = 2M + CO_2$$

Using iron as a specific example of the general process we find that the two principal ores of iron are hematite, Fe_2O_3, and magnetite, Fe_3O_4. The ore is crushed and placed in a "stove" for roasting to remove water, and sulfur. It also decomposes carbonates to carbon dioxide. It is then placed in a blast furnace with coke and limestone. The limestone (calcium oxide) reacts with sand to remove this impurity as slag.

$$CaO + SiO_2 = CaSiO_3$$

The slag is lighter than the iron and floats on top where it can be drawn off periodically. The coke is changed to carbon monoxide and reduces the iron ore.

$$Fe_2O_3 + 3CO = 2Fe + 3CO_2$$

The process for aluminum is slightly different. Bauxite, $Al_2O_3 \cdot 2H_2O$, is the most important ore of aluminum. It is first crushed to a fine powder and

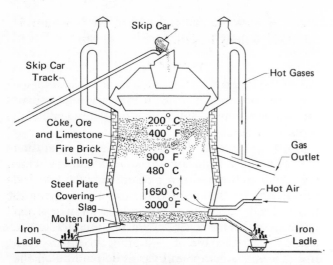

Figure XXV-1. Blast Furnace.

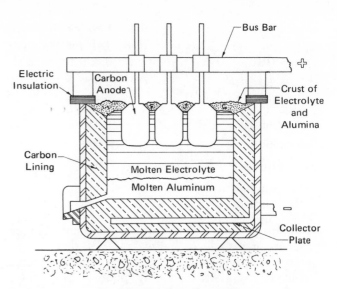

Figure XXV-2. Aluminum Cell.

dried. Then it is treated with steam and sodium hydroxide (lye) which dissolves the aluminum ore, but does not dissolve the impurities of iron and silicone oxides. It is separated by filtering from the impurities and is then dried and reconverted to the oxide. A flux is added to help melt the aluminum oxide and it is placed in an electric furnace. The electric current electrolytically produces aluminum by the method developed by Charles Hall (1863–1914) while he was a student at Oberlin College.

Besides the pure metals are many alloys with special properties. We have seen where technical developments, such as the automobile and airplane have had to wait metallurgical developments to get strength or heat resistance or weight specifications needed for an innovation. Today's metallurgist can frequently tailor a metal for a particular use.

The table below illustrates that the properties of steel can be changed by the addition of small amounts of other metals.

Name	Properties	Uses
Manganese	Hardness and resistance to wear.	Railroad rails, armor plate.
Invar (Nickel)	Expands slightly.	Metering and measuring devices.
Stainless steel Chromium and nickel	Resists corrosion.	Knives, hatchets, restaurant-ware.
High-speed steels	Keeps strength at high temperature.	High speed cutting tools, e.g., lathes.

Although metals are the backbones of industry there are many almost intangible materials used by the chemists to make our world.

CASTLES MADE OF AIR, GAS, AND WATER

The basic starting materials for any chemical synthesis have to be readily available and inexpensive. The most expensive starting materials are those we have already discussed, coal and petroleum. These provide the materials for many of the chemical substances that are produced. Sulfur, mined by forcing superheated water underground, is used for the manufacture of sulfuric acid. Sulfuric acid is used in making many other chemicals. So much so that it has been said that the degree of civilization of a country can be measured by the amount of sulfuric acid that they consume.

All of the above are a little more expensive than industry likes. What they prefer is something as free as the air or dirt cheap. This they really dig or pump as the case may be.

THE SOLVAY PROCESS

Ernest Solvay (1838–1922) was the son of a French salt refiner. He had but little formal education and learned some chemistry and physics by reading elementary books on the subjects. In 1863, he developed the process which bears his name and from which he became wealthy. He used his money philanthropically and opposed inheriting money as "unearned."

The spur for the development of a cheap process for washing soda and caustic soda came from the textile industries of the industrial revolution of over a century before. Wool and cotton needed to be cleaned and washed free of oils several times in the manufacture process. The earliest method was burning plants which gave a low grade caustic soda and in small quantities. Other methods replaced this to give greater amounts but the costs were still too high. This is the problem that Solvay solved.

The basic materials used in the Solvay process are salt and limestone. At the beginning a small amount of ammonia must be used but since this is regenerated in the process little additional is needed.

The limestone is heated to produce calcium oxide and carbon dioxide. Water is added to carbon dioxide to produce carbonic acid (the stuff that gives carbonated beverages their sharp taste). The carbonic acid is reacted with ammonia to produce ammonium bicarbonate. The ammonium bicarbonate is reacted with salt (sodium chloride) to make sodium bicarbonate. The ammonium chloride which is also produced is reacted with the calcium oxide from the first step to give ammonia which can thus be reused and calcium chloride. The useful materials are produced in the next two steps by heating. Sodium bicarbonate can be used as is or it can be heated to make sodium carbonate (also called washing soda). Carbonic acid is also produced which can be reacted with the recycled ammonia to start over. If the sodium carbonate is heated further it produces sodium oxide which can be reacted with water to produce sodium hydroxide.

Sodium carbonate is used in large quantities to make sodium hydroxide (caustic soda), glass, soap, paper and pulp, cleansers of different kinds, water softeners, and in the petroleum refining process.

The chief waste product in the process, calcium chloride, is used for a road dressing, in refrigerating brines, in making coal dustproof, and as a dehumidifier for naturally moist places such as basements.

Both the ingredients for the process are cheap and easily available. The entire central portion of the United States has thick beds of limestone and can be dug from the surface. The salt that is used comes from beneath the ground and there are a number of extensive deposits in this country. A bed about 450 ft. thick underlies parts of Kansas, Oklahoma, and Texas. The International Salt Co. has a mine covering 200 acres with 25 miles of passages 1/4 mile below the city of Detroit.

Salt is not only used for the Solvay process but can be electrolyzed into other products. If salt is heated to its melting point and electricity passed through it the metal sodium and gas chlorine are produced. If salt water is used and an electric current is passed through then the products are chlorine, sodium hydroxide, and hydrogen.

SALT WATER TAFFY?

More can be made from salt water than taffy. Each cubic mile of ocean water contains about 5 million tons of magnesium and every million pounds of sea water contains 70 pounds of bromine.

In order to get the magnesium it is necessary to have cheap starting materials. We need sea water, oyster shells, fresh water, and natural gas. A sea side location with an off-shore oil drilling well for the natural gas, an oyster fishery, and a river or stream nearby will provide all our starting materials. Each time you consume a can of oyster soup you are aiding in the manufacture of magnesium. This is not the old shell game for the oyster shells are calcium carbonate. They are heated to produce calcium oxide which is reacted with the fresh water to produce calcium hydroxide. The calcium hydroxide is added to the sea water. The magnesium forms a solid precipitate of magnesium hydroxide. The ocean is filtered away leaving a sludge. The sludge is treated with hydrochloric acid to produce magnesium chloride. The hydrochloric acid is produced later and is thus not a starting ingredient. The magnesium chloride is placed in a steel tank and it acts as one pole while a graphite rod lowered into the tank and it acts as the other. Electricity is passed through the tank and the products are chlorine and magnesium. The chlorine is burned with the natural gas to give hydrogen chloride which in a water solution is hydrochloric acid.

It is from sea shells from the sea shore with gas and water that go to make magnesium which with its alloys make cars, planes, and speedboats. They make pistons, crankcases, and propellers. Each Volkswagen has 42 pounds of magnesium in it.

Some of the chlorine produced in the above process is used to liberate bromine from sea water. The bromine is then blown out of the sea water by compressed air.

Bromine is used principally in the manufacture of ethylene dibromide which is a part of antiknock gasoline and reacts with lead tetraethyl so that the lead doesn't accumulate in the cylinder—it gets the lead out. It is also used in making some organic dyes, a light sensitive silver bromide for photographic film, and sedatives.

From the chemicals made from water we go to those that are mostly air.

IT LOOKS LIKE AIR TO ME

Only in those regions of the world where rain was scarce or in caves could supplies of natural nitrates be found. There were, in general, three limited supply areas. Animal wastes from the dirt of barnyards from which nitrogen compounds could be extracted with a poor yield and much labor was possible in all rural areas. There were a few scattered sources such as the natural saltpeter deposits in India and in Mammoth Cave in the United States. The main source was found in Chile where deposits were worked starting about 1825. Nitrates were vital to wars, both in the form of explosives manufacture and in fertilizers, for the production of food. Many attempts had been made to duplicate in the laboratory what nature performed in her operations.

In 1900, Friedrich Wilhelm Ostwald (1853–1932) discovered a way to change ammonia to nitric acid by the action of oxygen. Nitric acid is the material needed for most explosives. However, this did not really solve the problem for where was the ammonia to be obtained? This question remained unanswered until 1908 when Fritz Haber (1868–1934) discovered a way to make ammonia from nitrogen and hydrogen.

Nitrogen was obtained from the air through a liquefaction process and was readily available. The hydrogen was produced by the action of steam on red hot coke. The carbon dioxide that was produced at the same time was dissolved in water and so a relatively cheap source of hydrogen was available. The nitrogen and hydrogen were put under pressure and heated in the presence of a platinum catalyst and ammonia was produced. While this was not the method of nature it was an effective way.

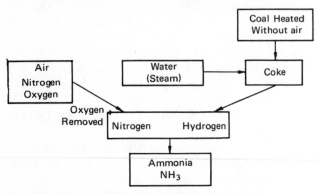

Figure XXV-3. Haber Process.

One of the reasons that England felt that World War I would be a short war was that Germany needed to import all of her nitrates from Chile, the chief source for nitrates, and she felt that her mastery of the sea would keep Germany from getting these vital war materials. To her dismay Germany was in less trouble than she for they could produce their nitrates while England had to import them with the

submarine peril which grew. Many have expressed the feeling that the Ostwald-Haber processes extended the length of World War I and II.

The ramifications of such a technique may transcend time and space. Chile, the chief supplier of nitrates, levied an export tax on nitrates that was the key income for the government of that country. Between 1880–1930, this income amounted to $1,000,000,000 which was 43% of the total government revenue for that period. The effects of synthetic nitrates on Chilean economy can easily be seen from the following: in 1889, Chile supplied almost all the world's nitrates; in 1910—64%; 1913—50%; 1918—33%; 1929—25%; 1938—8%; 1953—5%.

Following World War I the allies acquired all of the German patent rights as part of the spoils of war. Among these were important chemical patents such as the Ostwald-Haber processes. No longer did Germany have exclusive use of the process. It spread over the globe. By 1924, the Haber process was supplying 30% of the world's nitrates.

The opening of the Panama Canal in 1914 increased trade between the United States and Chile. The United States encouraged Chile to borrow money on its export tax income for public works such as fancy public buildings. In a few years Chile had a large foreign debt and as the tax revenues declined her debts, which she could not pay, built up. By 1926 she faced a deficit of about 150,000,000 pesos and by 1931 the world depression was affecting her. She owed money to creditors in the United States, France, Belgium, and Great Britain.

The Chilean government forced all of the nitrate companies including those foreign owned, to sell through the government which kept 25% of the profits. In addition, Chile forced the foreign debtors to accept a settlement on the bond issues much below par value. These measures were adopted in a response to widespread popular feeling against foreign "economic imperialism." Threats by foreign governments concerning the use of force to collect debts did little to make for good relations.

While the Ostwald-Haber processes made a country independent of nitrate deposits, it prolonged wars, upset economies, and disturbed diplomatic relations.

On the positive side are the many uses for ammonia in our industrial world. The largest consumer of ammonia is the explosives industry. The ammonia is converted into nitric acid which is used to make trinitrotoluene, nitroglycerine, nitrocellulose, nitrostarch, pentaethritol, and ammonium nitrate.

Perhaps the second largest consumer is the fertilizer industry where the ammonia may be used directly or compounds such as urea, ammonium nitrate, and calcium cyanamide can be formed.

The textile industry uses ammonia in several ways: directly in the synthesis cuprammonium rayon and nylon; before dyeing, fabrics such as cotton, wool, rayon, and silk are cleaned in an ammonia soap solution; in dyeing the ammonia is used to precipitate metallic mordants; to deluster acetate rayons and to give sheer, transparent, and crisp effects to lightweight fabrics; to mildewproof fabrics.

Ammonia is used as a catalyst and to reduce acidity in the production of certain synthetic-resins such as phenol-formaldehyde, and urea-formaldehyde.

A direct use of ammonia is in the manufacture of sulfa drugs including sulfanilamide, sulfathiazole, and sulfapyridine; while indirectly it is used in the manufacture of vitamins and anti-malarials.

The rubber industry uses ammonia to prevent coagulation of the raw latex and also as an atmosphere in vulcanization.

For large industrial refrigeration, ammonia is the most frequently used refrigerant. e.g., ice production, cold-storage plants, quick freezing units, air conditioning, and de-waxing lubricating oils.

Inorganic chemical producers use large amounts of ammonia to form ammonia salts. e.g., nitric acid + ammonia = ammonium nitrate. Any acid reacted with ammonia will form the salt of that acid. Also used in the lead-chamber process for the manufacture of sulfuric acid.

In the organic chemical production ammonia is used for the direct production of certain chemical types such as amines, amides, and nitriles as well as providing a liquid (i.e., liquified ammonia) in which some chemical reactions can be conducted.

The metals industries use ammonia in a solution for leaching some ores such as copper, molybdenum, and nickel. In addition, some industries break down the ammonia for use as a protective atmosphere for annealing and sintering, reduction of metal oxides, and atomic hydrogen welding.

In producing industrial alcohol from black-strap molasses ammonia is used to give nitrogen to the yeast cells and to keep the fermentation from becoming acid.

Ammonia is frequently used with chlorine in the purification of water to give a better taste, control algae growth, and keep the chlorine active longer to better kill the bacteria.

By substituting ammonia for calcium in the bisulfite process of the pulp industry better pulp is obtained with the possibility of greater reduction of stream pollution. In coating paper ammonia is used as a solvent.

SYNTHETICS BOUNCE JUST AS WELL

During the latter days of World War I, Germany had been faced with a critical shortage of rubber necessary for motor transport, gas masks, surgical gloves, airplane tires, and storage batteries. It was realized that in any future conflict rubber would always be in short supply because the rubber could only be grown in the tropics and Germany could not maintain control of the sea. A part of the idea of "geopolitic" was to produce synthetically any raw material not contained within the natural boundaries.

Following World War I when the United States was expanding its auto industry, Great Britain limited the production of her Far East plantations. Congress approved the expense of $500,000 to survey possible production areas in the Phillipines and South America. It takes years to get rubber into production so that Edison joined experts from the Department of Agriculture in attempting to find plants that would produce rubber faster than the rubber trees and in 1925 du Pont was stimulated to begin searching for a synthetic product.

In the early 30's Germany was successful in producing two synthetic rubbers—Buna-N which was a low grade material which could be used for hose and Buna-S a high grade product suitable for tire treads.

Du Pont was successful in 1931 in producing a synthetic rubber. Limestone is heated to produce calcium oxide and carbon dioxide. The calcium oxide is placed with coke into an electric furnace and heated to 2000° to produce calcium carbide and calcium hydroxide. The calcium carbide is reacted with water to produce acetylene (the gas used in cutting torches).

J. A. Nieuwland (1878–1936), a Catholic priest and professor of chemistry at Notre Dame, had discovered that if acetylene is treated with a water solution of copper chloride and ammonium chloride and then with hydrochloric acid, vinylacetylene is formed. Vinylacetylene is then used to make a rubber with the trade mark of neoprene.

The 1961 production was 265 million pounds. In many respects it was better than natural rubber. It was less affected by heat and resisted action by oils and gasoline better. During World War II when we were cut off from supplies of natural rubber neoprene along with Buna rubbers made from butadiene, a cracking product from petroleum, enabled us to keep military vehicles on the road although there was little for civilian use.

Buna-S, which was also used during World War II, was a combination of butadiene from petroleum cracking and styrene which is an important product. Benzene and ethylene, both from petroleum fractions, are put together to form ethylbenzene. The ethylbenzene is treated over a mixed metallic oxide catalyst to form styrene. It can be linked together to form polystyrene called styrofoam which is a lightweight packaging and insulating material or it can be reacted with butadiene to form rubber.

YOU CAN SAW IT JUST LIKE WOOD

Rubber and the synthetic forms of rubber are plastics of a type. That is they flow when warm and do not have a crystalline structure. The story of man-made plastics begins after the Civil War when $10,-000 was offered to anyone who could produce a new material for billiard balls. John Wesley Hyatt (1837–1920), a print shop employee, started to try and find such a substance. He tried many and none of them worked. One day he noticed that a bottle of collodion had overturned and some had run out and hardened into a smooth, ivory-like stuff. Collodion was an ether-alcohol solution of nitrated cellulose (cotton treated with nitric acid) and used as a protective film over wounds. He used heat and pressure instead of the alcohol and ether, which would increase costs too much. It was called celluloid and patented in 1870. Celluloid, not only made billiard balls, but piano keys, collars, ornamental boxes and cases, dressing table accessories, etc. The chief disadvantage was the flammability of celluloid. As cellulose nitrate it was used for years in the motion picture industry.

While casein, a milk product, could be reacted with formaldehyde to make a hardened plastic, Leo H. Baekeland (1863–1944) found in 1909 that phenol reacted with formaldehyde gave a molded type plastic. When World War I ended over 20,000 tons of phenol was surplus. Most of this was made into the new plastic—Bakelite and used as cases for the expanding production of radio receivers.

Newer plastics after World War I either centered on reactions of formaldehyde with urea (which is made synthetically) or they are of the styrene, ethylene type which have developed from the petroleum industries after World War II. Besides the growing list of things made from plastics—e.g., football helmets, telephones, toilet seats, wall tile, records, aircraft windows, unbreakable toys, records, electrical insulators, garden hose, hearing aids, frames for glasses—almost everything available for purchase is wrapped in a plastic film. As part of the solution to the building boom is the manufacture of a plastic bathroom complete in one unit with plastic pipes and all that can be set into a new house for an instant bathroom. Tomorrow's house may be all plastic with you sitting in an inflated plastic chair.

FABRICS OF WHOLE CLOTH

Both the fields of man-made plastics and man-made fabrics have their beginning at nearly the same time from the same cause—a problem. A silkworm disease was threatening one of France's largest industries. Hilaire de Chardonnet (1839–1924) sought, where others had looked in vain, for some way to replace the silkworm. He was attracted by collodion and constructed a machine which would turn and by centrifugal force push out the contents through small holes or spinnerettes and used hot air to evaporate the ether-alcohol solvent. The cellulose material came out in the form of light fluffy filaments but the cost of production was too high.

The essential idea for artificial fabrics was born in the mid 1880's. Find some way to put an inexpensive material into solution that will, when it is taken out of solution, form into a long filament. The fibers that man had used historically were linen, cotton, silk, and wool. These natural fabrics had the advantage of being cheaper than the artificial at first but had the disadvantage of being based on a growing and collecting systems. It would seem that the first to yield would be the most expensive—silk.

Regenerated cellulose can be made in several ways. Louis Henri Despeissis patented a technique in 1890 in which the cellulose was dissolved in a cuprammonium solution and then precipitated with sulfuric acid after it had gone through the spinnerettes. The viscose method of preparing rayon makes use of wood pulp. It was developed by C. F. Cross (1855–1935) and E. J. Beven (1856–1921) and patented in 1892. The pulp is treated with sodium hydroxide to form sodium cellulose. This is reacted with carbon disulfide to form cellulose xanthate which is dissolved in sodium hydroxide and forms a thick viscous liquid which is forced through the spinnerettes into an acid bath where the cellulose is regenerated into fibers. Cellophane is made in the same way except that the regenerated cellulose is run through rollers instead of the spinnerette.

In 1910, the American Viscose Corporation set up a plant at Marcus Hook, Pennsylvania and began the production of rayon. Synthetic fibers for this country are a product of the twentieth century.

Twenty years later, Wallace H. Carothers (1896–1937), a chemist at du Pont corporation, developed a new class of synthetic fibers called nylon. Chemically nylons are related to silk and wool—that is they are protein-like. One nylon is made from hexa-

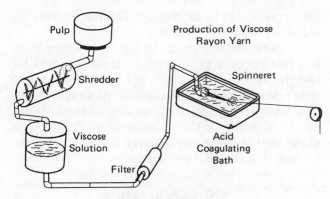

Figure XXV-4. Spinnerette.

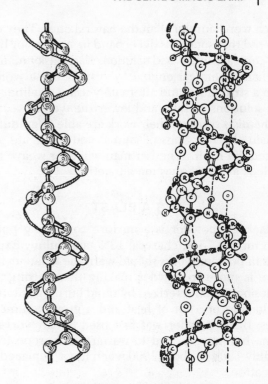

Figure XXV-5. Fiber and Protein.

methylenediamine reacted with adipic acid and the basic starting materials can be made from coal, air, and water. Nylon can be used as a plastic as well as a fiber but it was first marketed in 1940 as a fiber.

The same reactions of ethylene and acetylene that go to make plastics also make a variety of fabrics. There are the polyester fibers marketed under the name of Dacron and the acrylic fibers such as Orlon, Acrilan, and Dynel. Much of what we wear during the day, sleep under and in at night, walk on and sit on are products of the laboratory and made in industries no longer associated with plants or animals.[2]

The relation of life to fabric is more than a superficial one. There are about 25 different ways in lengths of from 250 to thousands of units long enough to make different kinds of proteins. Your body has over a hundred thousand different proteins in it. While your proteins are similar to other persons they are at the same time different from every other person's in the world. Yet they are more like another person's than they are like a different animal. If we eat pork or beef our digestive tract breaks down that protein to amino acids. The amino acids pass into our blood stream and go to a growing tissue or place where replacement is needed. The amino acids are then rebuilt into our protein—that way we have nothing to beef about.

The fiber cases we have dealt with in synthetics are similar in that they too are made of small units put together; they differ, however, in that while they are put together by the hundreds of units, proteins are put together by the thousands. Besides the difference of the units themselves there is a difference in the way proteins are put together.

The most typical shape found in the proteins' structure is a spiral or helix, each turn made up of one or more amino-acid units, linked together, like steps on a spiral staircase. The twists and turns of the spiral may be folded or coiled in various ways in the space occupied by the molecule. There may be short sidechains attached to the spiral at regular intervals. And there may be compound spirals, one helix entwined within another, bundles of such helices twined around one another and around a central helix, like twisted strands of yarn or rope or cable. . . . the real structure and texture of life is a fabric unlike any other fabric, soft as hair, hard as bone, or tough as a cable-like tendon.[3]

Proteins can be grouped somewhat on the above general characteristics. The keratins are structures like our outer skin, nails, hair, a chicken's feathers, and a turtle's shell; while the collagens are such things as bone, tendons, cartilage, and the corneas of the eyes. The myosins seem to be muscles, nerve fibers, blood cells or a kind of gel like protein. Even the control and active agents of the body, the enzymes and hormones, are protein. These control and direct the chemical reactions that go on in the cells and in the body as a whole. Finally there is the board of directors, the nucleoproteins, that make up structures such as genes and chromosomes. They lay the blueprint for each new cell and each new organism.

The coils of muscle protein are gel like and bear a charge on parts of the molecules. It seems to be that the repulsion and attraction of these charges account for the contraction and expansion of the muscle tissue caused by chemical reactions in the cells. These conclusions are based on work done by Aharon Katchalsky (1914–   ) and Wilhelm Fridrich Kuhne (1837–1900) who made synthetic polyelectrolyte gels which were simpler than the natural ones. They discovered that such gels did expand in an alkali or base and contract in an acid solution. Katchalsky made a mechano-chemical engine by suspending a weight from a strip of gel and alternately using alkaline and acid solutions to raise and lower the weight.

Chemists through their work are able to gradually explain the functions of animal and plant life and indeed life itself. Whether man will ever weave this fabric is another question altogether.

IT'S A BLAST

Black gunpowder is a mixture containing potassium nitrate 75%, charcoal 13%, and sulfur. Explosions in powder works should warn the amateur that there is more to powder making than mixing. An explosion is characterized by rapid burning with the production of a lot of heat and a great volume of gases. Black powder was first used for fireworks in China but was applied to warfare in Europe from roughly 1300 to about 1900 when it was replaced by guncotton.

The tales of synthetic fibers and explosives overlap for guncotton is a nitrated cellulose (most explosives are made by nitrating compounds i.e., by treatment with nitric acid) which is made liquid by a solvent and extruded in strands that are chopped to make pellets for smokeless powder. It was first made by Christian Friedrich Schönbein (1799–1868) in 1845. In the same year Ascanio Sobrero (1812–1888) nitrated glycerin to make nitroglycerin.

Called "soup" by members of the underworld it is a very unstable compound and many persons have been killed by just moving it from one place to another. Besides being used in the unlawful opening of safes it was used to bring in gushers in oil fields and to blast rock for railway cuts.

Alfred Bernhard Nobel (1833–1896) in 1866, found that a diatomaceous earth called kieselguhr if mixed with nitroglycerin would absorb three times its weight and would stabilize the "nitro." This was called dynamite and made it possible to safely handle and store "nitro." So concerned was Nobel about the use of explosives in warfare, he set up some of his profits in a fund to create the Nobel prize for peace and to encourage man in peaceful pursuits.

While the du Pont company had its origins in the manufacture of gunpowder about 1800, it also made dynamite and "nitro" by 1880. Later it moved into the manufacture of smokeless powder and from there into the fields of plastics and synthetic fibers. Explosives themselves have peacetime uses of hunting, stump removal for farming, mining, making highway cuts. In large projects, like the Panama canal, it saves a lot of digging.

The same technology can be used to produce fibers and billiard balls as to produce smokeless powder and the smokeless powder can be used for hunting either deer or man. In this case, the application of the product is not a question of morality for the scientist, the technologist, or the manufacturer. The decision of the way in which the material will be used is not his and it has several applications.

SIN AND SCIENCE

The real problem of "sin" begins when the applications of a technology or scientific development has only one use—the destruction of man or creating a situation that is harmful to human life.[4] Scientists are fond of saying that science is amoral and manufacturers argue that if we don't make the product someone else will. It would appear, however, that while science may well be amoral that scientists are not and that the argument of the manufacturers could be used by the murderer. He could say that statistically so many people are killed each year by homicide and that he is just keeping up the statistical figure for if he doesn't kill someone somebody else will.

There are a number of cases of clear "sin" and it has not always been sin accepted. Miguel de Cervantes (1547–1616) who wrote *Don Quixote,* had his left arm shot off in the bloody battle of Lepanto recognized sin and technology when he wished that the man who invented artillery roast in hell.

On the other hand, when Sobrero first prepared "nitro" he was so horrified that he refused to have anything else to do with it.

There is a story that during the Crimean War of 1853–1856, [between Russia and the allied powers of Turkey, England, France, Sardinia, located in the Ukraine on the north shore of the Black Sea] the British government asked Michael Faraday, the greatest living scientist of the day, two questions: 1) was it possible to develop poison gas in quantities sufficient to use on the battlefield? and 2) would Faraday head a project to accomplish the task?

Faraday said "yes" to the first and an emphatic "No" to the second. He did not consider patriotism excuse enough. During World War I, Ernest Rutherford of Great Britain refused to involve himself in war work, maintaining that his research was more important.[5]

Fritz Haber was extremely patriotic, he had already contributed mightily to the war effort with the process that bears his name, and developed methods of producing quantities of poison gas. He also supervised its use at Ypres in 1915 against the French.

It is perhaps gentle irony that Haber was a Jew and that the Germans selected poison gas as the method of solving the "Jewish Problem." It is perhaps more ironic that Haber lived long enough to need to be rescued from Hitler. Rutherford did much to effect the escape of Jewish scientists from Germany and even met the planes and shook hands with the rescued but he did not shake hands with Haber.

STUDY AND DISCUSSION QUESTIONS

1. What is a giant molecule?
2. What are the basic units of protein that make up living tissue of fish, fowl, and man?
3. What is the difference between "living" and "dead" protein?
4. Can the chemist get a protein when he puts the basic units together?
5. How did chemistry relate to the German idea of "Geopolitic"?
6. How do synthetics differ from natural fibers?
7. What is polyethylene?
8. What are the advantages of synthetic rubbers? Why were they developed?
9. What are aerosols? Purpose and composition?
10. What did Katchalsky discover? What is its significance?
11. What is the basic structure of life?
12. Why is the number of proteins almost endless? How are they classified?
13. Are all life processes merely chemical reactions?
14. Why is black beautiful?
15. What arts are prechemistry? Why are they arts?
16. How is petroleum formed?
17. What is the geology of petroleum?
18. What methods are used for petroleum exploration?
19. What pressures (social, economic, others) forced changes in the refining of petroleum?
20. What is the octane rating scale?
21. Why is gasoline colored?
22. What is the basic process for refining metals?
23. Why weren't all metals occurring free in nature known to the ancients?
24. What are the basic materials for synthesis? What factors determine if materials are suitable for synthesis?
25. What are the chief products of synthesis?
26. Explain the implications in the development of the Ostwald-Haber processes.
27. What are sins of science?
28. How can science be a double-edged weapon?

FOOTNOTES

1. The information that follows is modified from a chart in George W. Watt and Lewis F. Hatch, *The Science of Chemistry* (McGraw-Hill Book Co., 1949), p. 441.
2. For a general history of man made fibers and detailed information see George E. Linton, *Natural and Manmade Textile Fibers* (Duell, Sloan and Pearce), 1966.
3. Lawrence P. Lessing, *Understanding Chemistry* (The New American Library, 1959), p. 158.
4. An elaboration of some of the points that follow appear in Isaac Asimov's "The Sin of the Scientist" *Fantasy and Science Fiction*, (Nov. 1969), pp. 80-90.
5. *Ibid*, p. 89.

Suggested Readings:

A rather technical but somewhat useful study is A. I. Levorsen, *Geology of Petroleum*, (W. H. Freeman and Company, 1958).

Some idea of the number of chemicals derived from oil may be gained by looking at Sachanen, A. N., *The Chemical Constituents of Petroleum* (Reinhold Publishing Co., 1945).

Laurence P. Lessing, *Understanding Chemistry* (New American Library, 1959) is very good.

Other than standard chemistry texts which may prove useful for reference Louis Vaczek, *The Enjoyment of Chemistry*, (Viking Press, 1964) may be interesting to some readers.

Also see Helen Miles Davis, *The Chemical Elements* (Ballantine Books, 1959).

XXVI

Poisons, People, and Pollutants

Unlike the problems of "sin" and science the problems of daily living are more difficult to place in clear cut categories. Many of the beneficial, or at least what began as beneficial, aspects from science and technology have turned on their creators, like Frankenstein monsters, and created problems. Old questions rise about the "new" morality and many solutions for problems contain their own problems in turn. With World War I and following, wars attacked the civilian populations. There seemed a change in international morality or amorality from the usual rape and pillage of civilians to their slaughter. The major countries have developed and continue to develop chemical and biological weapons. A single drop of one of the nerve gases developed during World War II is ample to cause death. All of the plant and animal plagues have been made more deadly so that no only the population can be killed but also their ability to produce food so that more will starve.

The United States has recently renounced the use of biological warfare but will this hold during an all-out war? Gas was not used by either side during World War II but this may have been because the other side also had it or because gas can blow back on your own troops. Will banning the "bomb" work in a war between major powers?

Most of the chemical biological warfare weapons are not too new in concept. Greek Fire was used to burn ships and men and cities in the seventh century —today we use napalm. In antiquity the conquered land had salt sown to keep anything from growing— today defoliating agents. When we were conquering the Indians we sold them smallpox infected blankets —today we use more sophisticated varieties.

Does it matter to the man who is dead that he was shot through the heart or cooked in napalm? The real horror is death. The effect on family and friends. The fact that man cannot seem to rule himself without recourse to violence. What is the morality of war? Today, our allies are yesterday's enemies and our enemies are yesterday's allies. There has been no victory of morality and right but only a realignment of the power structure, dead heroes and famous politicians. Even war which promised beneficial results in the liberation of mankind has created new problems rather than curing the old.

The story of drugs is one of promised beneficial results and the creation of a whole new set of problems. The use of narcotics, particularly opiates, has occurred throughout the history of western man. Opium appears in the 9th century in Greece and reference is made to it in Homer's *Iliad*. The benefits of such drugs are only too clear if one reads descriptions of amputations or operations performed without pain killers. They are also useful in inducing sleep and relieving worry. On the other side is the horrible picture of addiction to the drug when man becomes a slave and loses his will.[1]

Opium can be taken internally as an alcohol solution, laudanum, or inhaled in a smoke. In the east it is more frequently smoked whereas in the west it was more frequently drunk as laudanum. At the beginning of the industrial revolution and following the poor sometimes used laudanum on small children to keep them quiet while their mother was at work. The parents found that gin was cheaper to ease the pain of hunger and day to day work. Before aspirin it was the only relief to be had.

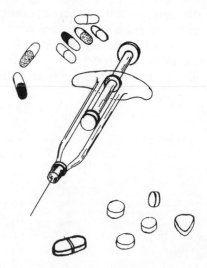

Figure XXVI-1. Drugs.

Morphine and codeine can be extracted from the raw opium and are used medicinally. Codeine is frequently used as a cough remedy and morphine is injected to bring almost immediate relief to severe pain. On the other side of the coin is the addictive properties of the drug. During the last century, attempts were made to find a synthetic compound or to so change morphine that it would retain the pain relieving qualities while eliminating the addictive properties.

One of the workers so engaged was Heinrich Dreser (1860–1924), who around the turn of the century produced a compound from morphine which met the qualifications. It bore the trademark heroin after the Greek god Heros (the offspring of a god and a mortal and were mortal but after death they became gods) because after taking the drug one felt like a god. The use of heroin spread quickly in this country and around the world.

> The invention of the hypodermic needle, the importation of Chinese laborers who smoked opium, and the wide spread use of morphine in the Civil War contributed to making American soil fertile for heroin.[2]

About 1900, there was an outbreak of teen age heroin addiction that led to the passage of the Harrison Act which outlawed the possession or sale of narcotic drugs.

There seems little to recommend the use of heroin and much that can be said against it. Addiction occurs readily usually by sniffing or having it mixed in with marijuana as those who push the sale of the drug try to do. The addict increases his dosage and frequently begins to inject it under the skin or into a vein. The dangers are death from an overdose or from tetanus or other infections because the addict does not keep his needle sterile. The pain killing properties of drugs also mask illnesses and infections until it is too late.

It is estimated that there are 100,000 heroin addicts in the city of New York and that for the age group 15 to 35 drug abuse is the leading cause of death. In 1969, 900 persons died due to drugs.[3] The rate of use and the number of deaths makes addiction a major problem.

A new drug, methadone, has been created which can be used to take the heroin user off the habit. The only catch is that he then becomes addicted to methadone. Methadone was developed by the Germans during World War II as a cheap pain killer. The difference in the addiction is that the drug is taken orally and avoids the dangers of infection and overdose. There is also no high and the use of methadone is legal so that the addict can move out of the crimi-

nal class. About 80% of heroin addicts who have the usual treatment of complete drug withdrawal return to heroin when they are released from treatment. The new drug allows 83% of former heroin addicts to become useful members of society and not return to heroin. It leaves the question of the morality of substituting one habit forming drug for another.

Amphetamines, first synthesized in 1927, were designed as a decongestant in bronchial asthma to give man breath. Today, they are used in a variety of uses: for reducing, to relieve mild depression, and to treat certain cases of brain-damaged children. When abused, used regularly or in large doses, it is called "speed" or "uppers." The drugs are energizers that remove the desire to eat and sleep, they speed up the heart, and stimulate the brain. They are abused commonly by housewives trying to reduce and by truckdrivers trying to stay awake on the road. They also increase irritability, ease of anger, and aggressiveness. Abuse with these drugs causes hallucinations, colds, nausea, extreme fatigue, muscle tremors, heart problems, malnutrition, and hepatitis if a needle or excessive methedrine is used.[4]

To the other extreme are the "downers" which include the barbiturates and tranquilizers. Barbituric acid was synthesized by Adolf von Baeyer (1835–1917) from urea and a compound of malic acid. His main work was with dyes and urea, you may recall, was used in making one of the plastics so there is a relatively close relation between medicine and dyes and plastics. Barbiturates are useful in inducing sleep and relaxing tensions when a person becomes overwrought. The dangers with barbiturates is that the person using them loses perception of the number taken and may accidently take an overdose. They also induce depression and may cause suicidal tendencies. In addition, the person using them tends to be fuzzy in his thinking and they reduce his effectiveness in being able to operate in society.

The tranquilizers, part of a group classed as psychotropic drugs, do not cause confusion and do not affect consciousness. One of these, compounds derived from the rauwolfia root, had been used for centuries in India to treat hypertension, and insanity but it has only been in the last twenty years that they have been used by the west.

The west has characteristically been harsh in its treatment of the mentally disturbed. Historically from Bedlam to the *Snakepit* was a slight distance. The insane or lunatics, as they were and are called, were kept under forcible restraint. Even the beginnings of psychoanalysis made little dent in those confined. Sedatives and opiates had been used to keep the patients quiet but did little towards helping them. In the 1930's, insulin and electric shock were used to force some out of their dream world but the treatment was harsh. The use of psychotropic drugs in about 1956 for the first time reduced significantly the number of persons hospitalized for mental illness. About 10% of the population suffer from some mental disturbance and 54% of those in hospitals are there for mental disorders. In 1955, there were 559,-000 mental patients while five years later in 1960, there were 536,000. The number may not seem to be much different, but 1956 was the first year to have a decrease in the number of mental patients. Many mental patients can now sit by windows instead of being tied down or staying in locked rooms. This effect cannot be measured in mere numbers as any that have suffered constraint can testify. They are valuable in treating schizophrenics and can control excitement in manic and alcoholic patients. They make possible, for the first time in many cases, for mental patients to be treated and by reducing their excitability enable some mental patients to return to society while they are under treatment. They enable neurotics to function better. The major danger with tranquilizers is that large portions of the population are not feeling any emotions. They are gradually cutting themselves off from those characteristics that make them human. They bring *1984* ever closer.

During World War II our sources of hemp were cut off and farmers were encouraged to grow a plant which would enable rope to be made. It was from this beneficial approach that large numbers of the plant that produces marihuana are to be found in the middle west. The drug or drugs derived from the plant has no medicinal value but it is used by an estimated 200 to 300 million people in the world. The drug was earliest used by an assassin cult in the east as a reward for services. Smoking marihuana has become prevalent in middle-class youth in North America in recent years and its use constitutes a problem. Its use at present is illegal but proponents of its use urge that it be made legal. There have been no reliable studies made as to the effects of marihuana. It does seem to upset the body's sensory systems and deceive them as to the length of time, shapes, colors, sounds and odors. If these can be verified as consistent results from the use of the drug then its use has no place for a person who wants to experience life for himself and it has less place in a mechanized society where the actions of one in a ton of steel may snuff out the lives of others.

The same Dr. Baeyer who made barbituric acid had a graduate student, Othmar Zeidler, who synthesized a compound as part of his doctoral research thesis in 1874. The compound, 2,2-bis-(p-chlorophenyl)-1,1,1-trichloroethane, became known popularly as DDT in the 1940's when a Swiss chemist, Paul H. Mueller (1899–1965) discovered that the chemical was an excellent insecticide. It is effective against a number of insects that carry diseases detrimental to man such as the fly, mosquito, tick, and louse as well as a large number of insects which attack food crops that man needs for his growing population.

DDT has some dangers associated with its use. One of the unpleasant results was heralded in Rachel Carson's *Silent Spring* of 1962 in which it was noted that DDT kills the food supply of birds and in addition is toxic to birds. In addition to losing birds, the whole ecological system or balance of nature may be upset resulting in the loss of other precious forms of life. DDT is not broken down in nature but remains and keeps its poisonous properties. Various forms of biological life concentrate this poison and it is passed from one form of life to another. The drug has been found at the North Pole in the ice and at different depths in the ocean indicating a growing concentration. A new DDT compound has been made to break down and become non-toxic but perhaps the damage has already been done and too much may be loose.

In parts of the world DDT has meant a new lease on life for millions would have died from diseases carried by insects. Many of those whose lives were saved at an early age by innoculation, water purification, penicillin or other drugs, are now faced with death through starvation. The area in which they live cannot always produce enough food to feed them. Pestilence and plague, two of the regulators of population, have been reduced in their effects but then the population that was saved dies of famine. There is no technological reason for famine at present because of the rapid forms of transportation available and the food surpluses in some parts of the world such as the United States.

The time is quickly approaching when the earth will not be able to support its growing population. Chemistry can delay the day by improved fertilizers. Biology by crops that give a higher yield of usable food and farming the sea. The work done by the use of chemicals to grow a form of algae high in protein is an example of the combination of the two fields toward a common problem. Another step in the food production process is the indication that it will be possible to produce a form of protein from petroleum.

All these procedures, however, bypass the basic question of the regulation of the population. The question has been around in the civilized world ever since Thomas Robert Malthus (1766–1834) pointed out in 1798 in his *An Essay on the Principle of Population* that poverty is inevitable because population increases by a geometrical ratio while food by an arithmetical ratio. He believed that the only ways to prevent the growth of population were war, famine, disease, and moral considerations. In the Eskimo society when the food supplies grow short, the old expose themselves to the cold. We have progressed in other means to control the population. In the United States the automobile kills more than war but disease and famine have been largely eliminated. There have existed for a long time various mechanical and chemical means for the prevention of conception. Most of these, however, require some foresight in use and the sexual act is not always a preplanned affair.

It has been contended that woman gained political freedom through suffrage acts, physical freedom from the household chores through electricity, and sexual freedom through the "pill." In 1961, two oral contraceptive pills were marketed. Conception could be prevented by taking one pill starting the fifth day after the start of the menstrual period and continuing for twenty days. Their effectiveness was demonstrated by 1963–64 and a new means was found for controlling population. It was not the discovery of any new drug rather it was the application of certain hormones in combination and their relative safety that was new. Today, an oral drug for inducing sterility in males is being sought.

Most birth control has not been exercised in those countries that most need to use it with the notable exception of Japan. India has been trying but there the amount of pride in a new born child coupled with a belief in the demonstration of male virility and the belief that if you die you will return again does not do much to force to the attention of the populace the immediate problems of starvation. The solution is at hand technically, but not the will and the means of educating the masses.

R. C. Gesteland, a biologist at Northwestern University, has forecast massive famines by 1975. About 1/3 of the world is well fed and 2/3 is not and in the United States a large percentage of the population is over fat and wastes food. The population problem is not simple for it is concerned not only with the birth rate in a country but also the infant mortality. The United States has a current rate of about 17 births per thousand population and an infant mortality of 22 per thousand live births, while Arab countries are nearly 40, more than double ours, but with an infant mortality of 80, nearly four times ours. Most European countries have rates near ours while Asia, Africa, and South America are near 40. If infant mortality is reduced through drugs and better medical attention the populations will increase faster and famine will stare at most of these because they have little agricultural surpluses.

The picture for life expectancy is of course reversed. In Europe and the United States a person can expect to live until his late sixties and early seventies while in Asia, Africa, and Latin America it is closer to forty and fifty or roughly twenty years less of life. Another way to solve part of the population problem would be to kill those over a certain age.

Certainly there is little to be said for ending one's life in a nursing home. There is no room in the homes of the Western world for the aged and they are committed to the final indignity of a mature person having another treat him as a child. There is no future and the look of despair while waiting the release of death is all too common. The best run nursing home or hospital has no humanity for the final days of a man. It is not the pain so much of a terminal disease as the indignities to which the sufferer is put to. There comes a time in the life of a person when death with dignity is much more meaningful and valuable than a few days more of life.

In this question, as with birth or with the hopelessly ill, who is to decide who is to live and who is to die or not be given the right to be born? What criteria can be used that will be meaningful and agreeable? If all are to live then it will have to be at a lower level. The west has the most to lose in terms of standards of living but if a selection is to be made, is it better to have lived and died or never to have lived at all?

Perhaps another solution for a man would be to stand on top of the dung heap that he has made of the earth and fling himself into the void in search of another world to befoul. This has been suggested by Eiseley who intimates that the drive that man has for exploring space is based on his nature to destroy and seek a new place.[5]

The idea of man as a polluter is not new. The aqueducts that Rome built to bring water was because the Tiber was too polluted to draw drinking water from. Civilization, from the word "city," has always been a polluter. As man congregates or as there becomes more of him and as he become more technological his wastes increase as does the disposal problem. In a pure biological culture of yeasts or bacteria if they have lots of nutrients they will continue to grow and reproduce until they are killed by their own waste. Perhaps this is the end of man—not to die with a bang or a whimper but with a gasp.

Man gradually makes his surrounding uninhabitable. He covers the top soil with asphalt, mine wastes, houses and buildings. Some he scrapes so that

Figure XXVI-2. *The Thinker* Used by Permission of Ray Osrin, *Cleveland Plain Dealer*.

water and wind can more easily remove it to the ocean where it cannot be used. He befouls the air with auto exhaust and industrial gases and in so doing may upset the climate and in crowded areas does prevent some from breathing. His body wastes and phosphate detergents that keep him clean pollute the waters so that he cannot use it to drink and so that fish and plants cannot live in it. The more "civilized" man becomes the more waste he produces. This he spreads over the countryside destroying beauty and making a health hazard to himself. He defiles natural places of beauty with graffiti and waste products as a trip to any National Park will show.

In this country we have reached an all time high in waste produced—we lead the world in waste production. With the Civil War we entered a packaged age. Canned food became possible and cans have covered the world. The age of packaging was combined with mass production and mass retailing so that after World War II almost all food products was packaged or wrapped in plastic. Following World War I the automobile made its appearance in large numbers and its rusting body has become a familiar scene all over the land. Every man, woman and child in this land produces a ton of waste per year and the crisis of waste disposal is upon us in the U.S. Every

year we use 60 billion cans and 30 billion bottles that adds to the disposal problem. Our own affluence makes waste disposal more difficult for it adds to the labor cost which is the major cost in waste disposal.

The cost of pick-up and disposal of refuse is about $30 a ton. If a landfill is used the cost is about $5 a ton or if burned $7 a ton. Once it could have been expected that most of the waste would eventually decompose and disappear with time. Iron from tin cans would rust, paper would rot. Today, however, materials are used which will not decompose because they were never a part of nature. Aluminum forms an oxide layer which is harder than the metal and sticks close to the metal protecting it from further attack while plastics will remain long after even clay tablets have disappeared.

Waste can be recycled. Emphasis has been placed on the reusable deposit bottle and the deposit costs have risen to encourage their return. Aluminum cans are being purchased by some canning companies to be melted and reused. Deposit charges for automobile pick-up and disposal are being considered to get the iron of cars back to smelters. Paper used to be recycled as it is in Germany and Japan where 90% of paper is put back to work but in this nation of six carbon copies of everything labor costs make recycling unprofitable.

Other approaches are possible. Plastics are now being made with dyes in them which will react with the ultraviolet light from the sun and become a powder. Garbage has been shredded and made into a lightweight building material which also solves its disposal problem. Glass has been crushed and made into a paving material called "glassphalt" and is presently undergoing tests. It is possible to make sewage into fertilizer and to use gas from the treatment plants for home consumption but there are the original costs of such methods and none of them solve the high cost of pick-up.

If we were willing to accept a less material way of life pollution could be reduced drastically. Automobiles could cease to be polluters while running and junk afterward if we were willing to develop a public transit program and to use it. Less garbage would be possible if we would reduce the amount of packaging and return to an unpackaged produce. Retaining newspapers separately and using paper drives could reduce part of our problem. If we want less power for air conditioners and appliances, electric needs could be cut and hence, the amount of pollution from burning the coal for the power. In the last 50 years the amount of refuse per capita has doubled and is expected to double again in the next 20. Where will we put it?

STUDY AND DISCUSSION QUESTIONS

1. What is a scientific "sin"? Would the same apply to manufacturers? Laborers?
2. How were drugs good? In what sense evil?
3. What is wrong with addiction?
4. Why isn't tobacco an unlawful narcotic?
5. What are some of the benefits of amphetamines? Dangers?
6. What is the value of DDT? Dangers?
7. What are psychotropic drugs and how have they been beneficial?
8. Should anyone have the right of life or death over another? Does the unborn have the right to be born? If there is a limited space on earth with a limited ability to feed the population, then how should the population be limited?
9. Does man destroy by nature?
10. In what ways does man pollute the earth?
11. What are the basic solutions for pollution?
12. What factors have made pollution a major problem?
13. How would you evaluate chemistry overall as a benefactor of mankind? Or has it been a benefactor?

FOOTNOTES

1. For excellent descriptive material on the effects of drugs written by users, see David Ebin (ed.), *Drug Experience* (Grove Press, 1961).
2. Lee Edson "$C_{21}H_{23}NO_5$—A Primer For Parents and Children" *The New York Times Magazine* (, 1970).
3. Philip H. Abeeson "Death From Heroin" *Science 168* editorial (June 12, 1970).
4. See Jonathan Block's "The 'Speed' that kills—or Worse" *The New York Times Magazine* (June 21, 1970).
5. Loren Eiseley, *The Invisible Pyramid* (Scribners, 1970).

Bibliography

Adams, Ansel and Nancy Newhall. *This is the American Earth.* (New York: Ballantine, 1960).

Aylesworth, Thomas G. *This Vital Air; This Vital Water* (Chicago: Rand McNally, 1968).

Behrman, A. S. *Water is Everybody's Business.* (New York: Doubleday, 1968).

Berryhill, John. "Our Troubled Urban Environment," *Nation's Cities* 5:5-7 (August, 1967).

Bracher, Marjory L. *SRO.* (Philadelphia: Fortress, 1966).

Brodine, Virginia, Peter P. Gasper and Albert Pallmann, "The Wind From Dugway," *Environment* 11:2-9 (January-February, 1969).

Carr, Donald E. *The Breath of Life.* (New York: Norton Co., 1965).

Carr, Donald E. *Death of the Sweet Waters.* (New York: Norton Co. 1966).

Carson, Rachel. *Silent Spring.* (Conn: Fawcett World Library, 1962).

Caydill, Harry M. *Night Comes to the Cumberlands.* (Boston: Little, Brown, Co., 1963).

Cleaning Our Environment. American Chemical Society. (Washington: 1969).

Commoner, Barry, "Pollution: Time to Face the Consequences," *Think* 34:24-27 (Winter, 1969).

Commoner, Barry, "The Social Significance of Environmental Pollution," *The Explorer* 11:17-20 (Winter, 1969).

Conway, Gordon and others, "DDT On Balance," *Environment* 11:2-5, (September, 1969).

Curtis, Richard and Elizabeth Hogan. *Perils of the Peaceful Atom* (New York: Doubleday, 1969).

Davies, Delwyn. *Fresh Water* (New York: Natural History, 1967).

Dasmann, Raymond F. *The Last Horizon,* (New York: Macmillan, 1963).

Edelson, Edward and Fred Warshofsky. *Poisons in the Air.* (New York: Pocket Books, 1966).

Evans, David M. and Albert Bradford. "No Deposit, No Return." *Environment* 11:17-23 (November, 1969).

"Fighting to Save The Earth From Man," *Time* 95:56-63 (February 2, 1970).

Farb, Peter. *Ecology,* 1963.

Fowler, John M., *Fallout.* 1960.

Frost, Justin. "Earth, Air, Water," *Environment* 11:14-29 (July, 1969).

Graham, Frank. *Since Silent Spring,* 1970.

Graham, Frank. *Disaster by Default,* 1966.

Grava, Sigurd. *Urban Planning Aspects of Water Pollution Control.* (New York: Columbia University Press, 1969.)

Grant, Neville. "Legacy of the Mad Hatter" *Environment* 11:18-19 (May, 1969).

Henkin, Herman. "DDT On Trial." *Environment* 11:14-17 (March, 1969).

Herber, Lewis. *Crisis in Our Cities.* (New Jersey: Prentice-Hall Publishers, 1965).

Herber, Lewis. *Our Synthetic Environment,* (New York: Knopf Publishers, 1962).

Kilgore, Bruce M. (ed.) *Wilderness in a Changing World* (San Francisco: Sierra Club, 1966).

Leopold, Luna B. *Water* (New York: Time-Life, 1966).

Lewis, Howard R. *With Every Breath You Take* (New York: Crown, 1965).

Lillard, Richard G. *Eden in Jeopardy* (New York: Knopf Publishers, 1966).

Lofroth, Goran with Margaret E. Duffy. "Birds Give Warning." *Environment* 11:10-17 (May, 1969).

Mannix, Daniel P. *Troubled Waters.* (New York: Athenium, 1964).

Marine, Gene. *America the Raped* (New York: Simon and Schuster, 1969).

McCaull, Julian. "The Black Tide." *Environment* 11:2-16 (November, 1969).

McHarg, Jan L. *Design With Nature.* (New York: Natural History, 1969).

"Menace in the Skies," *Time* 89:48-52 (January 27, 1967).

Milne, Lorus and Margery. *Water and Life* (New York: Athenium, 1964).

Moss, Frank E. *The Water Crisis,* (New York: Praeger, 1967).

Novich, Sheldon. "A New Pollution Problem." Environment 11:2-9 (May, 1969).

Overman, Michael. *Water: Solutions To A Problem of Supply and Demand* (New York: Doubleday, 1969).

Perry, John. *Our Polluted World.* (New York: Franklin Watts, 1967).

"Pollution: Is This The Air You Want To Breathe?" *Life* 66:38-50 (February, 1969).

Popkin, Roy. *Desalination.* (New York: Praeger, 1967).

Radford, Edward P. and others. "Statement of Concern." *Environment* 11:18-27 (September, 1969).

Roosevelt, Nicholas. *Conservation: Now or Never* (New York: Dodd, Mead, 1970).

Roueche, Berton. *What's Left: Reports on a Diminishing America.* (Boston: Little, Brown Publishing Co., 1969).

A symposium, "Search for Survival," *The Explorer* 11:9-27 (Fall, 1969).

Shaffer, Helen B. "Air Pollution: Rising Threat." *Editorial Research Report,* Vol. 1, pp. 301-340, 1969.

Shea, Kevin P. "Unwanted Harvest." *Environment* 11:-12-16. (September, 1969).

Selikoff, J. J. "Asbestos." *Environment* 11:3-7 (March, 1969).

Sears, Paul B. *Living Landscape.* (New York: Basic Books, 1966).

Simpson, Charles H. *Chemicals From the Atmosphere.* (New York: Doubleday, 1969).

Stewart, George R. *Not So Rich As You Think* (Boston: Houghton Mifflin, 1967).

Still, Henry. *The Dirty Animal.* (New York: Hawthorn, 1967).

Teal, John and Mildred. *Life and Death of the Salt Marsh.* (Boston: Little, Brown and Company, 1969).

"The Crisis In Water: Its Sources, Pollution, and Depletion," *Saturday Review* 48:23-44 (October 23, 1965).

"The Environment: A National Mission for the Seventies." *Fortune* 81:103-151 (February, 1970).

"The Ravaged Environment." *Newsweek* 75:30-47 (January 26, 1970).

Udall, Stewart. *The Quiet Crisis.* (New York: Harcourt, Brace and World, 1963).

U.S. Department of the Interior. *The Third Wave: Conservation Yearbook,* No. 3. (Washington, D.C., 1967).

Wise, William. *Killer Smog.* (Chicago: Rand McNally Co., 1968).

Wolozin, Harold. *The Economics of Air Pollution.* (New York: W. W. Norton Co., 1966).

Wright, Jim. *The Coming Water Famine.* (New York: Coward-McCann Publishers, 1968).

XXVII

When Electricity Was a Game

In almost every instance of a discovery or invention in science and technology, the device plays a role as an experimental toy to a full range of applications by the amateur before it has a practical use. In the popularization of the device or discovery, charlatans apply it to their own uses and frequently frauds are perpetrated on a gullible public.

As we have already seen with the steam engine, discoveries are the end result of the efforts of a number of men frequently spread over centuries of time and of many nationalities. This is certainly the case with electricity. Man's first experience with electricity must have been as with the other animals who experienced the lightning in thunderstorms and showed fear. This emotion is still displayed by many. The first man to net an electric eel or a torpedo, a kind of flat fish, must have had a shocking experience. As primitive man settled around the shores of the Baltic he became aware of the attractive power of amber.

Amber is a fossilized resin from pine trees, now extinct, which grew along the Baltic sea. It is found in beds of lignite and alluvial soils. It is yellow in color and is transparent to translucent. These properties made it attractive in the making of beads and hence it was once an important item of trade both in prehistoric times and in antiquity. This fact probably accounts for a knowledge of the properties of amber preceding the development of language. The Persians called it *karabe,* attractor of straws; while the Phoenicians made spindles of amber and as the spindle whirled it drew dust and hence was called *harpaga* or clutcher. We are more familiar with the Greek name *elektron* which they applied to both gold and amber, meaning children of the sun, because of their color.[1]

There is a strong possibility based on recent archeological evidence that the ancient Parthians around Baghdad about 250 B.C. to 224 A.D. had developed batteries and possibly electroplating. William Konig, a German archeologist, made the discovery about thirty years ago. He found what looked like vases, copper cylinders, iron and bronze rods, and asphalt stoppers. Apparently they were made as follows: thin copper sheet was soldered into a cylinder about four inches long and an inch in diameter. Into the bottom of the cylinder was crimped a copper disc insulated with asphaltum and the top was closed with an asphalt stopper through which an iron rod was stuck. The vase held it upright and contained the electrolyte.[2]

Other than magic, amusement and the above possible isolated use the only general use for electricity was in medicine. Claudius Galen (c. 130– c. 200) almost undisputed in medicine from his lifetime until the sixteenth century, recommended electric shock for curing certain illnesses. In the eighteenth century electric shocks from the torpedo were used to cure headaches and gout.[3] Electric shock is still used in the twentieth century in the treatment of schizophrenia.

These sporadic experiences with electricity were not regarded as being related nor was their nature known or understood. Indeed, it was not until the sixteenth century that formal experimental studies were undertaken in the Western world.

William Gilbert (1540–1603) entered St. John's College, Cambridge in 1558 where he took his bachelor's, Master of Arts, and a doctorate of medicine. He took further work on the continent leading to the degree of Doctor of Physics and then settled in London in 1573. He made a reputation as a physician but he also studied chemistry, physics, and cosmology. He was the first advocate of the Copernican system in England and held that the fixed stars were not all the same distance from the earth. He spent a good deal of time at the docks learning about ships and navigation. From these studies he invented two navigation instruments. He also spent time at the forge and foundry learning about the making of iron. In Gilbert we see the merging of the man of intellect respecting and learning from the craftsman.

Gilbert's greatest work, *De Magnete Magneticisque Corporibus et de Magno Magnete Tellure Physiologia Nova,* was published in 1600 after eighteen years of study and experimentation. It was the first important book on physical science published in England. For this study he became famous throughout Europe. He influenced both Galileo and Kepler and founded a scientific discussion group which was the forerunner of the Royal Society. In 1601, he was appointed physician to Queen Elizabeth and later to James I.

Although it was characteristic of the Renaissance to write in the vernacular this book was published in Latin. Gilbert first examines what was known of magnets from antiquity and some of the fallacies. He thus shows his own knowledge of Greek and of the ancient authorities. His proofs, however, are not those of authority and a prior logic but rather they are based on experiments. He venerates the ancients as also befits a man of the Renaissance but experiment and the evidence of the senses overcomes authority. Parts of the book are quite interesting and it is available in English in the *Great Books of the Western World,* volume 28.

Gilbert, at one point, compares the lodestone and magnetic attraction to amber, or electrics, and electrical movements. He states, "A loadstone attracts only magnetic bodies; electrics attract everything. A loadstone lifts great weights; a strong one weighing two ounces lifts half an ounce or one ounce. Electrics attract only light weights; e.g. a piece of amber three ounces in weight lifts only one-fourth of a barleycorn's weight." He notes that moisture or a moist breath destroys the power of electrics. He finds that a variety of substances when rubbed are electrics— amber, jet, diamond, sapphire, carbuncle, iris stone, opal, amethyst, vincentina, English gem, beryl, rock crystal, glass, fluor-spars, sulphur, mastich, sealing wax, and hard resin. Of weak electrics, he finds— rock salt, mica, and rock alum. Some substances are totally non-electric such as—emerald, agate, carnelian, pearls, jasper, chalcedony, alabaster, porphyry, coral, marble, lapis lydius, flint, bloodstone, corundum, bone, ivory, ebony, cedar, juniper, cypress, and metals.

Using any of the electrics and a bit of wool or fur it took a lot of rubbing to get a little charge; but it did attract a lot of attention. Further advances in experimenting had to wait until the idea of rubbing a finite length of electric was made into an infinite length by using a sphere of material. This step was taken by Otto von Guericke (1602–1686) about 1660 when he melted sulfur and poured it into a glass sphere. He let the sulfur solidify, broke the glass sphere, and mounted the sulfur sphere on a wooden frame. A crank handle was attached to the sphere and thus the first machine was made that generated electricity.

When a dry hand was held against the sphere and it was rotated in a dark room light was produced.

The age of electricity may be said to date from von Guericke yet he is largely ignored by historians and they have committed a crime of omission by so doing. He studied law at Leipzig and Jena, and mathematics and mechanics at Leiden. After this he returned to his native Magdeburg where he carried out his most important researches while serving as Burgomaster. He was aware of the experiments of Evangelista Torricelli and Blaise Pascal with air pressure. About 1654 he succeeded in designing and having constructed a copper cylinder and a piston which together with a stopcock enabled him to produce a vacuum. A vacuum is nothing. Yet without this discovery the age of electronics could never have been. Von Guericke used two copper hemispheres from which he removed the air to show the power of atmospheric pressure. It is reasonable to assume that his work aided in the development of the steam engine. In addition this technique could be applied in many areas to see what would happen to something acting in a vacuum. e.g., Galileo's statement that two objects of different weight will fall at the same rate in a vacuum could be tested; Boyle's law concerning the pressure and volume of a gas could be developed. But in this particular strand we are following the important step was to put together the two inventions of von Guericke's.

The results are described by the experimenter:

I took a glass globe of about 9 inches diameter, and exhausted the air out of it; then (having turn'd a cock, which prevented the return of the air) I took it from the pump. The globe being thus secur'd, I fix'd it to a machine, which gave it a swift motion with its axis perpendicular to the horizon: and then applying my naked hand (expanded) to the surface of it, the result was, that in a very little time a considerable light was produc'd. And as I mov'd my hand from one place to another (that the moist effluvia, which very readily condense on the glass, might, as near as I could, be thrown off from every part of it), by this means the light improv'd; and so continued to increase, till words in Capital letters became legible by it: (as has been observed by spectators). Nay, I have found the light produc'd to be so great, that a large print might without much difficulty be read by it: and at the same time, the room, which was large and wide, became sensibly enlightned, and the wall was visible at the remotest distance, which was at least 10 foot. The light was of a curious purple colour . . .⁴

Thus writes Francis Hawksbee (or Hauksbee) (?–1713(?)) curator of experiments for the Royal Society. He is either unaware of von Guericke's work or chooses to ignore it for he gives credit for the invention of the air pump to Boyle. Hawksbee tries a number of things in his glass globe such as: amber, woolen, oyster shells, glass, flint and steel, sealing wax, rosin, sulfur, etc. In one part he uses mercury vapour and notes the light produced. The mercury vapour lamp was made commercial by P. Cooper Hewitt in 1901. A technological lag of nearly 200 years.

Electrical phenomena could be produced but it was a fleeting spark. In damp conditions it was almost impossible to make and there could be little consistency in work that was so weather dependent. One day in 1745 Ewald Georg Kleist (?–1748), dean of the cathedral in Carmin, Pomerania, had thrust a wire from an electric machine into a jar of water in an attempt to charge water. The machine was being cranked and he was holding the jar in one hand. His other hand accidentally touched the wire and he received a tremendous jar. The jolt was from the stored electricity. The following year Pieter Van Musschenbroeck (1692–1761), professor at University of Leyden, devised a similar jar.

With the Leyden jar it was possible to build up charges and to store these increased charges for several days. This was particularly true after the improvements made by Sir William Watson (1715–1787) c. 1747. He lined the inside and outside of the jar with a metal foil, used an insulated stopper through which a metal rod was stuck. On the bottom of the rod connecting it to the bottom of the jar was a metal chain while at the top of the rod was placed a metal sphere.

Many experimenters tried numerous things with these jars. One such person was Abbet Gene Antoine Nollet (1700–1770); he had 700 monks join hands and then connected them to a Leyden jar for the amusement of the French court. They all jumped at the same time in inverse proportion to their size and age. He also tried shocking experiments on birds, worms, and insects. Some of these he electrocuted. He noted that when the bodies of birds were dissected they showed the same discoloration and ruptured blood vessels as in people struck by lightning, but did not see any connection.

The use of electricity for the specific purpose of electrocution of persons does not occur until Elbridge T. Gerry, appointed to chair a committee to find a more humane way of execution by Governor Hill of New York State, recommended alternating current. Experiments were conducted for the state by the Edison Laboratory and a dynamo was purchased for the state by Westinghouse. The Chair was given its trial run on William Kemmler August 7, 1890 who had been convicted of murdering a woman in a drunken frenzy. Needless to say he was poor—no man of wealth has ever been executed. The electric current killed and cooked Kemmler. Early executions were botched so frequently that the state made it mandatory to saw the top of the head off and remove the brain immediately after removal from the execution chamber. (This was, of course, before Westinghouse had its present slogan.) There was no doubt that the person was then dead.

In spite of the fact that various animals had been killed by the Leyden jar and the lethal effects of lightning and the electricity of the Leyden jar were one and the same. The experiment was conducted in a cow-pasture near what is now the corner of Fourth and Vine Streets in Philadelphia. The kite was a thin silk handkerchief attached to a frame of light cedar sticks from which a pointed wire projected. It was flown from twine fastened to a piece of silk ribbon which Franklin used as a handle. At the junction of the twine and silk was attached a brass key. Franklin stood in a cow shed. As the thunderhead came overhead, the loose filaments of the twine stood out from the twine and bringing his hand near caused them to move back and forth. He then put his finger near the key and a spark jumped in the same manner that sparks jumped from a static machine or a Leyden jar. Some of those attempting to repeat Ben's experiment have been killed. This experiment showed the oneness of electrical phenomena.

He postulated a one fluid theory for electricity and used the terms positive and negative to replace the two kinds of electricity. He also saw that the essential parts of the Leyden jar were two sheets of metal separated by an insulator and so used a pane of glass with metal foil on either side. Today such a device is called a condenser.

A practical outcome of Franklin's studies was the lightning rod which would "... draw the electrical fire silently out of a cloud before it came nigh enough to strike, and thereby secure us from that most sudden and terrible mischief?" During the American Revolution there was a controversy in England as to the relative merits of points or balls as terminals for lightning rods. Franklin advocated points. "Balls," cried the King and instructed the president of the Royal Society to support him. The president of the Royal Society intimated that the processes of nature could hardly be reversed at royal pleasure. He was told that a president of the Royal Society that had such a view should resign and he did. This is not the first nor the last time that politics attempts to bend scientific thought.

Others in the eighteenth century continued investigations into the nature of static electricity. Joseph Priestly (1733–1804) English theologian and natural philosopher was able to demonstrate in 1767 that the electrical charge is only on the surface. His device was made of silk shaped like a dunce cap with a thread through the peak. It was supported horizontally by a wire rim which was on an insulated base. He rubbed a rod and transferred the charge to the surface. He tested the charge with Gilbert's electroscope which was a needle device similar to a compass (a versorium). He then pulled on the thread making the outside the inside and tested again. The charge remained on the outside. (The gold-leaf electroscope was not developed for another 20 years.)

The fact that charge was a surface phenomena was important to the formulation of a quantified relationship. This relationship was recognized by Henry Cavendish (1731–1810) before 1772 but he did not publish his studies so that credit goes to Charles Augustin Coulomb (1736–1806), French military engineer. Coulomb constructed what he believed to be the first torsion balance about 1784. Actually the torsion balance was invented by Rev. John Michell (1724–1793), who was a true image of the Renaissance man. He lectured in Hebrew, Greek, arithmetic and geometry; was a theological censor, a violinist of considerable accomplishment, an astronomer and had observed the comet of January, 1760. In addition to these some of his work preceded Coulomb's, he formulated a theory as to the cause of earthquakes based on the Lisbon quake of 1755 and is regarded by many as the father of seismology for his method of locating the center of an earthquake.

The device was used by Cavendish to weigh the earth and establish the gravitational constant in 1798. Using the torsion balance, Coulomb showed both that the repulsive electrical force between two like charged spheres and the repulsive magnetic force between two like poles was inversely proportional to the square of the distance between the centers of the two spheres or poles.[5] (Newton had speculated that the Magnetic force varied as the cube of the distance.) Since there were no electrical units in 1784, Coulomb was prevented from putting the data into completely quantitative form.

The movement toward current electricity began with the studies of Luigi Galvani (1737–98), Professor of Anatomy at the University of Bologna about 1780. The laboratory where he did his work was also occupied with individuals doing electrical experimentation.

It had long been known that direct electric discharges could cause muscular contractions in dead animals and involuntary muscular movement in living bodies. Many of Galvani's later observations had been recorded by Jan Swammerdam (1637–1680) in his *Biblia Naturae* over a century before. What galvanized Galvani was the fact that frog legs twitched when near the electrical machine but without direct contact. (This was electrostatic induction which had been described about sixty years previously.) In order to see the effect of lightning, Galvani had some brass hooks fashioned which were hooked into the spines of frogs and hung outside on an iron trellis. Every time the lightning flashed the legs twitched but sometimes this happened when the lightning did not flash. He found that when the leg was pressed against the trellis the twitch occurred. Even when he took the leg and hook inside and pressed them against an iron hook that the twitch occurred. One of two conclusions was possible: either the phenomena was due to some kind of animal electricity or it was an electrical process depending on the metals. Galvani chose the first and thus muffed it but his discoveries which he published about 1791 had far reaching influences.

Many persons repeated Galvani's experiments using frog legs and dissimilar metals. Among these was Alessandro Giuseppe Antonio Anastasio Volta (1745–1827), Professor of Physics at the neighboring University of Pavia. At first Volta agreed with Galvani in his interpretation but as he worked further he began to support the second idea that the phenomena was related to the metals. He repeated the experiments of Johann Georg Sulzer (1720–1779), German Professor of Mathematics, who noted in 1750 that the contact of two different metals on the tongue gave a pungent sensation which Sulzer did not connect to electricity. Volta was able to stimulate flashes of light and of sound by connecting conductors in the mouth to those of the forehead or ear. From such experiments he became convinced that the metals were more than mere conductors of electricity—they were the generators and that the nerves were passive. He found that not all metal pairs had the same effect and about 1794 constructed a list—zinc, tin, lead, iron, copper, platinum, gold, silver, graphite, charcoal—showing the relative strengths. The further that two substances were apart on the list, the stronger the effects, i.e., zinc and charcoal coupled would be the strongest pair on the list. This was shown by the use of a gold leaf calibrated electroscope as well as the fact that of a couple the earliest metal on the list was charged positively and the latter one negatively.

Volta wrote a letter in 1800 to the president of the Royal Society in which he describes his batteries—

the first which came to be called a voltaic cell and the second the crown of cups:

> After a long silence, for which I do not attempt to excuse myself, I have the pleasure, Sir, to communicate to you, and through you to the Royal Society, some striking results which I have just obtained, in carrying on my experiments on the electricity excited by the simple mutual contact of different kinds of metals. . . . I obtain several dozen small plates or disks of copper, brass, or better, of silver, an inch in diameter, more or less; for example coins, and an equal number of plates of tin, or what is still better, of zinc, of the same shape and size approximately. . . . I prepare besides a sufficiently great number of disks of cardboard, or cloth, or of some other spongy material capable of imbibing and retaining considerable water or other liquid; for it is necessary for the success of the experiment that they should be well moistened. . . . I place horizontally on a table . . . one of the silver ones, and on this first plate I place a second plate of zinc; on this second plate I lay one of the moistened disks; then another plate of silver, followed immediately by one of zinc, on which I again place a moistened disk. I thus continue . . . to form from a number of these steps a column as high as can be formed without falling.

The problem with the discs was to keep the absorbent material between moist and this was overcome by the use of cups.

> We set up a row of several cups or bowls (small drinking glasses or goblets are very suitable) half-full of pure water, or better of brine or of lye; and we join them all together in a sort of chain by means of metallic arcs of which one arm which is placed in one of the goblets is of copper and the other, which is placed in the next goblet is of tin or better of zinc. . . . The two metals of which each arc is composed are soldered together somewhere above the part which is immersed in the liquid.

Here then we have a consistent means of producing electricity of rather constant strength any time and which can be increased by merely increasing the number of cells. For his invention, Volta was invited to Paris by Napoleon in 1801 and given a battery of honors.

Galvani's work stimulated quite a different sphere than just science. The Romantic poets George Gordon Noel Byron 6th Baron (1788–1824) and Percy Bysshe Shelly (1792–1822), ". . . were interested in contemporary science, particularly galvanism, which seemed to give a semblance of life to dead limbs, and there was speculation about the possibility of reviving the dead with electricity."[6] This possibility was extended by Mary Shelly (1797–1851) in her novel Frankenstein (1818). This strand has continued until the present day and as a matter of fact the technique has been used to revive the "dead."

One of the popular entertainments for the upper and upper middle classes of both the eighteenth and nineteenth centuries was the public lecture and one of the more popular topics was natural philosophy, which was what science was called prior to the mid-nineteenth century. Electrical phenomena had been a popular topic for some time. As a matter of fact Franklin's interest in electricity was sparked by such a public lecture. The development of Volta's battery made it possible to multiply the voltage by using a larger number of Voltaic cells. In 1802 in Paris the first electric light was produced from current electricity using a voltaic cell of 120 units and two pieces of charcoal attached to a wire at each end of the pile. Later the same year this was done in England by Sir Humphry Davy (1778–1829) who used 150 unit battery and astonished an audience with the brilliant light from the arc. So great was the public interest that when Davy wanted to make a battery of 2,000 cells, he was easily able to raise the needed funds. Electric arc lamps are still used in some motion picture projectors and for welding.

In the same year that Volta introduced his crown-of-cups battery a young man earned his degree of Doctor of Philosophy in Denmark. Hans Christian Oersted (1770–1851), who should not be confused with Hans Christian Anderson who was another great Dane and who was a good friend of the Oersted's, was interested in many things, e.g. poetry, and tried to relate some of these. Two concepts that man had long sought to relate were electricity and magnetism and it fell to Oersted to make this relation. It has long been debated if the discovery was by chance. According to several of his students the dis-

Figure XXVII-1. Oersted.

covery was made one day in lecture when he placed a compass needle parallel to the wire which was connected to a battery and closed the switch the needle swung until it was at right angles to the wire thus it showed that a magnetic force was produced at right angles to the current flow.

MAGIC, MEDICINE, MAGNETS, AND ELECTRICITY

Myths regarding magnets, lodestone as it was called, go back to the time of Thales and earlier. Onions and garlic were supposed to be able to take the power away from the lodestone so that the helmsman on a ship had to watch what he ate. Further if a lodestone were carried by a person it would cure various diseases, e.g. gout, headaches, cramps, enable him to be a gracious, persuasive and elegant conversationalist. It would also help a quarreling husband and wife to make up, would draw poisons from wounds and if it were ground to a powder and sprinkled on water it would cry like a baby. In the middle ages this same powder could draw iron from the body.

Phillippus Aureolus Paracelsus (1483–1541), his real name was Theophrastus Bombastus von Hohenheim, a great physician for his time advocated and praised the occult and miraculous powers of magnetism in the treatment of disease.

In the 15th century a group who called themselves Rosicrucians (brethren of the Rosy Cross) supposedly founded by a German Knight, Christian Rosenkreutz, (such a group exists today), pretended to use magnetism to draw or cure diseases by the use of a magnet. Comte Alessandro Cagliostro (1743–95) an alchemist of occult powers was supposed to have been one of the leaders of the European wide organization. Benjamin Franklin was also supposed to be one of its members. One of the group, John Baptist Van Helmont, wrote a book titled, *A Ternary of Paradoxes of the Magnetic Cure of Wounds* (London, 1650) in which the word "electricity" appears for the first time. They also claimed that their magnetic cure would work at a distance and in general preyed on the imagination of the nervous, the invalid, and other suckers. (A sucker is one who thinks he will get something for nothing.)

Once the ideas of Galvani about animal electricity and the battery of Volta became widely known many hucksters used the new electricity to "treat" diseases. Electrodes were attached to the patient and hooked to a battery. Once the galvanic currents had passed through his body he was to feel better at once.

From an advertisement in the Police Gazette in early 1894:

DR. SANDEN'S ELECTRIC BELT
The illustration showed a paunchy middle-ager in underdrawers. Around his midriff he wore a contraption that resembled a champion strong man's victory belt attached to a truss. Instead of the usual American eagle belt buckle there was a device that resembled a fancy doorbell.

The trusting reader was advised that Dr. Sanden's Electric Belt was, however, 'invented solely for the cure of all weaknesses of men.' Wars? Speculative investments? Their neighbors' wives? No, but the belt was 'guaranteed to cure all forms of nervous debility, impotency, spermatorrhea, shrunken parts, nervousness, forgetfulness, confusion, languor, dyspepsia, rheumatism, and the many evils resulting from bad habits in youth and the excesses of later years.' Dr. Sanden stated (in quotes) that the belt was 'an absolute positive cure.' Electricity 'which is nerve force' was the secret. The doctor added —it would seem rather daringly—that five thousand dollars would be forfeited if the wearer failed 'to feel the electricity.'[7]

The Sears, Roebuck catalog (the company was organized in 1893) for 1903 listed an electric belt that was claimed to relieve headaches and backache while the catalog two years later had electric insoles for shoes to relieve foot ailments.[8]

Following World War I magazines carried advertisements of the "Ohio Electric Works" which manufactured a number of devices whose main function was removing money from the ignorant and the gullible. Sticking your nose into the wondrous Medico Electric jar, it was claimed, would cure a wide range of ailments including: asthma, neuralgia, hay fever, and catarrh. This same company also manufactured devices which consisted of alternating disks of copper and zinc—sort of a self powered battery. The "Ohio Electric Belt," worn around the waist was advertised as a cure for rheumatism and guaranteed to give pep—"do you want to wake up tired blood?"— while the "Ohio Electric Insoles" (worn in the shoes) would cure perspiring feet, aches, pains, and cramps and would also provide "pep." According to one source, over 3,750,000 pairs of these soles were sold.

In the 1890's following the discovery of X-rays the Purity League of America introduced legislature which would prohibit the use of X-rays in opera glasses. The new is not always readily understandable to the public as this verse of the times shows:

I am full of daze
Shock and amaze,
For nowadays
I hear they'll gaze
Through cloak and gown—and even stays,
These naughty, naughty Roentgen rays.[9]

Needless to say companies were soon making and selling X-ray proof corsets and clothing. But then perhaps it's better to be X-rayed than ultra-violated. As short a time ago as World War II shoe stores were using X-rays so that the customer could see his feet in shoes and determine how well they fit. They were made illegal when the dangers of radioactivity were more fully understood and known. X-rays too were used by the charlatan in the treatment of a number of diseases and conditions.

Hertzian waves (today called radio waves) were and are still used by the quack. The apparatus usually has a series of dials and lights and sometimes a buzzer that all signal in a most convincing manner. Such devices are used for the "diagnosis" and "cures" of an almost unlimited variety of diseases—falling hair to fallen arches with a cancer in the middle. Electrodes are attached to the patient and by a reading on the dials the quack is able to tell that patient what is wrong with him. Furthermore, by adjusting the input current to the patient he may treat with the electric waves which set up sympathetic vibrations in the afflicted part of the patient.

One of the latest ruses of the medical huckster is radioactivity. The patient is told that all he has to do is place the "radioactive" packet where it hurts and the mysterious radiation will relieve the pain and cure the disease.

By no means should any of the preceding be misunderstood. Many of these devices were and are used in the treatment of disease. Some men were accused of being charlatans when they first tried a new technique and only after ridicule, dishonor and sometimes death were their discoveries really recognized for what they were.

Such a case was Franz Anton Mesmer (1734–1815). He studied medicine in German Universities and became interested in the ideas of Paracelsus which included that the stars affect the health (also in the writings of Hippocrates) of humans by means of an invisible fluid. He became convinced that there was such a power in his own hands which he called "animal magnetism." The regular doctors forced him to leave Austria and his problems continued in Paris where a commission was appointed to determine if his cures were genuine. Benjamin Franklin was a member of the commission that returned an unfavorable report. Mesmer was forced into retirement and died without receiving the honors due him as the father of hypnotism. Although still used by the charlatan, hypnotism has found its rightful place in medicine, psychiatry, and dentistry.

For a good, but somewhat dated, history of electricity see Alfred P. Morgan *The Pageant of Electricity,* New York, 1939

STUDY AND DISCUSSION QUESTIONS

1. What sort of materials lent themselves to electrical studies?
2. Who were some of the persons who contributed to early electrical studies and what did they do?
3. To what uses was early electricity placed?
4. Give examples of which illustrate the general sequence of invention and innovation.

FOOTNOTES

1. Alfred P. Morgan *The Pageant of Electricity,* Appleton-Century-Crofts, N.Y. 1939 pp. 8-10
2. Harry M. Schwalb "Electric Batteries of 2,000 Years Ago" *Science Digest,* April 1957 pp. 17-19
3. Alfred Still *Soul of Amber* Murray Hill Books, Inc., N.Y. 1944 p. 155
4. F. Hauksbee *Physico-Mechanical Experiments on Various Subjects Containing an Account of Several Surprising Phenomena Touching Light and Electricity, Producible on the Attrition of Bodies,* London 1709 p. 36. It is noteworthy that the same discovery was made at the same time by Pierre Poliniere. See *Isis 59* 1968 pp. 402-413
5. For experimental details see Morris H. Shamos *Great Experiments in Physics* Holt, Rinehart & Winston, Inc., N.Y. 1959 pp. 59-92
6. E. F. Bleiler *Three Gothic Novels* p. xxxiii
7. Theodore Roscoe *Only in New England* Charles Scribner's Sons, N.Y. 1959 pp. 78-79
8. Melvin Kranzberg and Carroll W. Pursell *Technology in Western Civilization* Vol. II Oxford University Press, 1967 p. 83
9. Ritchie Calder *Science in our Lives* New American Library, N.Y. 1955, p. 69

XXVIII

The Electron Rolls Up Its Sleeves

Another individual who suffered some because he was ahead of his time was Andre Marie Ampere (1775–1836) a French mathematician. He noted, within two weeks of Oersted's experiment, that when current flowed through a coil of wire that the coil acted as if it was a magnet. Up to this point electricity had been largely qualitative but Ampere applied mathematics. In this he was too far ahead of his time and was thus under attack by the establishment. It was apparent too that conservatism had already possessed electricity.

It may be conjectured that we have carried the power of the instrument to the utmost extent of which it admits; and it does not appear that we are at present making any important additions to our knowledge of its effects, or of obtaining any new light upon the theory of its action.

Thus wrote John Bostock in his *Account of the History and Present State of Galvanism* in 1818. This was, you will note, published just two years before the work of Oersted and Ampere.

The "power of the instrument" was considerably enhanced when two men working independently discovered the principle of electromagnetism. William Sturgeon (1783–1850) an English physicist, in 1825, bent and varnished a soft iron bar. He wrapped a copper wire around it to from a coil and attached the ends to a battery. When the current flowed through the wire the iron acted like a magnet and when the current was cut off it no longer was a magnet. The other man was Joseph Henry (1797–1878), head of the Smithsonian Institution, did the same at about the same time and a year later made a magnet by insulating the wire that could lift a ton. Henry's insulating material was silk cloth made by cutting his wife's petticoats into strips. Her reaction has not been recorded.

Both Henry and Michael Faraday (1791–1867), an English chemist, were hot on the track of something regarding coils and magnets. It happened that Faraday published first and is given credit for the discovery of the principle of the electric motor and

Figure XXVIII-1. Henry.

generator. (April, 1832.) If a magnet is moved into or out of a coil of wire an electric current is produced.

A poetic rendition of the discovery was made by Herbert Mays, a contemporary of Faraday's:

Around the magnet Faraday
Is sure that Volta's lightnings play;
But how to draw them from the wire?
He took a lesson from the heart;
'Tis when we meet—'tis when we part,
Breaks forth the electric fire.

Henry in 1831 had demonstrated an electric bell to his classes and to others which operated at distances over a mile. We shall shortly return to Henry's ding-a-ling.

Over the centuries man had used limited audio-visual methods of communications; drums, hollow logs, mirrors, smoke, signal fires, flags, semaphores and other devices. Each of these methods had decided limitations and for the common man during most of these centuries did not apply. His method was to give someone a written note or verbal message which went at the speed of the traveler i.e. about 3-8 miles/hour.

Figure XXVIII-2. Faraday.

A number of persons had attempted to make a telegraph work as early as the 18th century using static electricity but none of these could have been made practical except over very short distances. The first use of current electricity was made in Munich using 35 wires (one for each letter and number) and decomposing water to indicate which letter or number was meant. It worked for over a two mile distance. By 1811 a code had been worked out and the number of wires reduced to two. Either government ignorance, indifference, or opposition—the establishment—has blocked many an invention.

An attempt in England in 1816 was blocked by the official statement: "Telegraphs of any kind are now wholly unnecessary and no other than the kind now in use (a row of semaphore towers within sight of each other) will be adopted." This type had been developed by Claude Chappe (1763–1805) in 1794 with towers 6 to 10 miles apart and could transmit a message between Toulon and Paris (475 miles) in 10 to 12 minutes. Dr. Burbeck, of the Mechanics Institute, England, made this statement in 1837. "The electric telegraph, if successful, would be an unmixed evil to society; would be used only by stock jobbers and speculators—and the present Post Office is all the public utility requires."

In spite of such comments and indifference work went on. Ampere suggested in 1820 a needle electric telegraph using a wire for each letter and this technique was demonstrated three years later using five needles by Paul von Schillin-Cannstadt (1786–1837). In 1836 William Fothergill Cooke (1806–1879), an Englishman, constructed a magnetic type needle telegraph. Later he formed a partnership with Charles Wheatstone (1802–1875) and constructed a mile long telegraph line in 1837.

As a sculptor and portrait painter trained at Yale, Samuel Finley Breese Morse (1791–1872) had already made a name for himself not only in this country but also in England. His paintings hung in the Royal Academy and he had been elected first president of the National Galleries. Shortly after the death of his wife Morse returned to Europe to work and study classic art in Florence, Rome, and Paris.

At the age of forty-one and after three years abroad, Morse was on his return voyage from Le Havre to New York when a chance conversation with a fellow passenger changed his whole life. A Dr. Jackson was seated at the same dining table and was regaling those at the table with wonders of the new electromagnet and the recent experiments and theories of Andre Ampere. Morse became captivated by the idea of a telegraph and spent the rest of the trip filling his notebooks with drawings and ideas of the new telegraph instead of figures and landscapes.

He had almost no knowledge of electricity and even less mechanical skill and after three years of work nearly all of his savings were gone. Fortunately the New University of New York was founded in 1835 and Morse was offered the professorship of art and design which he took to make money. This is an example of what W. H. Cowley calls the "haven" function of the university—i.e. to provide an income to a person for research in perhaps another field.

He did become acquainted with Leonard Gale, professor of chemistry, who made a number of helpful suggestions to Morse. He finally made a receiver from a pendulum with a pencil on the bottom of it and a magnet which pulled the pendulum toward it. The pencil made marks on a paper tape. The current from batteries was much too weak to send over very long distances so that Morse had to invent a relay which in this case was an electromagnet which closed the circuit to another set of batteries thus adding more current to send the message further along.

At this point he was not only penniless but his reputation as an artist had been forgotten since the public had a short memory. In September of 1837 he made a demonstration of his equipment and sent a message over 1700 feet of wire. This demonstration was witnessed by Stephen Alfred Vail who offered his mechanical skill and $2,000 for a partnership with Morse. Some claim that Vail and not Morse devised Morse code.

January of 1838 saw a working telegraph and Morse persuading Congress to introduce a bill to appropriate $30,000 to build a telegraph line from

Washington to Baltimore. It lay on the shelf until February 3, 1843. In the meantime Dr. Jackson was trying to cut himself in for a claim on the patent and later there was a triangular fight among Morse, Wheatstone, and Henry. This situation repeated itself with radio and the telephone with the judgement going to the person or persons with the most economic backing—justice is always more easily swayed by the buck which is why she carries scales.

When Congress finally did act in the face of ridicule the measure only passed by eight votes. Success seemed assured with $30,000—keep in mind that in those days wages for skilled workers was apt to be less than a dollar a day. They had decided to put the wire underground and when they had spent 2/3 of the money found that it wouldn't work and had to start over stringing the wires on cross beams on posts spaced 100 feet apart and using bottles for insulators.

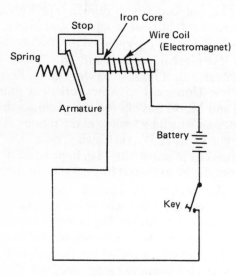

Figure XXVIII-3. Telegraph.

The line was completed and a demonstration given on May 24, 1844. Morse sent the message from Washington to Vail in Baltimore—"What hath God wrought?" which sounds about as artificial as the Genesis bit from the lunatics. Most who witnessed it refused to believe it. Belief was instilled from the fact that the Democratic National Convention was in session in Baltimore. James K. Polk was brought out as a "dark horse" candidate and the delegates nominated Senator Silas Wright as Vice-President. This was telegraphed to Wright who refused and sent the message back to Baltimore. The delegates didn't believe the massage and sent a committee to Washington the next day. When the message was confirmed suddenly everyone began to see the possibilities.

Hiram Sibley joined Morse and Vail to form Western Union in 1856 and poles were seen throughout the land.

Obstacles were not totally overcome. There was a southern Kentucky community that destroyed a telegraph line in 1849 on the grounds that "... it robbed the air of its electricity, the rains are hendered, and ther' ain't been a good crop sence the wire was put up." The buffaloes used the poles to scratch their backs and the Indians used pieces of copper wire for decoration. It took railroad officials seven years after Morse's patent to realize that this might be a good way to dispatch trains then every train station had a telegraph office. Even as late as 1885 the War Department was on record as stating that they had no need for electrical communication—this after the Civil War?

The spread of the telegraph to the west coast signaled the demise of a thrilling part of American history—the pony express. It had carried the mail between St. Joseph, Missouri, the western terminus of the railroad, to San Francisco. There were horse changes every twenty-five miles and a rider could cover seventy-five miles a day. They had set a record when they carried President Lincoln's inaugural address to the coast in seven days and seventeen hours. In less than 100 years, Western Union established 20,000 stations and strung 1,600,000 miles of wire.

In a Parisian paper, *L'Illustration,* for August 26, 1854 Charles Bourseul outlined what was almost a practical description of a working telephone.

> I have asked myself . . . if the spoken word itself could not be transmitted by electricity; in a word if what was spoken in Vienna may not be heard in Paris? The thing is practical in this way:

> Suppose that a man speaks near a movable disk, sufficiently flexible to lose none of the vibrations of the voice; that this disk alternately makes and breaks the connection from a battery; you may have at a distance another disk which will simultaneously execute the same vibrations.

Such a telephone was constructed by Johann Phillip Reis (1834–1874) a professor of Physics at Friedrichsdorf in Germany. In 1860, he stretched a sausage skin over a beer barrel bung which he had hollowed out and mounted it in an opening in a wooden box. He fastened a piece of platinum on the center of the skin near which he fastened a heavy brass spring which made and broke contact as the skin vibrated. The circuit also included a battery and a receiver which was made of a knitting needle wrapped with insulated wire and mounted to a cigar box which acted as a sounding board. Perhaps the honor for the invention of the telephone should go to

Reis and this has been argued. See S. P. Thompson *Phillip Reis, Inventor of the Telephone* London 1833 pp. 36-38 and 112 ff.

Credit is given to the Scot born teacher of deaf-mutes, Alexander Graham Bell (1847–1922). He was the third generation of teachers of the deaf and had taught a dog to growl a sentence with Bell holding the throat and mouth. When Bell was nineteen he was reading Helmholtz's book *Sensations of Tone* which was in German. Due to the fact that he did not read German well he got the mistaken idea that Helmholtz was writing of electrical transmission of tones and the seed was planted in his mind. He also had the good fortune to see Reis' instrument at the Smithsonian Institute and have it explained to him by its director, Joseph Henry. This may have given him the idea of substituting a stretched membrane instead of a vibrating reed. Bell and Elisha Gray both filed patent papers on the same day—February, 1876 —and the hour of filing had to be determined in order to establish priority. Bell was given a patent March 7, 1876.

Three days later while conducting an experiment Bell spilled some dilute acid on himself and in exasperation said: "Mr. Watson, come here. I want you." Almost instantly Watson was there shouting that he had heard Bell clearly over the telephone. In 1915 Bell repeated this sentence on the occasion of the first transcontinental telephone line—he was in New York, Watson in San Francisco. Watson, replied, "I would be glad to come, but it would take a week."

The first telephones used the receiver and speaker interchangeably and were membranes which vibrated and made or broke electrical circuits. Before long carbon granules were used in the speaker and the amount of spacing between them permitted more or less electric current to flow and vibrated a metal diaphram in the receiver through a permanent magnet.

England and Europe regarded the telephone with amusement as a silly American toy. Even in this country Bell had trouble raising the funds to make the telephone a commercial reality. In 15 years the tenth interest that Bell offered for $100 was worth $1,500,000.

Both the telegraph and the telephone are based on the simple electromagnet—a piece of iron wrapped with coils of insulated wire. Each time that electricity flows the iron becomes magnetic and each time the flow stops the iron ceases to attract. The current can go on or off in fractions of a second and the magnetism appears and disappears just as rapidly. The slightest change in the intensity of the current will

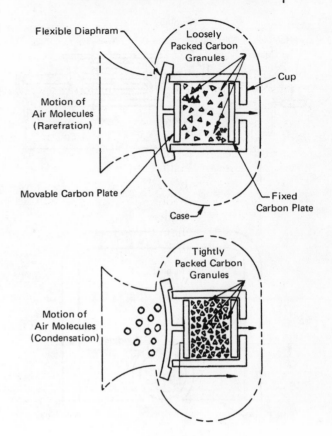

Figure XXVIII-4. Telephone.

increase or decrease the magnetic field so as to mechanically reproduce the quality and tone of a person's voice on the telephone. The electromagnet is used in a number of different devices—the teletype, dial telephones, the electric bell, elevator controls, traffic lights, railway signals, switches, generators and electric motors.

If instead of moving a wire in and out of a magnetic field of a permanent magnet as Faraday did, an electromagnet is used then the amount of electricity produced in the moving wire is increased and if the wire be increased by coiling it then this also increases the flow of current. By 1871 this principle had been developed into an efficient generator by Zenobe Theophile Gramme (1826–1901). It was operated by steam engines which turned the coils within a series of electromagnets. It was used to produce electricity for arc lighting in 1873. The arc began to replace gas lights in public lighting and in public buildings. More efficient and cheaper power was to come by using the fall of water to turn the coils of wire or by using wind to provide light and power for isolated farms all over the country.

A child was born in Milan, Ohio in 1847 who was to help shed light on the world. Thomas Alva Edison

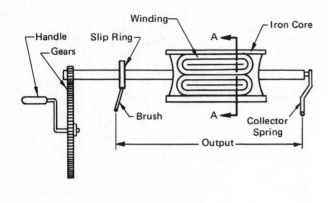

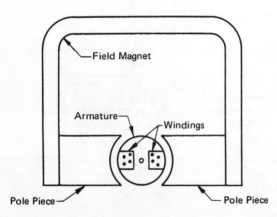

Figure XXVIII-5. Generator.

Figure XXVIII-6. Edison.

(1847–1931), a man already famous for his improvements in the telegraph, the stock ticker and the inventor of the phonograph, in the 1870's turned his attention to the electric light. From the time of the early electrostatic generators and the first air pump, various men had tackled the problem of producing light from the electrical fluid. The arc lamp was already established but the carbons had to be replaced frequently as they burned out. The first of these installed for illuminating a public street were twelve 2,000 candle-power (light intensity was and still is measured in candle-power—you can't hold a candle to the technique) arc-lamps set up on ornamental poles in Cleveland's Public Square in 1879.

Edison's work in the incandescent lamp was no accident. It was carefully planned and painstakingly exercised. The first patent issued on an incandescent lamp was to Frederick De Moleyns an Englishman in 1841. The problem was that the material in the partially evacuated tube vaporized and gave a black coating to the lamp thus cutting out the lamp. Edison did not invent the incandescent bulb but he invented the first practical one. After many experiments that failed he tried using a cotton thread in an evacuated bulb. (He spent eight hours evacuating

the bulb.) Success hung by a thread as the current was applied and the thread burned to create a carbon filament. The soft glow continued for 40 hours after it was started October 21, 1879 and would have burned longer if the current had not been increased.

In the same year in Menlo Park, New Jersey there was a public demonstration of Edison's electric bulb lighted by power from his own generator. The first American power plant was the Pearl Street station in New York in 1882. Its first use was lighting and the first electric bill was $50.40—but about 1940 the annual bill was $2,000,000,000. Other applications came swiftly as the electric motor was developed—it runs in just the reverse manner as a generator. (As a matter of fact a motor can be used as a generator if it is turned by some force.) Railways were electrified in 1907 and these were extended to suburban traffic and to street cars. With motors factories could be located far from water and could be more independent in location. At the Columbia Exposition of 1893 there were demonstrations of cooking and ironing using electricity. Soon everyone was using electricity or will.

MENLO PARK

On this site, half a century since, the searching brain,
The patient hand, began their work to find and wield
The latent powers Nature still held in store.
Today, a million men at rise of sun
March to the workshops of the world, their tread
A pean to the Chief, whose wizardry
Made live the whirring wheels, the vibrant wires
And woke new powers of service for the world.
RICHARD ROGERS BOWKER

Although Edison was the wizard of Menlo Park, he missed several opportunities in being the first in radio. November 22, 1875 Edison made observations reported in the January, 1876 issue of *The Operator* —the discovery of a new force which he christened "etheric." It reports that he noted that a metallic substance brought near a charged electromagnet caused sparks which were unlike any electricity static or inductive. He placed this force as "... the direct offspring of electricity and magnetism." He recognized that the new force performs independently of insulators. By 1885 he had recognized also that there was no need for a conductor either. May 23 of that year he filed for a patent (No. 465,971). In this application he had drawings of aerials and antennas and stated:

I have discovered that if sufficient elevation be obtained to overcome the curvature of the earth's surface and to reduce to a minimum the earth's absorption, electric telegraphing or signaling may be carried on by induction, without the use of wires connecting such distant points.

This discovery is especially applicable to telegraphing across bodies of water, thus avoiding the use of submarine cables, or for communicating between vessels at sea, or between vessels at sea and points on land.

In 1883, while trying to increase the efficiency of the carbon filament incandescent bulb by keeping the bulb from blackening, Edison placed a metal plate close to the filament and connected it to a galvanometer (a device that shows current flow), a battery and the base of the lamp. He was amazed to find that there was a current flow recorded on the galvanometer even while the circuit was open since there was a gap between the filament and metal plate. He also found that the only time a current appeared to flow was when the metal plate was connected to the positive pole of the battery. He saw no practical use for the device but he applied for a patent anyway.

Joseph Henry (1797–1878) had discovered in 1842 the existence of electrical oscillations produced by the spark discharge of a condenser and even guessed that there was some connection between them and light.

As reported in Oliver Lodge's book *Talks About Radio,* Henry said:

It would appear that a single spark is sufficient to disturb perceptibly the electricity of space throughout a cube of 400 feet capacity, and ... it may be further inferred that the diffusion of motion in this case is almost comparable with that of a spark from flint and steel in the case of light.

In the 1850's Faraday had worked with and described magnetic fields and electricity. In this world which could not be seen Faraday felt intuitively and worked out logically that there must be a necessary connection among all forms of energy—light, electricity, heat, gravity. One of the chapters in John Tyndall's book *Faraday As A Discoverer* is titled "Unity and Convertibility of Natural Forces ..." which demonstrates the extent of Faraday's speculations. Yet Faraday had been poor and while he was brilliant, he lacked the mathematical education to make his dreams more substantive. Others including Einstein and Heisenberg have had Faraday's dream and a part of this linkage was supplied by James Clerk Maxwell (1831–1879), a Scotch physicist.

Maxwell's father was a Scottish laird and a bit of a nut who was interested in technology and also de-

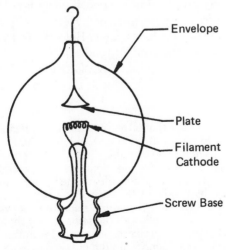

Figure XXVIII-7. Diode.

Figure XXVIII-8. Maxwell.

signed his son's clothing when he went to college. Maxwell's nonconformity and his peculiar clothing earned him the nickname "Dafty." His classmates did not make him conform but did provoke him to fight from time to time. After his college days he spent some eight years studying Faraday's ideas concerning the magnetic field produced by moving electrons. A current flow is an electron in motion and the faster the motion the stronger becomes the magnetic field. Maxwell reasoned that if one knew the amount of the charge and the strength of the magnetic field then it would be possible to calculate the speed of the field's movement. He found by devising an apparatus to give comparative measurements that this speed was that of light. As the wave, moving with the speed of light, spreads out, its two forces (magnetic and electrostatic), are at right angles to each other and to the direction of travel. He published this in four neat little equations in 1865 in his paper "A Dynamical Theory of the Electromagnetic Field."

Needless to say, Maxwell's paper did not go unchallenged. One outstanding scientist shortly published a paper titled, "On the Impossibility of Originating Wave Disturbances in the Ether by Means of Electric Forces." Maxwell had been dead nine years when the experimental evidence of Heinrich Rudolph Hertz (1857–1894) vindicated him by producing the waves that he had predicted.

In the late 1870's Hertz was attending the University of Berlin and in one of the lectures by the great Hermann Ludwig Ferdinand von Helmholtz (1821–

1894) he and his classmates were exhorted to verify the electromagnetic theory of Maxwell. Hertz did not begin immediately on the problem—he didn't know how to start. He began his studies in 1883 and two years later he was appointed Professor of Physics at the Technical High School at Karlsruhe. It was in his laboratory there that he made and recognized the waves that Maxwell had predicted. Three years previously Elihu Thomson (1853–1937), an eighteen-year old science teacher at Philadelphia High School had confirmed Maxwell's theory when he drew sparks from metal pipes on the side of the room across from an oscillatory circuit.

Hertz made use of an induction coil (a means of producing high voltage) attached in an incompleted circuit to two brass spheres. When a spark was produced between the two spheres an electromagnetic wave was produced which traveled through space. For a receiver he used another incompleted circuit with two brass spheres and a coil of wire whose length could be changed by sliding a piece of wire along the other two wires. The great von Helmholtz in 1887 at the Physical Society Meeting in Berlin described the work of his former student.

This phenomena Hertz had discovered was soon making waves all over the world. Hertz had shown that the waves possessed many of the same properties as light—they could be reflected, refracted (bent), and showed interference (the waves reinforced or cancelled). Other investigators were also at work—Oliver Lodge in England, Alexander Popoff in Russia, and Edouard Branley (1844–1940) in France.

In February 1892 Sir William Crookes' writing in the London Fortnightly Review predicted the possibility of using electromagnetic waves for wireless telegraphy and tuning to special wave lengths. Two years later, the year in which Hertz died, a young Italian, son of a businessman of Bologna and a Scotch-Irish mother, chanced to read an article which described the work of Hertz. Guglielmo Marchese Marconi (1874–1937) caught the signal. He later wrote:

> It seemed to me that if the radiation could be increased, developed and controlled it would be possible to signal across space for considerable distances. My chief trouble was that the idea was so elementary, so simple in its logic, that it seemed difficult for me to believe that no one else had thought to put it into practice. I argued, there must be more mature scientists (he was then twenty) who had followed the same line of thought and arrived at almost similar conclusions. From the first the idea was so real to me that I did not realize that to others the theory might appear quite fantastic.

Figure XXVIII-9. Hertz.

Marconi began his experiments in a loft at his father's country home, the Villa Griffone, at Pontecchio near Bologna in 1895. In reality, what he did was to have the genius to combine the efforts of others into a practical and workable arrangement. He consolidated the idea of an antenna (aerial), first used by Edison and Popoff; with Branly's coherer, a device with filings to detect the waves; and Hertz's electric waves to create the Marconi Wireless Telegraph Company. Being a true patriot Marconi offered his ideas to the Italian Navy and was refused. He then turned to the homeland of his mother in 1897.

He had a letter of introduction to Sir William Preece who was the technical director of the British Post Office and upon his arrival in England applied for a patent—British patent number 12,039 granted July 2, 1897. Within a short time Sir William issued a classic example of British understatement:

> The distance to which signals have been sent is remarkable. On Salisbury Plain, Mr. Marconi covered a distance of four miles. In Bristol Channel this has been extended to over eight miles. And we have by no means reached the limit.

In 1901 signals had been sent across the Atlantic. In 1909 Marconi shared the Nobel prize in physics for his development of wireless telegraphy.

The British government had given Marconi its full support and cooperation. This support included the gathering and lending of experts. Three men of special interest to the development of radio were made available to Marconi by the government—John Ambrose Fleming (1849–1945), Dr. Lee DeForest (1873–1961), and Oliver Joseph Lodge (1851–1940).

In 1901, some did not believe that Marconi had heard the three dots that signaled trans-Atlantic communication. A man who had narrowly missed being the discoverer of radio waves himself did believe and lent his moral support—Edison. The Edison organization not only contributed ideas, and inventions but also proved the training ground for future workers in electricity and radio.

Reginald Aubrey Fessenden (1865–1932) had served his apprentice as inspecting engineer for the Edison Machine Works and chief chemist of the Edison Laboratory (1886–1890). He taught at Purdue and then while professor of electrical engineering at Western University of Pennsylvania (now University of Pennsylvania) he began researches into radio. All the basic equipment for the sending and receiving of Hertzian waves had been patented by Marconi but Fessenden had an idea that perhaps it might be possible to send without a spark by simply connecting a generator to an antenna. The problem was one of frequency. The frequency of his alternating current was the number of times the generator's armature (center coil) turned each minute. To get the frequency needed for wireless without having the generator fly into pieces was the next problem. Dr. Ernest F. W. Alexanderson, (1878–   ), of General Electric, perfected an alternator named for him which produced a frequency of 100,000 cycles per second. It was this carrier wave which when controlled made possible the first radio broadcast from the United States. On Christmas Eve, 1906, wireless operators were amazed to hear sounds other than dots and dashes coming from their earphones. They heard music—Handel's "Largo," a violin solo of "O Holy Night,"—and a Bible reading. It was not the second coming but a radio broadcast from Brant Rock near Boston.

When the Marconi Company was first organized, Fleming was the chief technician and later he worked for the Edison and Swan Lighting Company in London. It was here that he had available some of Edison's lamps with the extra electrode built in and these he used for his experiments which were to lead to the vacuum tube or valve as it is called in England. In 1904 he devised from the Edison Effect the Fleming valve which was essential to detect weak radio waves and also to boost power from the broadcasting end.

In 1906, Dr. DeForest added a third electrode to the Edison Effect to produce what he called an Audion tube. Deforest formed a company and sold stock to produce his tube. He was brought to trial in 1912 for using the U.S. mail to defraud.

> DeForest has said in many newspapers and over his signature that it would be possible to transmit the human voice across the Atlantic before many years. Based on these absurd and deliberately misleading statements, the misguided public, Your Honor, has been persuaded to purchase stock in his company.

This was extracted from the trial record. Two company officials were convicted. In 1914 the first transatlantic telephone message was sent using a development of DeForest's vacuum tube.

When regular radio broadcasting started in 1920, Edison put a sign on his office door which read: "I WILL NOT TALK RADIO TO ANYONE." He had missed the discovery of both radio waves and the vacuum tube.

It is possible that both television and motion pictures have the same father. At least they are based on the same physiological principle—the persistence of vision—which was described before the Royal Society in 1824 by Peter Mark Roget (1779–1869).

Professor Joseph Plateau (1801–1883) of the University of Ghent in Belgium attempted to apply theory to practice by cutting slits at regular intervals in a disk which could be mounted over another disk containing pictures in progressively varying acts of motion. He held the disks in front of a mirror and slowly rotated the outer one. He had confirmed Roget's theory and had produced the first motion picture.

In 1875, the first television set was designed by G. R. Carey. It was based on the photoelectric effect of certain metals. If light shines on some metals electrons flow and produce a current. Carey's device consisted of a mosaic of selenium cells and each cell was connected by a separate wire to a small electric lamp. A set using such a principle would require millions of wires and electric lamps. But this idea of photoelectric metals and mosaics has value later.

Dr. Paul Nipkow (1860–1940) of Berlin put together the ideas of Plateau and Carey by devising a rapidly rotating metal disk with holes in it. This device was actually used by John L. Baird (1888–1946) in 1926 to demonstrate a complete television system. In other early mechanical scanning systems President Harding's picture was transmitted by telegraph line from Washington, D.C. to Philadelphia in 1923 and a transatlantic picture was transmitted in 1928.

Motion pictures as an industry had their beginnings in a little horseplay. Leland Stanford, founder of Stanford University, bet some of his horse racing enthusiasts that there were times when a running horse had all four feet off the ground and hired Eadweard Muybridge to prove it. Muybridge set up a series of cameras with their shutters connected to strings which were tied across the track. As the horse moved along it tripped the shutters and took successive pictures. The photographs proved that Stanford was right. In 1888 Muybridge went to Edison to see if he could not project the pictures and Edison nursed the industry into existence. He first produced the kinetoscope which was a peep show of a few hundred photographs attached to a wheel which turned. Each photo was lit by an electric spark for seven-thousands of a second.

Here Edison missed the boat's departure again, for the motion picture machine was invented by a pair of men—Charles Francis Jenkins (1867–1934) and Thomas Armat (1867–1948). Their device was called the Vitascope and was first exhibited in 1895 in New York. The first shows only lasted two or three minutes and were shown as part of the show at vaudeville houses all over the country in 1903–1904.

In the year Baird first demonstrated TV the first talkies were shown. They made use of the photoelectric cell and the vacuum tube of DeForest.

A slight digression seems in order to pick up several important technological discoveries—important not only to the development of television but also to pure science, the electric incandescent lamp, and electronics in general. Heinrich Geissler (1814–1879) devised a new type of vacuum pump and developed a technique for putting metal electrodes through glass. An evacuated tube with metal electrodes in each end is called a "cathode ray" tube. Three years after Geissler's invention of the vacuum pump, Julius Pluecker (1801–1868), professor of physics at the University of Bonn, noted that the green fluorescent glow produced in the tube could be shifted by a magnet.

The marriage between craftsmanship and cerebral gyrations has produced many a lusty child but perhaps none as bright as the various forms of Crookes' tube. Conceptually from the mind of Sir William Crookes (1832–1919) but the handiwork of his assistant and glassblower G. C. Gimingham.

In 1911, A. A. Campbell-Swinton, an Englishman, stated that if television were to become practical and of good quality, it would contain the key device of an electronic camera tube based on the Crookes tube. Following up on this idea were Ferdinand Braum of Germany and Allen Balcon DuMont (1901–   ) in the United States. World War I intervened and it was in the years following that the race for a camera narrowed down to two Americans—Valdimir Kosma Zworykin (1889–   ) and Philo Taylor Farnsworth (1906–   ).

Both men won the race but at different speeds and with different techniques. Zworykin applied for a basic patent in December, 1923 and it was awarded fifteen years later. The device was the iconoscope. It used a lens and focused the light mosaic of millions of tiny dots of metal on an insulator backing. This mosaic is scanned by a beam of electrons and the amount of current flowing from each dot is dependent on the amount of light shining on it. It is just like the Crookes tube with a magnet. The signal produced is then amplified and controlled so that the image formed in the TV set is almost the reverse of the camera. In the receiver the signal is converted into a current which regulates the intensity of the electron beam that scans the screen. The screen which is covered with a material, like zinc sulfide, glows when struck by the electrons.

In the years that followed Farnsworth patented over 100 devices basic to TV. Most of these devices were purchased by RCA. In 1929 Zworykin had become director of research at RCA. TV became the bauble of industries. The cost was too high for an individual. Farnsworth had thought $5,000 would be enough. It took $1,000,000 and thirteen years to get a complete system of television with a picture of good quality. RCA (Radio Corporation of America) spent ten years and $4,000,000 perfecting Zworykin's camera.

The depression had delayed TV because investors were reluctant to put their money into any scheme but in 1939 the first home television set was marketed by Allen B. DuMont (1901–   ). The large screens and the better quality of motion pictures had also delayed the development as well as the high cost of the first sets. On top of all these factors World War II began and the materials and skills were needed for the war effort. It was then 1946 before television became today's practical reality and about another ten years for the market to grow to size with sets that most could afford.

It seems strange that by using a glass tube filled with nothing and electricity which no one fully understands that such things as television and radio have come. Yet when Sir William Crookes was lecturing in 1879 on the nature of cathode rays he became somewhat prophetic: "We have actually touched the borderland where matter and force seem to merge into one another. I venture to think that the greatest scientific problems of the future will find their solution in this borderland, and even beyond; here it seems to me, lie Ultimate Realities, subtle, far-reaching, wonderful."

Just as there seems to be no accounting for some tastes, today there is no accounting without a computer. In the dim past knots in ropes and stones were used to count until someone put the two together to form the abacus. We need to collect several basic ideas at this point before we can put them together.

Charles Babbage (1792–1871), an English mathematician, became interested in a calculating machine to eliminate errors which were current in tables of logarithmics and trigonometric tables and hence introduced errors in computing in other fields, such as astronomy, as well as removing the need for making the arithmetic calculations in preparing tables. Although he labored for some years on the ideas, they were too far in advance of the technology of that day but his description of the stored program for his analytical engine was the basis for the stored program of the modern computer.

Another mathematician, George Boole (1815–1864) in 1847 developed an algebra of logic where statements could be reduced to equations and could be handled in one of two ways—true or false. In Boolean algebra, false can be represented as (0) and true can be represented as (1). Complex statements or complex circuits to represent statements can be managed in truth tables constructed with few symbols.

An electric circuit only has two states—on or off—and it is for this reason that Boolean algebra and binary digits are used in today's computers. On is a (1) and off is a (0). This same concept applies to the next development.

One of the founders of automation and input data for computers was Joseph Marie Jacquard (1752–1834) who developed an improved loom about 1745. One of the features of this loom was the use of holes in cards to control the selection of threads in weaving. Babbage applied this idea to his proposed analytical machine.

While working with the U.S. 1880 census, the widest topic census up to then, Herman Hollerith became aware of the need for a mechanical method of counting data. He developed 80 column rectangular punch card in the 1880's and his method was selected for the 1890 census. He organized a company which later became International Business Machines Corporation (IBM). James Powers also connected with the census designed a 90 column circular punch card and founded what became Remington Rand. It was quickly found that you couldn't put data of a rectangular nature in a round hole.

A huge mechanical calculator designed by Howard Aiken of Harvard in 1937 would automatically carry out a sequence of arithmetic operations. Built by IBM and given to Harvard in 1944, the Mark I, as it was designated, contained 72 adding accumulators and 60 sets of switches for setting constants. It was told the series of operations it was to perform using buttons, switches, punched tape and wire plug boards.

A year later, J. Presper Eckert and John W. Mauchly of the Moore School of Engineering at the University of Pennsylvania used electronic components to design ENIAC (Electronic Numerical Integrator and Calculator) often called the first electronic computer. It was faster than the Mark I and the forerunner of today's computers.

In the 1950's a group, Bell Laboratories, found that crystals of certain metals can amplify signals and can be made to produce oscillations. Thus meant that not only computers but all electronic devices could be made smaller, more compact. Vacuum tubes produce heat and burn out while transistors are cool and less apt to burn out. This use of transistors is what the electronics industry calls solid state.

STUDY AND DISCUSSION QUESTIONS

1. Who was the wizard of Menlo Park?
2. What practical instrument did Faraday invent?

3. What form of algebra is used in computers?
4. Who was the first to use punched cards? For what purpose?
5. What was De Forest's invention? Why was he prosecuted?
6. Why wouldn't Edison talk about radio?
7. Why was the Marconi Co. English?
8. Who was Alexander Popoff?
9. Who first produced Hertzian waves?
10. Why did Edison put a plate in his light bulb?
11. Why didn't Watson come when Bell called in 1915?
12. To whom does the honor of an invention go?
13. What is the "haven" function?
14. Why would the telegraph be an evil?
15. What was the principle of the telegraph and telephone?
16. How is electricity generated?

XXIX

Measuring the Unseeable

How does the scientist work with that which he cannot see? If he cannot see it how does he "know" that there is anything there? We "know" that air exists because of the wind and because if we wave our hand we feel something. In other words, we know of something which we cannot see by what it does to things that we can see or feel. We say that the tree moves because of the wind which we say is air in motion. In a similar manner the scientist can work with what he cannot see by what it does. There are many things in this category that the scientist works with. We are going to select only one of these things to examine to see what methods of direct and indirect "proof" that he uses to work with the unseeable.

The object we have chosen does not even have a name when we take up the story of our investigation. Our tale in putting the jigsaw together, and in many ways science does resemble the working of a jigsaw puzzle or the work of a detective, begins with evacuated tubes through which electricity is running. Even to get this far in the story depended on the works of many others in a number of countries. The separate stories of electricity and of the investigations of air and the air pump merge.

Johann Wilhelm Hittorf (1824–1914) was professor of physics and chemistry at Münster from 1852 on. In the middle of the nineteenth century it was still possible to be unspecialized to an extent. He was interested in working with an evacuated glass tube with an electric discharge. He noted that when the negative end of the wire was a point source and the tube was nearly evacuated so that the glass was brought to fluorescence.

"If, in such conditions, any object is set in the space filled with the glow, it casts a sharp shadow on the fluorescing wall by cutting off from it the cone of light which goes out from the cathode as an origin."[1] He thus concludes that since the shadow is sharp the rays must travel in straight lines. This same argument is also used for light as we shall see later.

It is some ten years later that Sir William Crookes (1832–1919) published his studies of the glowing evacuated tube. He too noted the sharp shadows cast by objects in the tube and noticed that the fluorescence of the glass is dependant on the type of glass used. "Most of my apparatus are made of soft German glass, and this gives a phosphorescent light of a greenish-yellow colour. English glass phosphoresces of a blue colour; uranium glass becomes green; a diamond became brilliantly blue."[2]

Crookes' most important observation, for our interest here, was that the rays which came from the cathode seemed to be deflected by the action of a magnetic field and that the stronger the magnetic field the greater the deflection. The fact that the ray can be bent by a magnet is a useful property that can be used in further investigations. It would seem too that if the ray can be thus bent then it must be made of some kind of particles. The conclusions that he makes are not entirely in agreement with the views of today.

The phenomena in these exhausted tubes reveal to physical science a new world—a world where matter may exist in a fourth state, where the corpuscular theory of light may be true, and where light does not always move in straight lines, but where we can never enter, and with which we must be content to observe and experiment from the outside.[3]

Eugen Goldstein (1850–1930), who became noted for his investigation of the rays from the positive end of an evacuated tube, named the rays coming from the negative end "cathode" although he and others maintained that the radiation was wave like. This argument of wave vs. particle nature of the cathode ray was to continue in the same way as the arguments concerning the nature of light. It was settled for a time by the work of Thomson and revived again toward the middle of this century.

Since they didn't know what they were working with the matter of a name was a little vague. Goldstein used cathode rays while G. J. Stoney suggested the name "electron" to stand for the natural unit of charge after it was discovered that the ray had a charge. On the other hand, when investigations began on radioactivity the rays coming off were named

Alpha, Beta, and Gamma (the first three letters of the Greek alphabet). It was later discovered that beta rays were electrons. All three names are still confusingly used.

The fact that the cathode rays were negatively charged was shown in 1895 by Jean Baptiste Perrin (1870–1942). It was suggested to him by the effect of a magnet on the rays. By means of an ingenious apparatus he made a circuit in such a way that the charge of the rays could be determined when they entered a cylinder inside a cathode ray tube. When a magnet was used to keep the rays out of the cylinder, it did not become charged but when the rays were allowed to enter, it became charged negatively.

Using much the same idea concerning the effect of the magnetic field, Joseph J. Thomson (1856–1940) was able to measure the ratio of the mass to the charge of a cathode ray particle. He figured that since a tube, or anything placed at the focus of the rays, was heated up by the rays that the number could be found by measuring the amount of heat produced by the collisions of the particles in the ray against the tube. If he used a uniform magnetic field he could field out an idea of the mass (or weight) of the particle. Just as a stone on a rubber band moves further out on a curve if it is swung about the head if it is heavier than another or if more force is used. From the heat the kinetic energy of the particles could be calculated, and from that the velocity of the particle. By then using some basic equations and making substitutions of values that he was able to measure he arrived at a value for these particles of 10^{-7}. The path that the particles took was not a clear line and the values ran between 1.1×10^{-7} and 1.5×10^{-7}. He repeated these experiments a number of times to cross check and as an additional safeguard he devised a different method making use of the charge, rather than the magnetic field for the deflection of the particles and a different principle for his calculations. In scientific work, if a cross check can be obtained from two different approaches there is more reason to believe that the data is correct. This is particularly true if a numerical value is being obtained.

Coulomb had shown in 1788 that the force of attraction or repulsion between charged bodies acted on the inverse square law principle so that it had long been established. Certainly the knowledge that opposite charges attract and similar charges repel was well known.

Using these facts plus his own work, J. J. Thomson tried using the settling rate of charged clouds of water in an attempt to determine the charge of a cathode ray particle. It was realized that if the charge

Figure XXIX-1. Thomson.

could be determined then there was a simple calculation to determine the mass of these particles. The chief problem was in getting the same shaped cloud of water each time.

About 1907, Robert Andrews Millikan (1865–1953) became interested in the problem and was able to devise apparatus which enabled him to make these measurements. In Millikan's method, droplets from an atomizer were allowed to fall through a small pinhole into a chamber consisting of two brass plates. These plates could be charged with electricity. The drop under examination could have its rate of fall measured between two cross hairs. Its fall with just the influence of gravity or with the influence of the charged plates could be measured. The plates could

Figure XXIX-2. Millikan.

have their charge reversed which would speed up or decrease the rate of fall. The amount of charge on the oil drop could be changed by placing a small piece of radium salts nearby. Charge was either knocked off or added in this way. In this experiment the particle itself is not observed but only what effect the particles have on a falling droplet of oil. It is assumed that one or more charged particles are on the oil drop. Millikan repeated the experiment many times not only with the same drop, by reversing the charge on the plates and causing the oil drop to rise, but also on a large number of droplets.

In the table that follows, part of his observations are shown for one such observation.[4]

$e_n \times 10^{10}$	n	$e_1 \times 10^{10}$
34.47	7	4.923
39.45	8	4.931
44.42	9	4.936
49.41	10	4.941
59.12	12	4.927
53.92	11	4.902
68.65	14	4.904
83.22	17	4.894
78.34	16	4.897
24.60	5	4.920
29.62	6	4.937
19.66	4	4.915

The first column represents the amount of charge on the oil droplet, (e) stands for the electrical charge of the electrical particle and (n) for the number of such charges. To get the second column Millikan had to try different sets of values because he wanted the charge to be the same for each particle. In the third column what he expected seems to be borne out that the charge is the same for each particle.

[He wanted] to present direct and tangible demonstration . . . of the correctness of the view advanced many years ago and supported by evidence from many sources that all electrical charges, however produced, are exact multiples of one definite, elementary, electrical charge, or in other words, that an electrical charge instead of

being spread uniformly over the charged surface has a definite granular structure, consisting, in fact, of an exact number of specks, or atoms of electricity, all precisely alike, peppered over the surface of the charged body.[5]

The slight differences in the second and third decimal places are explained as experimental error and does not affect the general thesis of uniform and granular charge. This charge was further modified in other units as 1.60×10^{-19} coulomb for the size of the negative charge of every electron.

This experiment of Millikan's stimulated work to obtain values about other things which could not be seen. The mass of the electron could be obtained and this is usually given in terms of the weight of the hydrogen atom—it is 1/1837 times the weight of the hydrogen atom. It was also found that the electron has a diameter of 10^{-12} centimeters. From this approach it was realized that removing a charge from the hydrogen atom and thus giving it an opposite charge that it would be possible to bend a ray of such particles and by so doing to measure the weight of the hydrogen atom.

A question left by this experiment concerns the splitting of the realms of the very large from the very small. An important premise from the time of Newton was that what was discovered about nature in one size was applicable to all sizes. In other words, that nature was uniform and that rules which apply to atoms apply to stars. It is commonly known that no two things are exactly alike in the world that we can see. Fingerprints are all different, snowflakes all have different patterns—everything that we know of in the large world differs. In Millikan's work and subsequently it is important to electronics and to the structure of the sub-atomic world for all electrons to be exactly alike. If this is the case then there is no uniformity in nature and two sets, at least, of rules apply in nature. One set for the very large and one set for the very small. But the uniformity of nature is a basic assumption that the scientist makes in order for experimental work to make sense. It seems that the scientist is also caught up in a dilemma of whether to keep a uniform world as an assumption or to assume that there are two sets of rules as demanded by having all electrons and atoms alike.

STUDY AND DISCUSSION QUESTIONS

1. How can we know about what we cannot see?
2. What do we know about the electron from the experiments?
3. Why does Millikan use oil instead of water for his drops?

4. Why do we use cathode rays, beta rays and electrons as terms for the same thing?

5. What work does the determination of the charge of an electron stimulate?

6. What are the implications if the charge of the electron is the same for every electron?

FOOTNOTES

1. The quotation is a translation from "Keber die Elektricitätsleitung der Gase" which originally appeared in the *Annalen der Physik and Chemie,* Vol. 136, p. 1, 1869 and is cited in William Francis Magie (ed.) *A Source Book in Physics.* Harvard University Press, 1963, p. 563.

2. *Ibid,* p. 564. Although the original article is published in the *Philosophical Transactions, Part I,* p. 135, 1879.

3. *Ibid,* p. 576.

4. *Physical Review,* 32 (1911), p. 349.

5. *Ibid.*

Suggested Reading:

Morris H. Shamos (ed.) *Great Experiments in Physics.* Holt, Rinehart and Winston, 1959. pp. 238-249.

R. A. Millikan *Autobiography.* Prentice-Hall, 1950.

R. A. Millikan *The Electron.* University of Chicago Press, 1924.

XXX

We Get a Charge out of It

We have to an extent sized up the electron without taking its measure. In a larger sense, for the industrialized world of western Europe, Japan and the United States, the electron is Mr. Genie bottled in a Leyden jar to be aged and served in the twentieth century as a greater class leveler than gin or communism. It has proved to be both a liberator and captivator of mankind.

Any discovery is never an unmixed blessing as can be illustrated by one of the earliest comments about the earliest practical application of the electron. "The electric telegraph, if successful, would be an unmixed evil to society; would be used only by stock jobbers and speculators—and the present Post Office is all the public utility requires."[1] Ten years later lines had been established between Washington, D.C. and New York. By 1851 there were fifty telegraph companies in this country. Five years saw the consolidation of these and other companies into Western Union.

The Western Union killed one colorful part of the West—the Pony Express—but aided in reweaving the fabric of a nation torn asunder by conflict even as it had been used to bring about the reunion by fast communication of military information. The telegraph was used in the stock ticker by speculators as had been predicted and many people are still living who can recall that the telegraph was the harbinger of bad news. Death notices were frequently sent this way and the Defense Department still transmits news that, "We deeply regret to inform you . . ."

The telegraph was the life blood of the penny press and made possible the growth of the great newspaper kingdoms and the foundation of "yellow" journalism with the power to mold public opinion and to create wars. The reading public were drawn closer to national events—political, military, catastrophies— and they could react more quickly to them. After 1866 when Cyrus W. Field laid the Atlantic cable an umbilical cord was reestablished with England and was probably responsible for our entering WWI on the Allied side. It provided a means of rapid diffusion of propaganda as well as other government uses. The world was beginning to shrink and the wars of Europe were reported as they were happening instead of weeks later.

One of the earliest large users of the telegraph was the railroad in order to control traffic and to prevent wrecks. Intelligence about track conditions or hot boxes could be wired to a station and the affected trains flagged. Many trains were saved from collisions, derailments, and fires by such tactics.[2]

Still another use for the telegraph was to speed money, the financial life blood of the nation, from coast to coast. This made capital much more fluid and aided the growth of our country in the boom years following the Civil War. This device is still used—by great corporations and by individuals—when getting money with rapidity is necessary.

To us of the twentieth century it may seem merely a natural extension of the telegraph to the telephone but in 1876 when Bell received his patent the London *Times* reported it as the "latest American humbug." Few people had any faith in the idea and Bell was frequently hard pressed for money. In the first years after the birth of the telephone a half interest could have been bought for $2,500 which within fifteen years was worth $10,000,000. There were 800 phones in use in 1877 which grew to over a million by 1900. By 1940 telephones in the world numbered forty-five million with nearly half in this country. Today there are more telephones than families in this country and we have three quarters of the phones in the world.

The telephone placed instant communication into the hands of a large number of persons. In times of emergencies—fire, sickness, crimes—aid could be summoned rapidly. (The phone gradually replaced the fire box which was a form of telegraph.) It would, however, be an error to assume that the masses had a telephone available early.

In the Presidential election of 1948, a Gallup poll taken by telephone showed that Dewey would receive 49.5% and Truman 44.5% of the total vote. Such reliability was placed on polls that the *Chicago Tribune* ran a headline election night "Dewey Defeats Truman." History records not only Truman's victory but also his cackle. The error made was in assuming that telephone listings were a true cross section of the American public. The majority of telephones were of subscribers in the upper class and the upper middle class that are traditionally Republican so that the pollsters missed the large Democratic vote. Indeed many families did not own phones until after WWII when the family finances had recovered from the depression.

Before the days of the radio soap operas, real life dramas could be tuned in on the party line. Most of the subscribers were on party lines of eight to sixteen and a call for one was a call for all. The audio peeping Tom could be identified, at least to the number of them, by the number of clicks as receivers were lifted up and down the line.

Lifting your receiver and turning a crank put you in touch with the local operator (could this have been the origin of the phrase "crank calls"?) The local operator in a small town was the source of help, advice and frequently of heroism. Histories record the efforts of operators during catastrophies and personal emergencies that were unrewarded and beyond the call of duty. Today even where there is an operator the person may not be physically located in the city and only too frequently you may be talking to a recorded message.

The introduction of the telephone could also be an annoyance as well as an aid. Wrong numbers may interrupt bath or nap. The subscriber becomes a victim to commercial intrusion undreamed of even in the day of the door to door salesman. Only recently has answering services and automated answering devices been introduced to stand between subscriber and annoyance. With any luck the subscriber has a teenage daughter or wife who may prevent him from ever using the phone at home and he need not answer because it's not for him.

The institution of "calling" (when you dropped in on friends or they dropped in on you) has gradually been driven from our lives as today we are expected to use the telephone to arrange a call. This also applies more and more in a world of business that increasingly leaves the individual out in the cold who has not called for an appointment. Drop in and see us sometime has become synonymous with "don't call us we'll call you" and a friendly American tradition has fallen before the advance of technology.

Long distance calling is gradually replacing letter writing. In the 19th century, when London had eleven daily mail deliveries, letter writing was quite popular. Today the pen has become supplanted by the phone. The long range effects in terms of the lack

of material for biographers and historians of the future cannot be foreseen, but it's going to make a hard job harder. On the other hand, the use of dictaphones, tapes and records make it possible to preserve the spoken word which may give rise to a new kind of history.

Besides the telephonic invasion of one's home is the additional possibility of the further invasion of privacy by phone bugging. This is, of course, not new as has been previously noted with the party line or with a nosey operator but in neither of these cases is the intent to spy on you. They would spy indiscriminately on anyone while a person bugging the phone is after you. It may be the law (presumably with a court order) or an industrial spy, or a private detective seeking divorce information, an individual wanting information to use for blackmail or any other person.

In the early days of the telephone there were cases where you knew someone listened in. If you subscribed to one phone company and your friend across the road subscribed to another, the only way of communication was to call someone who was a subscriber to both companies and have your message relayed. This was frequently the general store in town which provided many services not directly concerned with profit.

The telephone generated a number of song hits on the new gramaphone: "Hello Central Give Me Heaven," (Central was the operator) "Mr. Cohen on the Telephone" which was a comedy skit—this device is still used on a number of records, TV shows, radio, etc., "All Alone by the Telephone,"[3] and many others. The last tune perhaps adding to the already noted observation of many that even in a crowd one can be all alone—perhaps more alone than on a desert island. The telephone connects the world yet cannot connect people. Technology cannot replace communication technics.

Both the telephone and the telegraph were new devices and did not replace items already in use. Such was not the case with the electric light. Homes had candles and kerosene lamps. Those in the cities had gas lights that were cheap and dependable. Some experiments had provided electric arc lamps but these were expensive and complex as well as requiring constant attention. What was needed was a steady light that was simple, of light weight, and flexible. One which would work under all conditions and in all positions (with a noiseless and inoffensive flame). Above all it must be cheap.

Edison began experiments with an incandescent lamp in 1877 and was successful within two years. No match was needed to light it nor was there blaze. It was whiter and steadier in its light. It did not give off obnoxious fumes nor smoke and thus will not blacken ceilings or furniture.[4]

In 1882 there was one power plant in the United States serving a few blocks of New York with arc lamps and a few motors (1883). In 1900 electricity provided 0.1 horse power per worker in this country while by 1927 it was three horse power which was three-fourths of all power used by industry.

Homes built about 1900 were supplied with both electricity and gas. The electricity was still a fad and used for show, but by 1927 about 17 million homes were equipped with electricity.

Jefferson had a dream which he expressed in a letter to George Fleming in 1815 for a small steam engine which would liberate the housewife from various onerous chores. In his vision it would "raise from an adjacent well the water necessary for daily use; to wash the linen, knead the bread, beat the hominy, churn the butter, turn the spit, and do all other household offices which require only a regular, mechanical motion."[5]

Oliver Evans, in a similar vein, stated that steam could be so applied that ". . . the very old women would do the common business of housewifery with it . . ." and indeed it would become ". . . as a part of its (the house's) furniture, to wash the clothes, scrub the floors, etc."[6] While by 1853 the *Scientific American* had gone further to state:

> The farmer wants them to thresh his grain and cut his straw, to saw his wood, and . . . to draw his plow. The mechanic wants them . . . and we may hope to see the day when they will become almost a necessity of the household. The world is growing wiser and lazier every day.[7]

Only a part of these dreams were to come to pass using steam. The remainder of these dreams had to wait for the harnessing of electricity for their fulfillment.

At the Columbia Exposition in Chicago, Illinois (1893) there was a demonstration of ironing and cooking electrically. The vacuum cleaner was devised in 1901 and the washing machine was electrified in 1907. In general the use of most of the devices did not spread until after WWI when electrification became more widespread. By 1928 there were four and a half million washing machines and they were third in popularity with the flatiron and vacuum cleaners.

Electricity, suffrage and the pill has produced the 20th Century emancipated female.

The question I would raise is whether woman has been liberated by all of the electric contrivances? Let us look at some of the aspects of the pre-electric middle class home at the turn of the century. The rural idea of a large family was still dominant when the family became urban. Although the woman of the house did not have everything electric, she did have help in the form of cheap labor provided by young immigrant girls to do the heavy work. She could also send out her washing and ironing and thus provided an honest living for widows who took in wash. Almost every home had a maiden aunt or a grandmother who helped with the work, supervised the hired girl, and had a ready ear for the woes of the children while she worked. However, as the home became increasingly urban a large family became a detriment and the move was towards smaller homes. No longer was there room at home for the maiden aunt or grandmother. After WWI it became increasingly acceptable for women to work in factories so that there was no longer a cheap source of hired girls for the home. In addition with all the handy electrical helps what need had the wife for hired help? If she is ill there is no electrical device, as yet, that runs the house. Even assuming that suddenly this electrical marvel was available, who is to care for the children? The suburban housewife of today has become a slave. Without extra hands to help, she acts as a taxi driver to take her husband to work, to take the children to their various activities, to shop. In the pre-electric days there were delivery boys who brought the ice, groceries, etc. and various door to door salesmen to save the wife.

Electric lights made a new world in both the home and factory, as indicated by this 1932 poem.

Through the long night, so still and dark,
  While all the world in silence slept,
  A man and boy their vigil kept,
By a new light in Menlo Park.

Glass formed its globe, and cotton thread
  By his own fingers shaped and bent,
  Served as an inner filament,
While wires from jars the current led.

Strongly it burned. So clear and bright
  It shed its rays, and on and on—
  Could it but last until the dawn,
And mark the conquest of the night?

Thus watched they on that upper floor,
  Silent, its fate yet unknown,
  Nor did they spend those hours alone—
A waiting world stood at the door.

A world whose day had just begun,
  Fifty and three long years ago,
  Nor guessed the blessings soon to flow
From that Lamp made by Edison.

Nor was the glow on that dim stage
  Revealed alone by his new light,
  A brighter radiance glowed that night,
The dawning of a nobler age.

WM. A. SIMONDS[8]

In many respects the housewife does not live better electrically. While release from some of the tasks has been brought about she is none the less a slave to the house and is without help. If she works, the children grow up unsupervised in a bedroom suburbia into which we have been lured by the promise of a better tomorrow.

Widespread electric service has liberated industry from the large urban centers. Just as the age of steam built factory towns belching black smoke so the age of electricity enabled industry to choose on the basis of other factors. They could locate based on markets, availability of transport, or of labor force and raw materials. Increasingly smaller towns were picked for the better living conditions.

Not only did electricity reduce costs by eliminating the boilers and engines but also the factories were cleaner and lighter places to work. Since a belt drive was not needed the buildings could be built of lighter materials which reduced initial cost. Cost was also reduced because the operation of one machine with steam cost almost as much as with the whole factory running while electricity could run one motor without the rest of the factory running.

Another application of the nobler age was to the wireless telegraph which was developed into radio. One of the first applications of the wireless was seen off the coast of South Foreland, Great Britain on March 3, 1899 when the lightship *East Goodwin* was run down by an off course steamer and those doomed to a watery grave were saved by sparks in the air. Many a home in New England still has an architectural feature called the "widow's walk" to bear witness to the fact that once a ship sailed they were not seen or heard from until they reached port. The wireless changed all this. Distress calls could be sent and received, daily time signals are sent (these are important for navigation), frequent weather reports and storm warnings could be sent or received, and the crew and passengers had a means to communicate with the shore. The poisoner Crippen was apprehended by the wireless when in an earlier age he might have got away. In 1912, news of the *Titanic* sinking was flashed almost as soon as the S.O.S.

The first radio broadcast of Christmas Eve 1906 was followed in 1910 by Lee De Forrest broadcasting a program carrying the voice of Enrico Caruso. Regular broadcasting began in 1920 with station KDKA of East Pittsburgh, Pa. By the end of 1921 there was a total of eight stations which led many to seek profits from the manufacture and sale of receiving apparatus; and others soon saw the value of radio advertising which soon became the chief means of support for broadcasting in the United States.

Radio is the miracle of the ages. Aladdin's Lamp, the Magic Carpet, the Seven League Boots of fable and every vision that mankind has ever entertained, since the world began, of laying hold upon the attributes of the Almighty, pale into insignificance beside the accomplished fact of radio. By its magic the human voice may be projected around the earth in less time than it takes to pronounce the word "radio."[9]

Just as religion in the Middle Ages promised a better life in the after life for the common man as an attraction, radio's soap operas provide an escape now from one's own problems into a far more active life. Never again would man be alone. Radio broke the Great Silence to provide contact with all the world for all mankind. One of the greatest hardships that the colonists, arctic explorers and pioneers suffered was loneliness. Perhaps too, radio may be the reason no one hears the small inner voice or the voice of God anymore. Silence has become rare. Radio may have been the reason that the twenties were roaring.

A man's radio image became important in advertising and politics. Radio provided a means of getting the news around the country and also of the importance of propaganda to sell anything to anyone. Man was captured by the little black box.

You little box, held to me when escaping
So that your valves [vacuum tubes] should not break,
Carried from house to ship from ship to train, [today, make it car and plane]
So that my enemies might go on talking to me
Near my bed, to my pain
The last thing at night, the first thing in the morning,
Of their victories and of my cares,
Promise me not to go silent all of a sudden.[10]

The phonograph and the radio are the great leveler in the realm of music for with them the common man can command any orchestra or vocalist living or dead to perform for them wherever they chose to be and at any time. Even the greatest Emperors of the past did not have this degree of control. A dramatic performance could have an audience which no amphitheater could hold.

Other services which radio could provide include bringing the classroom, the church, musical and sporting events to the individual. This service is, however, a two edged one. On the one hand it brings a service to the isolated individual, and on the other, it isolates the individual because he no longer needs to go where the action is but can have a second hand participation at a distance.

Now that we have had a word about instantaneous stupidity, let us scan visual vulgarity. Television has often been called the Great Wasteland and while I will not challenge the definition, I may modify it.

Mass television is a product of the last half of the twentieth century. Only following WWII did the forest of antennas rise over urban areas. From the first marketed sets in 1939 to the more than one set per family of today the industry has grown almost exponentially. Almost everything said of radio also applies to television with some exceptions that we will examine.

A cartoon showing a man and woman in front of a TV set and he turns to her and asks, "What did we look at when we listened to radio?" The similarity between a topless waitress and television is unmistakable—neither leaves anything to the viewer's imagination. Like the charge brought against many of the new toys of today, television, unlike radio, requires no imaginative skills. It tends to accent two dimensional thinking and stifles the mind in its ability to create. Too frequently the stories written by children in the early grades are merely carbon copies of the shallow formulas used in most television script writing.

As the home has increasingly become centered about the boob tube, as evidenced by TV trays, TV dinners, and snack pacs, conversation has languished. The various portions of the captive audience include: cartoons for children, sports for father and soap operas for mother. Wives complain to the sob sister columnists that they lose their husbands to football Saturdays and New Years, but perhaps television saves marriages by filling a conversation gap for two persons who have nothing to communicate or who fear to. It may stave off the concept of suburbia being a bedroom community, as created by the automobile, by drawing the family members home.

For many persons TV, like radio, fills the silence, provides background noise, or blocks unwanted sounds. For others it is an effortless escape. "... it is a tale told by an idiot, full of sound and fury, signifying nothing."

As the movies killed vaudeville and the legitimate theater, so television has killed the movies and revived vaudeville under the title of variety shows. As a part of economic democracy television gives the public what it wants and those who complain about either TV or politicians should keep this in mind. The man regarded by many as the father of television, Vladimire Zworykin, said in an interview that he seldom watched it nor did it live up to his expectations. But the creator and the critics do not make up the bulk of viewers. There are occasional signs that the public is becoming more sophisticated in its taste, or at least questioning. In an issue of *TV Guide* there is a series of questions by one watcher and responses by a critic. She raises questions such as why is "Peter Pan" good and "My Mother, the Car" is not?[11]

It might also be noted, for the benefit of critics, that one showing of one of Shakespeare's plays had a larger audience than the total number of people who have seen all of the plays since the time that they were written.[12]

When the vacuum tube was applied to radio, television, and other fields, the age of electronics was born. Automation, in the sense that I am going to use it, becomes possible only with electronics. The concept of automation is more than an automatic machine. The iron spheres on Watt's steam engine which controlled the rate at which the engine worked would contain the elements of the way in which some would define automation. The main idea is that a mechanical, chemical, business, or other pursuit of man is wholly or partially controlled within certain predetermined bounds by partially electronic apparatus without the need of man's interference once man has established the bounds within which the process will operate. In a number of instances it was the substitution of the transistor for the vacuum tube that made components small enough to fit the machine system.

The idea of automation, naturally, goes back to the Greeks when Aristotle said, "When looms will weave by themselves, then the slavery of mankind will end." It is interesting that one of the fields where automation was applied was in the textile industry where punch cards were used to indicate weave patterns.

Conversion or design towards automation is driven by desires to cut labor costs or to do jobs that man cannot do.[13] The crowning achievement for automation would be the automatic collection, digging, or production of raw materials which would then automatically be loaded on carriers, which when filled, would automatically proceed to the factories where they would be automatically unloaded and automatically machined, fabricated, or packaged for the consumer; which would then automatically be loaded on automated carriers to proceed to retail establishments; there to be automatically unloaded and placed into receptacles which could receive a signal generated by cash or credit card of the consumer; the signal would trigger a mechanism which would load the product automatically aboard a local automated delivery service. In other words, the consumer would receive goods which had been untouched by man.

From the worker's side, if he works in a semi-automated plant, he has all of the problems that the factory system entails plus a few more. The concept of mass production has a number of characteristics that create problems for the worker. These are that the machine sets the work pace and is uniform without variation; the skill level is low as is the amount of mental energy; there is a repetition of the same task using the same tools and techniques that the worker does not choose; only a small part of the total product is the responsibility of the individual worker.[14] These imply certain effects on the worker. Since the machine sets the pace, the individual worker must subordinate his own rhythm to the unvarying one of the machine. The worker makes no use of his ingenuity, personal choice, or his experience and judgment in his repetitive tasks. Even his bodily needs must be subordinated to a semi-mechanical and scheduled system of relief. Possibly the main effect which with automation tends to increase is the feeling of loneliness, lack of personal relation to the impersonal factory and the submergence of the individual in the crowd.[15] With automation too there is the substitution of increased tension or mental effort for muscular fatigue,[16] which is much more tiring and leads to cardiovascular problems and ulcers as well.

For industry and business automation means: "increased productivity, uniform quality, better control of the flow of production, and reduced running costs."[17] The office worker, since the application of automation, has become a worker in much the same sense as the factory worker and has many of the same problems.

Automation means, at least at the outset, the danger of overproduction, an increase in unemployed, fewer work hours with attendant increased leisure, increasing isolation for the worker, and an increase in required skilled labor for very complex machines to mention but a few of the areas in which problems have or are developing. We are still a long way from the realization of freeing man from slavery.[18]

Footnotes

1. Dr. Burbeck, of the Mechanics Institute, England, 1837 as quoted in Lloyd William Taylor *Physics The Pioneer Science* Houghton, Mifflin Co. 1941.
2. See also Marshall McLuhan *Understanding Media* McGraw-Hill 1964 pp. 217-227 (Signet ed.) and Lewis Mumford *Technics and Civilization* N.Y., Harcourt, Brace & World 1934 p. 199.
3. Additional effects of the telephone may be found in McLuhan *op. cit.* pp. 233-240.
   On phone bugging see John G. Burke *The New Technology and Human Values* Wadsworth Publishing 1966 pp. 253-26.
4. Francis Jehl *Menlo Park Reminiscences* Edison Institute Dearborn 1937 p. 217.
5. Cited in *Early Stationary Steam Engines* by Carroll W. Pursell, Jr. Smithsonian Institution Press, Washington, 1969, pp. 124-125.
6. *Ibid.*
7. *Ibid.*
8. Jehl *op. cit.* p. 358.
9. Gleason L. Archer *History of Radio to 1926* American Historical Society Inc., Chicago, 1938 p. 3.
10. Berthold Brecht as quoted by McLuhan *op. cit.* p. 260. Also see rest of section pp. 259-268.
11. *TV Guide,* March 19, 1966 pp. 6-9.
12. McLuhan *op. cit.* pp. 268-294 contains a number of views of TV.
13. Lewis Mumford maintains, in his book *The Myth of The Machine,* Harcourt, Brace and World 1966, that man is the central figure and that all aspects of man's relationship to the machine, whether a stone ax or automation comes from man's own characteristics and not the machine.
14. Robert H. Guest, "Men and Machine: An Assembly-Line Worker Looks at His Job," *Personnel,* 31:496-503, May, 1955 as cited in Charles R. Walker *Modern Technology and Civilization,* McGraw-Hill 1962 p. 98.
15. C. R. Walker et al., *The Foreman on the Assembly Line,* Harvard University Press 1956 pp. 121-125 as cited in Walker *op. cit.* p. 109.
16. C. R. Walker, "Life in the Automatic Factory," *Harvard Business Review,* 36:111-119 Jan-Feb 1958. Cited in Walker *op. cit.* p. 204.
17. Melvin L. Hurni, "Decision-Making in the Age of Automation," *Harvard Business Review,* 33:49-58 Sept-Oct 1955 as cited by Walker *op. cit.* p. 162.
18. For further insight in these problems see Burke *op. cit.* pp. 31-62, 105-208, 243-296, Delbert A. Snider, *Economic Myth* and *Reality* Prentice Hall, 1965 pp. 61-71, John T. Dunlop (ed.) *Automation and Technological Change* Prentice Hall, 1962 Lewis Mumford *Technics and Civilization op. cit.* pp. 321-435, and McLuhan *op. cit.* pp. 300-311.

XXXI

Light—A Duet?

I had thought of calling this the charge of the light brigade but then one should not make light of such an important topic. Many phenomena were observed by man about light almost from the time he became man. The sun was seen as the prime source of light and worshiped by different peoples. The beauties of the rainbow were long noted and the rather practical observation that one did not aim for a fish when spearing it. It was in early times too that light became associated with good and dark, its opposite, with evil or the dark powers.

It is indeed strange that those who would study light must always be in the dark. Is all knowledge merely a study of contrasts? Can good be known without evil, light without dark, sound without silence, motion without rest . . . ?

Again, as far as the Western World was concerned, it was the Greeks who began to formalize studies about light (optics). The basis of optics was the observation that light "rays" seemed to be straight coupled with the application of Euclidian geometry and Aristotelian logic. In the fourth century B.C. Aristotle had worked out a theory for rainbows, including the geometry, and had logically decided that the speed of light was finite.[1] That is that it took time to get from one place to another. Empedocles also taught this idea. There were even some, such as the Pythagoreans, who held that light was made up of small particles. Plato, on the other hand, believed that light came from the eyes. The Greeks were not able to cast much light on the subject.

Little of real worth was contributed by the Hellenistic Greeks and the accumulated knowledge passed to Islam scholars. In the tenth and eleventh centuries considerable advance was made especially by (Alhazen) ibn al-Haitham (965–1038) in his great work *The Treasury of Optics*.[2] These works were translated into Latin in the twelfth century.

Works in the late Middle Ages on optics were done largely by the Franciscans and we might deviate here for some reflection on the matter.

St. Francis of Assisi (1181/1182–1226) after a serious illness in his twenties decided to follow the footsteps of Christ. He denied himself everything and others, inspired by his way of life, followed his way so that he finally founded (1209) the Franciscan Order, the Grey Friars. This was approved by the pope in 1223. The poor and all of God's creatures were his friends. In 1224 he went up Mount Alverno in the Apennines with some of his disciples and after forty days of fasting and prayer on September 14 he had a vision. *In the warm rays of the rising sun* he beheld a seraph which flew towards him and filled him with pleasure. Suddenly a cross appeared and the seraph (a kind of angel with six wings) was nailed to it. After the vision was over Francis felt pain mixed with the pleasure and found the Stigmata—his body had the same marks as Christ's had had. The following year he composed *Cantio del Sol,* The Canticle of Brother Sun, which reads in part: "Praised by Thou, my Lord, with all Thy creatures, *especially* Sir Brother Sun, who is the day, and *Thou dost give us light through him."*

This passage is illuminating as to why it was the Franciscans who took the lead in the study of light. This urge is of course hooked to the translations of the works of Aristotle and Alhazen which provides the background for these studies.

Among these Franciscans were Robert Grosseteste (c.1175–1253), Bishop of Lincoln, and his pupil, Roger Bacon (1214–94) who experimented with mirrors and lenses and knew the nature of refraction.[3]

Between the thirteenth and seventeenth centuries little seems to have been done in the realm of light, or indeed in any of the scientific endeavors.

It is perhaps ironic that two of the leading figures in optical thought should have had that work suppressed in the modern mind by their work on laws of motion. Johann Kepler whose first work on optics, *Ad Vitellionem Paralipomena,* was published in 1604 in Frankfort and Isaac Newton whose *Opticks* was published exactly one hundred years later were both religious mystics who have been stripped of mysticism for an age of science.[4]

Johann Kepler is regarded by many as the father of optics after the publication of his second book on optics, *Dioptrice,* (1611), stimulated by the discovery of the telescope (1609). Kepler felt that the passage of light must be instantaneous[5] while Galileo, after his experiment in attempting to time a light flash between two mountains believed light required time for its passage.

Rene Descartes (1596–1650) in 1637 published his *La Dioptrique* in which he agrees with Kepler that light is instantaneous in transmission but makes the light particulate to agree with his ideas on vortices. He expects that light will obey mechanical laws and used a tennis ball as a mechanical proof for the laws of relection and refraction. In his analogies he is led to the statement that light travels more easily and faster in a dense material than in a rarer even though this contradicts the idea that light moves in an instant.[6]

A new phenomena with light was observed by Francesco Maria Grimaldi (1618–1663) which was reported in his collected studies published in 1665 as *Physico-Mathesis de lumine, coloribus, et iride.* In this work he reported his evidence which seemed to contradict the fact, accepted since the early Greeks, that light travels in straight lines and that it was a particle as Descartes had suggested. He let light enter a dark room through a small hole. Between the hole and the wall he placed an object and then examined the shadow that the object cast on the wall. He found that the shadow was wider than it should be if light traveled in straight lines and further that the border of the shadow had colored bands along its edge. He explained this by assuming that light was a fluid with a wave-like motion.

Thirty years after Galileo's death, Olaus Römer (1644–1719) was observing the eclipses of the innermost moon of Jupiter, Io, which revolves about the planet every 42 and a half hours. He found that the revolution of the moon varied over an earth year. How could this be? One way in which Römer could have explained the phenomena was by assuming that the velocity of Io varied. However, he chose to explain the variation by assuming that the speed of light was finite and that the difference in time was caused by the difference of the earth's distance from Jupiter during a year. From this assumption he calculated that the speed of light was 48,000 leagues per second, (this would be 193,120 km. or 120,000 miles). The view that the speed of light was finite was rejected by the Cartesians (followers of Descartes).

The observations of Grimaldi and the proof of the finite nature of light by Römer were synthesized into an elaborate theory of the wave motion of light by Christian Huygens (1629–95), a Dutch physicist and

Figure XXXI-1. Huygens.

son of a poet. Huygens published his theory in his *Traite de la Lumiere* in 1690.

The minute particles, which Huyens imagined, making up a light source transferred impulses to a medium which was everywhere. This medium, called the aether or ether, was made up of small, hard, totally elastic bodies which transmitted the impulses but remained itself unmoved. (In a sense these particles were very similar to Descartes concept of matter.) The light traveled in the form of waves and wave fronts. Each point on the wave-front acts as a center of a new small wave. (Huygens' principle) Using this theory he could explain all light phenomena then known.[7] This included light traveling in a straight line (rectilinear propagation), reflection, refraction (bending of light), diffraction (Grimaldi's observations), and double refraction in calcite which had just been discovered. His theory on refraction assumes that light travels slower in a transparent material medium than in a vacuum.

At about the same time that Huygens was doing his work in France, Isaac Newton became interested in improving telescopes (Newtonian Telescopes as they were later called) and had bought some prisms in hope of eliminating chromatic aberration (the colored fringe around objects). At first Newton agreed with the idea of wave propagation but later he found that he could not make it explain light traveling in a straight line and he therefore rejected it. I suspect that as Newton grew older he was also influenced by the idea of making all information fit the grand scheme of motion which idea would be best suited by a corpuscular theory of light. In his *Opticks* of 1704 and following editions, he denies the use of any theory but when explaining particular phenomena he refers to light in a corpuscular form. The great prestige and fame of Newton was enough to suppress the wave theory of light for a time.

Perhaps the first strong challenge to Newton's pontifications was that of Thomas Young (1773–1829) who at the turn of the century presented the evidence of his now famous double-slit experiment. In this experiment he allowed light of the same color (monochromatic) to fall upon two narrow slits. The two light sources thus created were allowed to project on a screen. A series of light and dark bands were produced on the screen. Young reasoned that the light was behaving like waves which reinforce where they meet in phase and cancel out where they meet out of phase. The out or in-phase is produced by the difference in distance between the two light sources and the screen.[8] This seemed to strongly support the undulatory theory of light.

Additional support in the argument was supplied within a few years by Augustin Jean Fresnel (1788–1827). Fresnel devised a set of equations against which elaborate experiments were tested. With these equations and experiments, which were published between 1816 and 1826, he established the undulatory theory of light.[9]

In 1849, Armand Hyppolyte Louis Fizeau (1819–1896) sent a beam of light a distance of 5.39 miles between two mirrors and a toothed wheel. By rotating the wheel at a fast and uniform rate he was able to chop the light beam into small pieces. In finding the point at which the light seemed to be at a maximum he could calculate its speed. (Somewhat the way in which a timing light is used in a car.) Using a rotating mirror Jean Bernard Leon Foucault (1819–1868) was able to reduce the distance 65.5 ft and could also place a tube of water between the reflecting mirror and the rotating mirror. In this way he established that the speed of light was less than in air.[10] This fact also supported the wave theory of light.

James Clerk Maxwell (1831–1879) in 1864 developed a mathematical theory which pulled together and explained all of the then known phenomena of light. The equations he developed also related light, electricity and magnetism—the whole electromagnetic spectrum as it was to be called and predicted the discovery of radio waves.

In 1887, Hertz confirmed the existence of the waves which Maxwell's equations had predicted. He did notice a slightly peculiar phenomena when he produced the waves. When the light from the flashing sparks of his transmitting apparatus shone on the open ends of his receiving loop, the sparks in the space between the spheres seemed to come slightly more readily.

With Newton's equations of motion to explain the behavior of matter and Maxwell's equations to account for the actions of energy it seemed that the pinnacle of man's power to reason over and eventually to control the forces of nature itself had been reached.

"Here indeed was reason to be proud. The mighty universe was controlled by known equations, its every motion theoretically predictable, its every action proceeding majestically by known laws from cause to effect."[11]

FOOTNOTES

1. George Sarton *A History of Science,* Vol. I, Harvard University Press 1959, p. 518. See also Aristotle *Meterology,* Book III Ch. 1-3.
2. For an expansion of Alhazen's work see Charles Singer *A Short History of Scientific Ideas to 1900,* Oxford Press 1959, pp. 152-153.
3. *Ibid,* p. 170-172. Also see A. C. Crombie *Robert Grossetests and the Origins of Experimental Science* 1100-1700, Oxford 1953, Chs. V, VI, & X.
4. For the effects of Newton's work on *Optics* on the 18th century poets see Marjorie Hope Nicolson *Newton Demands the Muse* Princeton University Press 1946 and Clark Emery "Optics and Beauty" *Modern Language Quarterly,* III 1942, pp. 45-50.
5. For an expansion of this and Kepler's other optical ideas see Abraham Wolf *A History of Science, Technology, and Philosophy in the 16th & 17th Centuries* Harper Torchbooks 1959, Vol I, pp. 245-250.
6. William Francis Magie *A Source Book in Physics* Harvard University Press, 1935, pp. 265-278.
7. *Ibid* pp. 283-298. See also Wolf, *Op. Cit.* pp. 260-264.
8. Morris H. Shamos *Great Experiments in Physics* Holt, Rinehart and Winston 1959, pp. 93-107.
9. *Ibid,* pp. 108-120 and Magie, *Op. Cit.* pp. 318-324.
10. For Fizeau's and Foncault's work see Magie *Op. Cit.* pp. 340-344 and Bernard Jaffe *Michelson and the Speed of Light* Doubleday & Co., 1960, pp. 29-34.
11. Banesh Hoffmann *The Strange Story of the Quantum* Dover 2nd Ed. 1959, p. 14.

XXXII

Is Bundling a Solution?

This side effect which Hertz had noticed was investigated by his student, Philipp Eduard Anton Lenard (1862–1947), who found that the emission from the negative terminal (cathode) had the same ratio of charge to mass as that found by Thompson for the cathode rays. Among the others who investigated the phenomena was Wilhelm Hallwachs (1859–1922) who showed that the emission was negative electricity.

In other words, light shining on certain metals in an electric circuit caused the electrons in the circuit to flow faster. (Photoelectric effect.) It was found that the stronger the light source, the more electrons flowed but that the energy of the electrons was related to the frequency of the light source and not its strength.

In the field of heat other phenomena were being studied. A hot body emits radiation. If the heat is great enough this radiation is in the form of visible light. At a steel mill the color of the metal is a crude indicator of temperature and most of you have seen metal that is at red or white heat. A hot body also emits radiation of the kind that cannot be seen including infrared and ultraviolet.

Investigations relating energies lost and temperatures were carried out in the last thirty years of the nineteenth century by Josef Stefan, John Tyndall, Ludwig Boltzmann (1844–1906), and Wilhelm Wien (1864–1928). The method developed for the study involved what was called a black body radiator. That is a device which would absorb all radiation falling on it and perfectly give off all radiation as it is heated. In practice the closest that could be achieved was to use a black box with a small hole leading into a hollow spherical interior. The radiation emitted from the hole would be a true black-body radiation.

The physics to deal with the problem included Newtonian mechanics, Maxwell's electromagnetic theory, and thermodynamics. (This is called classical physics.) The information from these areas indicated that as temperature increased and the frequencies increased toward the violet end of the spectrum that the energies involved should increase. It was found

that they, in fact, decreased. This is frequently referred to as the "ultraviolet catastrophe."

Wien proposed a solution but it only worked for short wavelengths. James Jeans (1877–1946) developed another (the Rayleigh-Jeans formula) but it only worked for long wavelengths. What was needed was apparently a new construction material.

Max Karl Ernst Ludwig Planck (1858–1947), during whose lifetime the German Empire was born, grew, and died and the Third Reich was born, grew, and was dissected, put forward a revolutionary concept to explain energy distributions. What Planck proposed in 1900 was that energy can only be emitted at certain levels and is emitted in bundles or quanta rather than continuously. To compare energy and sugar—it always comes in cubes rather than granulated. It seemed that all of nature was lumpy. Just as Millikan had shown that electrical charges were always whole numbers so Planck showed that solid matter can only radiate energy in whole quanta. The grainy nature of energy and charge.[1]

With Planck's quantum hypothesis available, Albert Einstein (1879–1955) proposed in 1905 that it be applied to the photoelectric effect. The quanta of light strikes the metal surface, which is a sea of elec-

Figure XXXII-1. Planck.

trons, and the energy is transferred from the light quanta (called photon) to an electron; just like one ball striking another in pool.[2] The observations were explained by the mathematics and this went far in causing reluctant scientists to accept Planck's work.

But in the solution of one problem a paradox had been created. The Young double slit experiment and Maxwell's equations called for light to be a wave; yet the photoelectric effect and light traveling in straight lines called in the quantum mathematics for light to be a particle. Particle or wave? That was the question. In the ensuing arguments someone suggested that perhaps light could be a wave on Tuesday, Thursday and Saturday; and particle on Monday, Wednesday, and Friday; with Sunday reserved for a day of rest. The nice system of physics of the nineteenth century was going to particles.

Matter and energy had always been considered separately but there were new developments coming about with the basis of matter too.

Just as working with cathode ray tubes had given rise to data about the electron, similar work began to give clues to the internal structure of the atom. Wilhelm Konrad Roentgen (1845–1923) found in 1896 that interposing a piece of metal at an angle between the electrodes of the cathode ray tube and using high voltages, invisible rays were given off which had the peculiar property of being able to penetrate opaque substances which would record differences of densities on photographic film. He called these rays the unknown X-rays.

The paper which he published describing the X-rays remarked on the fluorescence of the glass produced by the X-rays. This comment stimulated Henri

Figure XXII-2. Roentgen.

Becquerel (1852–1908) to experiment with various fluorescent materials, among those tried was Potassium uranyl sulfate (a salt of uranium). He discovered in the process of his investigations, almost by accident, that penetrating rays were given off by the substances. Becquerel's rays did not seem to give pictures of bones as did the Roentgen rays. So that further investigation was left to one of Becquerel's graduate students, Marie Sklodowska Curie (1867–1934). In the year 1898, she discovered two new elements from uranium ore. She named the first Polonium—for her native Poland, and the other Radium—from Latin meaning ray.

In the study of these various radioactive materials, three kinds of rays seemed to be given off: alpha, beta and gamma—named in the order that they were found. Ernest Rutherford (1871–1937), in 1902, experimented and discovered some of the radioactive decay series and also postulated a theory as to the structure of the atom based on his gold foil experiment which indicated that the mass of the atom was concentrated at the center of the atom. He suggested that the positively charged particles, or protons, were concentrated at the center, which is called a nucleus, and that the negatively charged particles, electrons, surround them. He drew the analogy that the nuclei are distributed amongst the electrons like plums in a pudding thus this is sometimes referred to as the plum pudding theory of the atom. However, since opposite charges attract, the question was raised as to how they are kept apart. Rutherford proposed that perhaps the electrons revolved about the nucleus in the same way that a planet goes about the sun and that its motion keeps it from the nucleus just as the planet's motion keeps it from falling into the sun.

Maxwell's equations predicted that the electron should spiral away from the nucleus as it takes on energy and should spiral in toward the nucleus as it radiated energy. What happened in experiments, however, did not verify this theory.

A student of Rutherford, Niels Henrik David Bohr (1885–1962), proposed in 1913 a solution to the problem which called for the electron to move in a discrete orbit and to jump from orbit to orbit when accepting or releasing energy. This idea was suggested to him by the work which Max Planck had been doing with black body radiators. Planck had proposed that energy was given off from a radiator only in discrete units or bundles. Bohr used three of his own assumptions and took the fourth from Planck.

The Bohr atom, as it was called, seemed to account for the spectral phenomena of the lighter elements,

Figure XXXII-3. Bohr.

but did not seem to work for other of the elements. As you may recall from an earlier chapter the prism spectroscope was invented by R. W. Bunsen and G. R. Kirchhoff in 1859 so that from then on it was possible to not only predict what kinds of elements were there from the line spectra of the flame but also to get some idea of the energy state of the electrons from the positioning of the lines. For example, using the equations which Bohr had developed it could be calculated that one of the spectral lines for hydrogen had a value of 3.25×10^{15} while the observed value was 3.290×10^{15}. The difference is well within the range of experimental error.[3]

Bohr was one of the most eminent scientists of this century. He had won a Nobel Prize and occupied Carlsberg Castle, which is reserved for Denmark's first citizen. He was an advisor to Presidents, Roosevelt and Truman, and entertained royalty. Yet to those he worked with and those who studied under him, he reacted well and could take a good joke.

Hail to Niels Bohr from the Worshipful Nations!
You are the master by whom we are led,
Awed by your cryptic and proud affirmations,
Each of us, driven half out of his head,
   Yet remains true to you,
   Wouldn't say boo to you,
Swallows your theories from alpha to zed,
   Even if—(Drink to him
   Tankards must clink to him!)
None of us Fathoms a word you have said![4]

But we must not linger for fearing of Bohring you.

In 1923 Arthur Holly Compton (1892–1962) set out to verify a theory which he had developed concerning X-rays. The discovery of X-rays in 1895 had set everyone working with them and one of the interesting phenomena which had been noticed was that the scattered beam of the radiation seemed to be softer (absorbed more easily). Compton assumed that this was the result of a billiard-ball-like collision between the photon and electron. In other words he was assuming mass for the photon and making it a particle. In his experiments, he reflected X-rays and found that the results confirmed equations which he predicted.[5] The phenomena itself became called the Compton effect. It verified Einstein's equations and predictions concerning the photoelectric effect because X-rays like light rays are part of the electromagnetic spectrum. But it only continued to fog up the question concerning waves and particles.

Following WWI during which he had been concerned with radio, Louis Victor, Duc de Broglie, (1892—   ) in 1924 turned his attention to the problem of dualism in light and extended the concept to matter. His interest in music may have assisted him in making the assumption that light is a wave field with aspects of a particle. If this were the case then perhaps what had been considered particles also had aspects of waves.

Such an experiment was designed and carried out by Sir George Paget Thomson (1892-   ), son of Sir Joseph John Thomson, who passed a beam of electrons through a crystal and demonstrated diffraction interference patterns which showed the wave characteristics of the electron. This established de Broglie's ideas.

In the Newtonian world picture all matter and energy could be considered as points which could be treated geometrically to predict their exact behavior. Now that light, electrons, protons, neutrons, and even atoms seemed to exhibit wave properties the world picture was changing. Instead of talking of matter as points the language changed to speaking of matter fields.

The considerations of de Broglie went something like this: particles of matter have mass; mass is energy, shown in Einstein's special theory of relativity in 1905; energy displays frequency; while frequency implies pulsation; pulsating particles behave very much like photons; and photons are related to light waves, therefore matter, or particles, should have a relation to matter waves and there should exist matter fields.[6] His solution was to create equations in which wave factors such as wave length, and frequency were included along with Planck's constant.

In 1925 the equations of de Broglie were combined with matrix algebra (today called quantum mechanics) to give the position and velocity of the electron in its atomic orbit.

Light or any form of energy if given to an atom causes the electrons in the atom to be at a higher energy level. When the atom releases the energy, the electron goes to a lower energy level. These changes in energy levels could be observed with a spectroscope. By examining the spectra produced it was possible to see if the mathematics successfully predicted what would happen in the way of energy levels.

Bohr tried with some others in 1924 to reconcile the wave and particle picture by suggesting that the electromagnetic waves should not be thought of as real waves but only as a probability wave. Max Born (1882–   ) accepted this idea in part and made the wave a part of a many-dimensional configuration space (instead of a three-dimensional one) and the idea became an abstract mathematical quantity.

Praise of de Broglie's wave equation by Einstein stimulated the thoughts of Erwin C. Schrödinger (1887–1961). He began by extending the ideas of de Broglie that if a fixed string vibrates, it always vibrates in nodes of whole numbers so that if this idea is extended to a circle (or metal ring), as a circular atomic orbit, it would be expected that the vibration would be in whole numbers. This would fit the atomic ideas and the ideas of Planck, and Bohr. Schrödinger also used classical mechanics from Newton as developed by William Rowan Hamilton in 1835. The picture which emerged was one which smeared the electron into a probability of position. It introduces the psi Ψ, or probability function into physics, and was to give a possibility of predicting the position of an electron, at least its probable position.

However, from purely philosophical considerations, which have been verified mathematically and to a certain extent experimentally, Werner Karl Heisenberg (1901–   ) in 1926 stated his indeterminancy principle that it will never be possible to determine both the velocity and the position of the electron at any time. The degree to which this can be a factor is determined by the nature of the distrubing observation, such as light or X-rays, the degree or lack of sharpness of the image (of the electron or whatever is being observed), and Planck's constant. The uncertainty of the measurement of the position may be expressed as delta p and the uncertainty of the momentum is delta mv (m is the mass and v is the velocity). Planck's constant is h. In an equation this would be expressed as (delta p) (delta mv) = h/2pi. The amount of uncertainty increases as the size decreases so that for the smallest body, the electron, the uncertainty is greatest.[7]

It is a well known fact in education that a class under observation behaves differently than one which is not being observed. It is also true that if one were observing a hummingbird around its nest the

Figure XXXII-4. Heisenberg.

presence of the observer would affect the flight pattern of the bird. In these two instances it would be possible to screen the observer and reduce his effects. In the case of the electron, or other small particles, in order for the observer to observe beams of energy having small wavelengths must be used and they have higher levels of energy which give the electron a kick thus changing its velocity.

Part of the problem also arises out of the grainy nature of things. In the same way that looking closely at a newspaper photograph doesn't give a clearer picture because the photograph is made up of little white, black and grey dots. Getting closer only allows the grainy nature to show. This is because energy and mass comes in packages of fixed sizes related to Planck's constant.

In basic principle the Heisenberg Principle of Indeterminancy states that the very act of measurement changes the amount of what is being measured. This implies that every process of measurement will be subject to the same limitation on its accuracy and indeed that this is a fundamental general law which operates throughout the whole of nature.

For additional information see:

Banesh Hoffmann, *The Strange Story of the Quantum,* Dover, 1959. pp. Preface + 1-140, 229-235.

Louise B. Young (ed.) *The Mystery of Matter,* Oxford University Press pp. 84-94.

Alfred Romer, *The Restless Atom,* Anchor Books, 1960.

Morris H. Shamos, *Great Experiments in Physics.*

Albert Einstein and Leopold Infeld, *The Evolution of Physics: The Growth of Ideas from Early Concepts to Relativity and Quanta,* Simon and Schuster, 1961.

Isaac Asimov, *Inside the Atom,* Abelard-Schuman, 1966.

Selig Hecht, *Explaining the Atom,* Viking, 1959, pp. 9-94.

David Park, *Contemporary Physics,* Harcourt, Brace and World, 1964.

Friedrich Hund "Paths to Quantum Theory Historically Viewed" *Physics Today* August, 1966, pp. 23-29.

Martin J. Klein "Einstein, Specific Heats, and the Early Quantum Theory" *Science* 148:173-180, 1965.

William T. Scott, *Erwin Schrödinger,* University of Massachusetts Press, 1967.

David Bohm, *Causality and Chance in Modern Physics,* Harper & Brothers, 1957.

A readable book on quantum chemistry in which the mathematics can be skipped is Audrey L. Companion, *Chemical Bonding,* McGraw-Hill, 1964, pp. 1-33.

Advanced books requiring considerable mathematical background are:

Robert L. Sproull, *Modern Physics,* John Wiley and Sons, 1963.

Behram Kursunoglu, *Modern Quantum Theory,* Freeman, 1962.

Edward V. Condon and Philip M. Morse, *Quantum Mechanics,* McGraw-Hill, 1929.

STUDY AND DISCUSSION QUESTIONS

1. What is the relation between black body radiation and light?
2. What proposal did Planck make?
3. What was the crux of the particle vs. wave argument?
4. What was the real meaning of Curie's discovery?
5. What was the Bohr atom?
6. What established de Broglie's ideas?
7. What was Heisenberg's principle?

FOOTNOTES

1. Morris H. Shamos, *Great Experiments in Physics,* Holt, Rinehart and Winston, 1959, pp. 301-314.
2. *Ibid,* pp. 232-237.
3. *Ibid,* pp. 333-347.
4. As cited by William H. Cropper, *The Quantum Physicists and an Introduction to Their Physics,* Oxford University Press, 1970, p. 38.
5. Shamos, *Op. Cit.* pp. 350-348. Also see Werner Heisenberg *Physics and Philosophy,* Harper & Brothers, 1958, pp. 36-37.
6. Banesh Hoffmann, *The Strange Story of the Quantum,* Dover, 2nd Ed., 1959, p. 74.
7. Heisenberg, *Op. Cit.* passim.

XXXIII

Let's Flip

It seems that any particle in motion has associated with it certain wave characteristics but it would be an oversimplification to say that the particle makes waves as it moves or that the path of the particle is a wiggly one. Rather it is stated that both energy and matter display both wave and particle characteristics. The most important factor, from our standpoint, is not the exact nature of light but the argument which rose from it.

The argument itself is subordinant because there are always disagreements in any field of endeavor. Many laymen and even those who are Roman Catholics are not aware of the amount of latitude in belief that is permitted among and within various orders of the church. There are many disagreements on matters of doctrine which have and are going on all the time. It is when the argument itself has widespread significance. Many monks had nailed their theses to church doors for the purpose of debating issues and many monks had been critical and called for reforms before. So that it was not the content nor the procedure of Luther that had the tremendous effects but rather that the argument had spread to a level of popular understanding and that the printing press could spread it more rapidly.

Such was the case in the argument concerning the nature of light. The evidence of either side was absolute, so it seemed, and conclusive allowing only one interpretation. The argument spread from the purely scientific community to the avant guard of the cultered and then to the popular press.

The overall effect was to create doubt as to the ability of science or the scientist. In the result that brought agreement to the scientists, it remained too abstract in the main for either the cultured intellectual or to the commoner. If science had replaced religion and the scientist replaced the priest, then to whom would man turn next with the downfall of belief in the ability of science?

There are a number of popular beliefs about science that have followed from the various attributes that have been given to science.

They are the ones that have struck the common imagination or that have received most attention from the popularizers. Thus there is the doctrine of the 'three great hurts' inflicted by science on the vanity of man: Copernicus (or Newton) displaced man from the center of the universe, Darwin thrust him back into the animal kingdom, and Freud put him at the mercy of his unconscious. These scientific blows are held to explain modern man's self-contempt and justify his giving up will and responsibility: 'scientifically speaking,' he is altogether conditioned and helpless.

It is also true that periodically Heisenberg's 'uncertainty principle' in physics is used to give new validity to moral commandments and religious faith. Similarly, the sophisticated interpretation of natural laws as being an expression of statistical probability rather than of causal necessity, offers an escape from the prison of matter and motion. Every time that a competent scientist produces a book in which science is made compatible with religion and metaphysics he is sure of a large appreciative audience. This shows how hard it is to say when and how much the public trusts the certainties of science and accepts its reduction of experience to particles and abstractions.[1]

Perhaps it doesn't matter a particle but the belief is more important than the reality. The same is true and with far greater consequences for Heinsenberg's principle.

Pierre Simon de Laplace, near the beginning of the 19th century, made the logical extension of the idea that had started with Newton and the mechanists when he stated, "If we knew the exact position and velocity of every particle in the universe at any one particular moment, then we could work out all the past and future of the universe." It was maintained that every particle in the universe obeyed Newton's laws of motion and so if there were a brain large enough to do the calculations it should thus be possible to calculate the motion of every particle throughout all time. This position leads to an extreme certainty and a belief in determinism. It caused scientists, and indeed others as well, to be filled with an optimism that soon all problems would be solved, or at least that all problems were subject to a solution. Every effect had a cause which could be known. This feeling of certainty also carries with it a sense of scientific predestination—i.e. if everything is determined by certain natural laws and the future can be forecast in great detail if enough is known then what happens to free-will? There are many who still retain this position since ideas do not change overnight but many have been influenced by the principle of indeterminacy.

The principle of indeterminacy, as stated by Heisenberg in 1927, shows that the dream of Laplace can never be even in theory. In addition it casts strong doubts on determinism or any exact laws which could make predictability certain. Since the time of Newton up to the 20th century it has been assumed that the macrocosm was reflected in the microcosm. In other words that the laws which apply to large bodies, like planets and stars, also apply to the very small, like atoms, molecules, electrons, etc. The converse has also been assumed—i.e. that the microcosm reflects the macrocosm. If one were to split the microcosm from the macrocosm then the uncertainty principle would have little effect on the predictions for large bodies and an amount of certainty might be retained for the macrocosm. To do so would however need to destroy another assumption that has and is still made—that of the uniformity of nature. It is basic in science that laws must apply uniformly and that there is no discontinuity in nature. But perhaps this position might be more tenable than the pessimistic view presented by uncertainty and the laws of chance.

If it is impossible to predict for the individual electron except in terms of probability for its position and only possible to predict if one has a large number of events to consider which would approach a valid probability of occurrence then one is left with a universe that is chaotic, meaningless, and purposeless.

Part of the restlessness of our time has been brought on by a growing feeling which is reinforced from a number of sources that there is no order in the universe. For nearly a thousand years, one of the strongest rational arguments that has been put forward by Christians for the existence of God has been based on the apparent order of the universe. A universe without order is a universe without God. I cannot, from the view of physical science, argue against the recent theories and evidence but I feel somewhat as Einstein must have felt when he said, "I cannot believe that God plays dice with the world."

Bohr, on the other hand, disagreed with Einstein and accepted the statistical view of nature. They had many arguments concerning this but neither could convince the other.

Physicists had held the general philosophical view of a deterministic mechanism ever since the time of Newton but following Heisenberg's indeterminacy principle a number of ideas had to be abandoned.

Among these was the idea of causality, at least in the realm of atomic sizes. Another concept that fell was that of continuity of motion. It had been assumed that the path of a particle could be traced by seeing where it made occasional tracks in an emulsion and filling in in between but this could not be supported any longer. Indeed some maintained that these ideas must be abandoned, not only in atomic domains, but at all levels. This view has seeped out and is another factor affecting public belief patterns.[2]

The principle also implies a mixture of the observed and observer. We are told that in the

nineteenth century natural science conceived of man as a detached spectator of an objective universe. It held spectator-spectacle polarity to be genuine and fundamental. During the present century, discoveries concerning the nature of the atom rendered this doctrine untenable. The nucleus of a new philosophy of nature emerged with Heisenberg's principle of uncertainty, whose basic meaning implies a partial fusion of the Knower and the Known. This theory led to a mathematical formalism which, in order to attain its purpose, namely lawful description of experience, has to speak of probabilities rather than unique events.[3]

Since scientists are persons they long to establish order the same as all men have done since the days of the ancient Greeks. They want to find in nature a logical and mathematical diagram which will reproduce on paper what other scientists find to be the case in their experimental work.[4]

Schrödinger, like Einstein, cannot accept and indeterminate view of the universe and argues against it and the moral and ethical implications. ". . . we are not prepared to regard our fate as the outcome of pure chance . . ."[5] He comments further,

It would thus appear that Bohr's considerations adduce a physical unpredictability of the behavior of a living body again precisely from the lack of strict causation, maintained by the quantum theory. Whether or no this physical indeterminacy plays any relevant role in organic life, we must, I think, sternly refuse to make it the physical counterpart of voluntary actions of living beings, . . .

The net result is that quantum physics has nothing to do with the free-will problem. If there is such a problem, it is not furthered a whit by the latest development in physics. To quote Ernst Cassirer again: 'Thus it is clear . . . that a possible change in the physical concept of causality can have no immediate bearing on ethics.'[6]

It matters not which view is correct. The important factor is one of belief. It is the extent to which belief in the idea that the principle of indeterminacy does play a role; and how far down the pyramid of society that the idea has permeated that makes the difference. It is also the degree to which the idea combines with other factors in our society, as will be discussed in the last chapter, that makes the effect total.

Physics had moved into another area of mathematics for a model. An area which had begun as a game. The father of the area of mathematics, which we call probability, was Jerome Cardan (1501–1576) who was a doctor and taught medicine and mathematics at the University of Pavia. Most of his books are on medicine or religion[7] but the one that attracts our notice was written of his chief pastime which was gambling. He studied the probabilities of throwing sevens and of cutting aces from the deck. In his book, *Liber De Ludo Aleae,* he puts together the results of his studies in a practical form. He even points out that the chance of drawing a certain card can be increased by the application of soap. It is not stated if this is soft-soaping. From these beginnings is founded an area of mathematics that is fundamental today in the business world of insurance and also to science in the theories of gases and atoms.[8]

Others have claimed that Chevalier de Méré (1610–1684) began the theory of probability with a dice problem in 1654, but Galileo had also asked a dice problem of why a throw of three dice more often turned up a sum of ten rather than nine. De Méré sent his problem to Blaise Pascal (1623–1662) who solved it but the problem gave a different problem to Pascal which is up again in the twentieth century.

With two dice there are thirty-six possible face combinations which could occur and assuming that the dice are not loaded the chances for each face is equal. The combinations do not give thirty-six face totals. A two or twelve will only appear once in thirty-six times while a seven will appear six so that the odds for throwing a two is 1/36 while that of throwing seven is 6/36 or 1 in 6.

Two terms tend to cause confusion, "probability" and "statistics," because they are sometimes used interchangeably. In the case just given we were dealing with probability because we were concerned with the instances of what could possibly occur. If the dice had actually been thrown and records kept then we would have been dealing with statistics or an accumulation of data based on what actually happens. In many cases, however, especially where prediction is concerned on the basis of accumulated statistics, the two terms are sometimes used in a fuzzy manner.[9]

The introduction of probability dominated physics has created confusion. Monotheism dictates a god that is purposeful and a universe that is orderly. Also, one of the fundamental assumptions of science is that the universe is orderly. Yet science has found that, at least for the very small, there is no order for the single event in nature. Conditions cannot be reproduced "exactly" in an experiment. When dealing with large numbers of events then there seems to be an order which we call probability or statistical. A number of persons have sought order and the determination of the existence of God. Pascal became very disturbed[10] for he calculated the probability for the existence of God was very small. This same theme is explored in Thornton Wilder's novel *The Bridge of San Luis Rey* (1927).

Those who wish to have order and determinism place an emphasis on the predictability of a large number of events while those who cannot see how any order applies, except the operation of the laws of chance, must then accept chaos for the individual with its attendant free will in a capricious universe ruled by a mad God, if there is one. Schrödinger and others have defended the former position very well but the popular interpretation and belief tends toward the second as reinforced by other events and ideas for other fields.

In physics the use of chance appears in statistical mechanics; it is our heritage from Boltzmann. First, we assume certain statistical laws for individual atomic collisions or other events. Then we apply the rules of counting and the law of large numbers. And behold, regular behavior for large bodies of gas or liquid and for large solid objects results. Furthermore, we not only come out with Newtonian physics as a good approximation for the macroscopic world, we can explain its failures for small objects, e.g., for the random Brownian motion of microscopic bacteria that are bombarded by water molecules on all sides but not quite uniformly.

Statistical mechanics provides an ordered description of several aspects of the physical world and involves characteristic features such as temperature, pressure, volume, number of molecules, compressibility, and other large-scale measures of material properties. Even without an underlying hypothesis of causality, statistical mechanics exemplifies the concept of intelligible chance.[11]

The fact that statistical methods were so successful in prediction of the probable events that creates for us the present problem. Laws of chance do not make most persons see an order around them even if the laws are "workable."

The very forces of nature that had been admired for their simplicity, order, and invariability included unexpected and unexplainable tidal waves, volcanic eruptions, and earthquakes. Nature suddenly appeared unpredictable, perverse, and capricious.

Thus the very same world which the eighteenth century regarded as rigidly determined and designed in accordance with immutable mathematical laws seemingly had to be viewed now as chaotic, lawless, and unpredictable. Reality appeared totally void of purpose, a 'tale told by and idiot, full of sound and fury, signifying nothing.' Man, in particular, was but an accident of the blind, fortuitous concourse of events. *The mathematical laws of science amounted to no more than convenient, usable summaries of the average effect of disorderly occurrences.* This attitude toward nature and its laws, which affirms that nature is chaotic and unpredictable and that its laws are not more than convenient, impermanent descriptions of average effects, is known as the *statistical view of nature.*[12]

The rigid laws of the past are gone leaving a haze that obscures the future. A feeling remains like lost virginity, which indicates that things will never quite be the same. In the dawn of a new approach by science that there is not quite enough light, though there has been enough heat, to see where this approach will lead us.

This uncertainty from science was multiplied in 1931 by uncertainty in mathematics. Kurt Gödel (1906–   ) showed that within a mathematical system there are certain statements which cannot be proven nor disproven so that certainty about the axioms of a system is impossible. "Perhaps doomed also, as a result, is the ideal of science—to devise a set of axioms from which all phenomena of the natural world can be deduced."[13]

STUDY AND DISCUSSION QUESTIONS

1. What guarantees to us the constancy of the laws of nature?
2. Where has doubt risen about the objective existence of the regularity of nature?
3. What exact laws exist for mass phenomena?
4. What determines the behavior of individual entities?
5. How can disorder be subject to orderly laws?
6. How are statistical data obtained by Insurance companies? The National Safety Council?
7. How can any order occur if the trend is towards disorder?
8. Is the causal principle necessary to our thought?
9. What is the conservative view? The revolutionary?
10. Why is philosophy becoming more closely linked to physics?
11. Before 1915, what was the fundamental dogma of practical physics? What is today's view? Why?
12. What is the limitation of determining the position and velocity of anything?
13. Why can't we get identical results from identical conditions?
14. Why is every observation discontinuous?
15. What does interpolate mean? Why doesn't it work with an atom?
16. What kind of a world did 18th Century thinkers envision? What did they argue from?
17. What occurred during the 19th and 20th Centuries to crumble the world of order?
18. What kind of "chance" phenomena seems to be "predictable" by the theory of probability and statistics?
19. What do scientific laws seem to be?
20. Does a statistical view deny the possibility of underlying determined behavior? What has Einstein said relative to this?
21. What is meant by the "statistical view of nature"?
22. What are the distinctions between the deterministic and statistical views?
23. Are the axioms in Euclid's geometry "true"?
24. Is the argument of simplicity a valid one?
25. If something works—does this make it "true"?
26. Does man make God his own image?

FOOTNOTES

1. Jacques Barzun, *Science: The Glorious Entertainment,* Harper & Row, 1964, pp. 62-63.
2. David Bohm, *Causality and Chance in Modern Physics,* Harper & Brothers, 1957, pp. 68-70, 84,89.
3. Henry Margenau, *Open Vistas,* New Haven, 1961, p. 201.
4. William T. Scott, *Erwin Schrödinger,* University of Massachusetts Press, 1967, p. 119.
5. Erwin Schrödinger, *Science and Humanism,* Cambridge University Press, 1951, p. 64.
6. *Ibid,* p. 67.
7. Jerome Cardan *The Book of My Life,* Dover, 1961, E. P. Dutton, 1930. For his life also see Jane Muir *Of Men and Numbers,* Dell, 1961, pp. 35-58.
8. Morris Kline, *Mathematics in Western Culture,* Oxford University Press, 1953, Chapters 23 and 24. Warren Weaver, *Lady Luck,* Doubleday & Co., 1963.
   William J. Youden, "How Mathematics Appraises Risks and Gambles" in Thomas L. Saaty and F. Joachim Weyl *The Spirit and the Uses of the Mathematical Sciences,* McGraw-Hill Book Co., 1969, pp. 167-187.
   Several articles in Readings from Scientific American *Mathematics in the Modern World,* Freeman, pp. 151-178.
9. Some statistical applications are discussed in O. G. Sutton, *Mathematics in Action,* Harper & Brothers, 1960, Ch. 6.

10. Kline, *Op. Cit.*, pp. 374-375.
11. Scott, *Op. Cit.*, p. 124. Also see Banesh Hoffman, *The Strange Story of the Quantum,* Dover, 1959, pp. 180-181.
12. Kline, *Op. Cit.*, p. 386.
13. Carl B. Boyer, *A History of Mathematics,* John Wiley & Sons, 1968, pp. 655-656.

Other Readings:

Erwin C. Schrödinger, "The Law of Chance" and "Indeterminism in Physics" in *Science Theory and Man,* Dover, 1935, pp. 39-80.

See also *Mystery of Matter,* Part 8, "Does order arise from Disorder." Werner Heisenberg, *Physics and Philosophy,* Harper and Brothers, 1958, pp. 187-206.

William H. Cropper, *The Quantum Physicists,* Oxford University Press, 1970.

John Herman Randall, Jr., *The Making of the Modern Mind,* Houghton Mifflin, 1940, Ch. 18.

XXXIV

Relativity—Can We Get to Yesterday?

Relativity, according to the seventh edition of Webster's New Collegiate Dictionary, is "a theory formulated by Albert Einstein and leading to the assertion of the equivalence of mass and energy and of the increase of the mass of a body with increased velocity and based on the two postulates that if two systems are in relative motion with uniform linear velocity it is impossible for observers in either system by observation and measurement of phenomena in the other to learn more about the motion than the fact that it is relative motion and that measurements of the velocity of light in either system always give the same numerical value." This is called the special theory of relativity. It was extended in the general theory of relativity to include a discussion of gravitation and related phenomena.

The two systems referred to in the above paragraph may also be called frames of reference. Each of us use ourselves as a frame of reference and the earth as the other. We say that we are in motion relative to the earth.

The idea about motion being hidden between two frames of reference that are in relative motion with uniform linear velocity is not new. It may be found in the works of Galileo. In his *Dialogue Concerning the Two Chief World Systems,* (published about 1630), Drake translation pp. 186–187, he states:

Shut yourself up with some friend in the main cabin below decks on some large ship, and have with you there some flies, butterflies, and other small flying animals. Have a large bowl of water with some fish in it; hang up a bottle that empties drop by drop into a wide vessel beneath it. With the ship standing still, observe carefully how the little animals fly with equal speed to all sides of the cabin. The fish swim indifferently in all directions; the drops fall into the vessel beneath; and, in throwing something to your friend, you need throw it no more strongly in one direction than another, the distances being equal; jumping with your feet together, you pass equal spaces in every direction. When you have observed all these things carefully (though there is no doubt that when the ship is standing still everything must happen in this way), have the ship proceed with any speed you like, so long as the motion is uniform and not fluctuating this way and that. You will discover not the least change in all the effects named, nor could you

tell from any of them whether the ship was moving or standing still.

The idea concerning relative motion has been about for at least three hundred years. It is possible that you have observed the same thing when you have stopped at a red light and you notice that your car seems to be moving. Only when you look at something that is motionless, e.g. a light pole, can you tell that it is the car next to you that is moving and not your own car.

Most writers would agree that the physical universe has not basically changed in the last two thousand years but that our concepts of the universe have and are changing considerably. Let us examine some of these changing concepts leading up to our own century.

In the Aristotelian-Ptolemaic system the center of the earth was a fixed point which provided an absolute frame of reference. Motion was of two types—natural and unnatural. There were two types of natural motion—sub-lunar and celestial. Sub-lunar motion, motion below the moon, was accounted for by the mixture of the four elements (earth, water, air, and fire) which were seeking their appropriate level relative to the earth's center. This type of motion was always perpendicular to the center of the earth, i.e. up and down. Celestial motion, on the other hand, which accounted for the motion of the heavenly bodies, was always perfectly circular and at a uniform rate about the earth's center. A deity supplied the force which keep these in motion. The force was transmitted through the concentric, transparent spheres which filled space—there was no vacuum. From this it follows that space was not isotropic (did not have the same properties in every direction) because there were different rules for motion depending on where the motion was located in space.

With the gradual acceptance of the Copernican theory the absolute reference point is shifted from the center of the earth to the center of the earth's orbit. The sun and stars are at rest relative to this point and the earth and the other planets travel in perfect circles about this center. There are still two forms of motion—natural and unnatural and two types of natural motion—earthy and heavenly. The force that moves the planets is supplied by God and transmitted through a filled space—no vacuum.

From the planetary studies of Kepler and the motion studies of Galileo, Newton formulates his ideas which provide only one type of motion whether natural or unnatural. All motion continues in a straight line. The planets move in ellipses because of an undefined force called gravity which somehow acts through empty space. The planets were put into motion in the beginning by God. Although it has not been proven there is a strong indication in Newton's letters that he thought the force of gravity was another manifestation of God. Newton makes a number of assumptions in formulating his system. He assumes that space is the absolute reference frame and that space and figures and bodies in space obey the theorems of Euclidian geometry. From this it follows that space is homogeneous (i.e., that space everywhere in the universe has the same geometrical properties). He also assumes an absolute time, i.e. that the measurement of time everywhere and under any conditions would be the same.

Since space itself is Newton's absolute frame of reference he does have the problem of where to place an observer in his system so that he is not on a moving frame relative to another moving frame of reference. Newton, himself, recognizes that an observer who can frame laws from a moving body, such as the earth, does not have an absolute position. The observer that Newton chooses who does have absolute observations of space and time is God.

Thus it can be seen that the great scientific ideas of Newton rested on a number of metaphysical assumptions including God, absolute space, absolute time, and absolute laws. Also included was the mysterious idea that gravity was a force which acted over a distance.

Scientists on the continent while accepting most of Newton's assumptions were very reluctant to accept the idea of a force which could operate over a distance without pushing or pulling through any physical matter. It seemed to be black magic and they wanted to divorce science from any concepts which seemed to have the stigma of the unknown.

As a force, gravity describes the motion between two bodies but it does not explain why bodies are drawn toward each other. The idea of gravity as a force gives every physical body two distinct properties—mass and weight. Mass is defined as the amount of resistance an object offers to a change in its speed or direction of motion (which is an inertial defini-

tion). It is also sometimes defined in some number to indicate that amount of matter that it contains. Weight, on the other hand, is defined as the force with which the earth attracts an object. Under usual conditions mass is a never changing constant but the weight of an object varies as the square of its distance from the center of the earth. The ratio of the weight to the mass of an object is always the same at a given distance from the center of the earth. The ratio is always the same at any given distance but the two properties are distinct. In the Newtonian explanation no reasons could be given for this fact.

Even while the shouts of acclaim for Newton still echoed there were a number of problems developing and it gradually became apparent that Newtonian physics was not the Philosopher's stone for physics. It would not solve all of the physical problems with one little neat set of equations.

The orbit of the planet Mercury did not lend itself to the analysis of the Newtonian formula. The planet did not follow the predictions of the equations of motion. Newton himself felt that God stepped in every once in awhile and straightened out the orbits of the planets if they did not follow His natural laws precisely. He introduced the idea of a *deus ex machina* to account for the irregularity of the orbit of Mercury. Other scientists, in order to save the phenomena—i.e. to make the facts fit the theory, introduced the idea that there was another planet on the opposite side of the sun from Mercury which being close to the sun so we could not see it would account for the irregularity of Mercury's orbit. They christened the undiscovered planet Vulcan.

Newton tried to include an explanation of light in his system by making light particulate. There was some phenomena of light, however, that was better explained by considering light a wave rather than a particle. Most scientists accepted the wave notion of light. In order to conceive of light as wave motion it became necessary to postulate a transmitting medium for the light just as water transmits waves or sound is transmitted by air. Thus an assumption was made that there existed a substance which completely filled space, called the ether (also aether), but could not be seen, tasted, smelled, weighed, nor touched. It was also fixed, i.e. had no motion of its own, but did not act frictionally on the planets.

This was more or less the way the situation was and remained up to the 19th century when a number of other problems began to arise which were not accounted for by Newtonian physics.

James Clerk Maxwell (1831–79) put together ideas of heat and light together with the work of Faraday in electricity and magnetism into a set of equations which related these seemingly different phenomena. He used Faraday's idea of a field and treated the units as being acted upon in this field by other units or elements. This remained in agreement with the ideas of the mechanists and the followers of Newton to explain all physical phenomena by purely mechanical means.

From Maxwell's equations (c. 1865) it was apparent that electricity, magnetism and light were all electromagnetic phenomena. All traveled with the velocity of light through the luminiferous ether. Maxwell's equations predicted the existence of waves from an electrical discharge which were demonstrated by Heinrich Rudolf Hertz (1857–94) to have all the properties of light waves including reflection and refraction. (These were the Hertzian waves or radio waves that we have already discussed.) Maxwell's equations seemed to work well if the phenomena being studied was wave in nature. But was electromagnetic phenomena wave or particle? Was nature continuous or discontinuous? The equations did not seem to work well if the phenomena was a moving particle—e.g. a moving electron. The space containing the field was Euclidian in Maxwell's considerations.

A number of investigators had been interested in investigating the nature of the ether over which these various waves were transmitted. In order to determine the amount of drag of the ether experiments were devised to measure the difference in the velocity of light as it came from a source moving toward the earth or moving away from the earth. As a rather crude example of the difference in velocities suppose I am walking toward the front of a train at 5 miles an hour while the train is going 50 miles an hour. Thus my velocity relative to the ground is 55 miles per hour. If, on the other hand, I am walking toward the rear of the same train at 5 miles an hour my velocity relative to the ground should be 45 miles per hour. A number of attempts to find this difference in the velocity of light were made by Albert Abraham Michelson (1852–1931) and Edward Williams Morley (1838–1923) on the Case-Western Reserve University campus between 1885 and 1900. All these attempts failed and this experiment is famous because it was not successful.[1]

As attempts to show the existence of the ether began to fail an attempt to save the phenomena was postulated by George Francis Fitzgerald (1851–1901) in 1892. He proposed that objects (which would include light) moving in one frame of reference would appear contracted to another moving frame of reference and that this contraction would just offset any attempts to determine the difference in velocity. This same idea was independently proposed by Henrik Antoon Lorentz (1853–1928). Experiments showed that the contraction could not be observed within the moving system itself and that indeed there was no way to show the absolute velocity of the frame of reference. Motion within one frame of reference could be converted to the other by means of what became called the Lorentz transformation which was a factor equal to the square root of one minus the square of the object's velocity divided by the square of the speed of light. This seemed to make problems work but was not satisfactory as an explanation.

Let us diverge from our main theme and discuss some developments in pure mathematics which are important a little further on.

You may recall that Newton in his *Principia* uses Euclidian geometry instead of the calculus which he had invented partly because he wished to avoid a proof of the calculus at this point but primarily because Euclidian geometry was well accepted as the model—physical and mathematically—from Neoplatonism and from its acceptance by Copernicus, Kepler, Galileo. It was also readily acceptable to Newton's contemporaries.

In 1829 Nikolai Ivanovich Lobachevsky (1793–1856) and in 1832 Johann Bolyai (1802–1860), working completely independently, each invented a system of geometry that was non-Euclidian. This was a radical departure from the traditional belief, held for over 2,000 years, that Euclidian geometry was the only conceivable one.

In Euclidian geometry the shortest distance between any two points is a straight line and parallel lines never meet. However, in non-Euclidian geometries such as developed by Lobachevsky, and have and continue to be developed by others, the shortest distance between two points (called a geodesic) is a curved line. See e.g. the pseudosphere of Bolyai and Lobachevsky.

Georg Friedrich Bernhard Rieman (1826–1866) in the 1850's developed a geometry for a spherical surface and showed that, on the surface of a sphere all lines are not parallel, and that the geodesic is the arc of a great circle. This is very important in an age of air travel—e.g. the polar route between New York and Europe.

The works of Bolyai, Lobachevsky, and Rieman liberated mathematical thought from the Euclidian captivity. Its importance for mathematics in the 19th century ranks with that of Darwin in biology, Freud in psychology, and the Romantic reaction in literature. Each of these revolutions against the classical

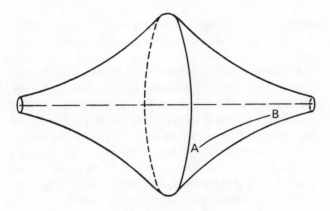

Figure XXXIV-1. Pseudosphere.

bonds which fixed man's thoughts into habit patterns. It is very difficult to break thought habits and all of us have them.

At about the same time a similar revolution was occurring in the field of algebra. Sir William Rowan Hamilton (1805–1865), an Irish mathematician, invented an algebra, called quaternions, which was non-commutative. Algebras up to this time were all commutative, that is, (A) (B) = (B) (A), but in a non-commutative algebra (A) (B) ≠ (B) (A). Hamilton's book, *The Elements of Quaternions,* was published in 1866. This property is important in describing rotations in a three-dimensional space. The order of rotation in a plane (or Euclidian space) is not important so that in a plane the commutative law holds.

Curbastro Gregorio Ricci (1853–1925) and Tullio Levi-Civita (1873–1941) at the turn of the century formulated a new algebra and analysis system which they called absolute differential calculus (tensor analysis). This system is used where more than three numbers are required to specify scalars (amounts which can be read from a scale such as a ruler) and vectors (quantities which have direction as well as amount).

We need one more mathematician before continuing with the main stream of our discussion. Herman Minkowski (1864–1909), who was one of Einstein's teachers, showed in 1908 that if time were made into a kind of distance by multiplying it by the speed of light, called space-time, then the space-time interval between two events is the same for all observers and is an absolute quantity in a four-dimension space-time, but that space and time are not absolutes separately. (In a three-dimensional world we have the quantities of length, width, and depth which are all quantities of extension—whatever extension might be. Try to define extension or length, depth, and width.) Another way of stating what Minkowski did

is to say that he created a new non-Euclidian geometry to deal with four-dimensions instead of three.

Albert Einstein (1879–1955) was aware of the mathematical discoveries, of the failure of the Michelson-Morley experiments, the flaws in Newton's system, and of the Galilian frame of reference. He had two facts to use. The first is the fact that there is no means of determining an absolute frame of reference. In Einstein's explanation he uses a freely falling elevator with constant acceleration instead of a ship as Galileo, with constant velocity. The second fact is the failure of the Michelson-Morley experiment.

Figure XXXIV-2. Einstein.

In 1905, at the age of 25 and while working as a clerk in the Swiss patent office, Einstein published his theory of Relativity. In the special theory he does away with the absolutes of space, time, and mass; but he assumes the velocity of light in a vacuum to be absolute. Using this assumption and the two facts from the paragraph above, he is able to reformulate into a field theory which will fit all frames of reference.

In the following equations: S stands for the special coordinates and contains values of x, y, and z; V stands for velocity; c stands for the velocity of light; t stands for time; m stands for mass in motion while m_0 stands for the mass at rest; and E stands for energy. The units are seconds for time, cm/sec. for velocities, grams for mass, and ergs for energy.

$$S^1 = \frac{S - Vt}{\sqrt{1 - \dfrac{V^2}{c^2}}}$$

$$t^1 = \frac{t - \dfrac{VS}{c^2}}{\sqrt{1 - \dfrac{V^2}{c^2}}}$$

$$m = \frac{m_o}{\sqrt{1 - c^2}}$$

$$E = mc^2$$

Compare Einstein's first three equations with the Lorentz transformation and you will note that each of Einstein's three equations contain this transformation. Einstein's equations explained how the Lorentz transformation worked and why Newton's equations did not work for systems containing motion near the speed of light or with very large masses. His postulates indicated that the Lorentz transformations of the Fitzgerald contraction was not a "real" physical change but an appearance which was due to the relative motions of frames of reference.

Consider the case of two parallel runways and a control tower. On one runway a Piper Cub is taxiing at a constant speed; while on the other there is a jet taking-off with an accelerating speed. It is night and only the Cub has a light. The light, seen from the stationary control tower, seems to be moving in a straight line but from the accelerating plane the light seems to be moving in a curved path, a parabola. In other words, what the observer sees is dependent on the motion of his frame of reference relative to the frame of reference in which the event is occurring. From these considerations it follows that simultaneous events do not exist if observers are in motion relative to each other nor is it possible to know the absolute velocity of ones own frame of reference without a fixed point for observation.

As we have already seen, Newton's system contained both weight and mass. Weight and mass seemed separate and yet were somehow related. Weight was defined in terms of a gravitational force while mass was defined in terms of acceleration. We see the latter today expressed as so many "G's" of acceleration in dives of planes and take-offs of rockets.

Einstein saw that he could explain both mass and weight in terms of geometry rather than by forces. His considerations were delayed—between the formulation of his special theory of relativity (1905) and the general theory of relativity (1915) by the lack of a proper coordinate system. This lack was made up when Einstein's former teacher Minkowski developed his four-dimensional space-time, (1908). The other mathematical ingredients were already present. Einstein selected for his model the non-Euclidian geometry invented by Bernard Riemann which curved the space-time described by Minkowski's equation. The analysis of a body moving through such a space-time could be handled by tensor analysis, developed by Ricci (et al) about 1900.

In the general theory of relativity, Einstein made the following postulates: (1) nature's laws are the same for all observers no matter how they are moving relative to each other, (2) the geometry of space-time was non-Euclidian, and (3) the degree of space-time curvature is a property of the amount of matter present in a given region. In this theory Einstein was able to unify all the laws of mechanics in physics into one set of equations.

The geodesic, the shortest distance between two points, is a curved line. Planets go around the sun in a particular manner because of the warping effect of space caused by the mass of the sun. It is as if they were in a trough or depression.

From this it followed in Einstein's theory that the motion about the sun should be in the form of a rotating ellipse (in Newton's theory the ellipse should be stationary). Urbain Jean Joseph Leverrier (1811–77) had shown that the orbit of Mercury did rotate about the sun. The amount of observed rotation agreed closely with Einstein's calculations. At the time that Einstein formulated his general theory of relativity this was the only confirming evidence.

His equations further predicted that light should follow the geodesic and should curve when passing a massive body. This was not verified until the solar eclipse of May 29, 1919 when measurements were made of the bending of light from stars passing by the sun. These agreed fairly well and the observations of the 1952 eclipse agreed very well with the theory.

The equations also predicted that light given off by hot atoms in a star should lose some of this energy if the light is moving away from the strong gravitational field of the star or if the star was moving away from the observer. The spectral lines should move toward the red end of the spectrum. This shift towards the red was observed between 1923 to 1928.

The experimental proof of the special theory of relativity came from a number of sources. Alfred Heinrich Bucherer (1863–1927) in 1909 was investigating the radiation from radium and found that the beta rays had different velocities and different masses. Placed into Einstein's equations these checked that as velocity increased so did the mass. This was the first experimental proof and the first verification of the special theory of relativity.[2]

In the world's largest particle accelerator, the European Particle Accelerator (CERN) near Geneva, protons were accelerated to within 100 miles/second of the speed of light (186,000 miles/sec.) and their mass increased 30 times.[3]

$E = mc^2$, the rather famous equation, which relates energy to mass and destroys them as separate entities was proven (twenty-seven years after it was written) by John Douglas Cockroft (1897–   ) and Ernest Thomas Sinton Walton (1903–   ) in England in 1932. They bombarded lithium with protons and the collision broke up the lithium into two parts releasing energy. Of course, it was proven explosively July 16, 1945 in the first atomic bomb.[4]

In 1938 Herbert Eugene Ives (1882–1953) proved time dilation by accelerating hydrogen atoms to 1100 miles/second and observing their spectra. He found that the frequency decreased with acceleration.[5] This was also shown in the famous Harvard clock tower experiment with atomic clocks at the top and bottom. The clock at the top ran slower since it was moving faster.

From 1915 until his death, Einstein was working on equations to pull together all physical phenomena in what he called a unified field theory—in this attempt he was not successful.

Others have continued in these efforts towards what is being called geodynamics. Werner Heisenberg (1901–   ) who claimed recently that he had developed a unified field theory but has not substantiated his claim, is one. John Archibald Wheeler (1911–   ) is another. These latter attempts try to do away with particles altogether and to explain all in purely geometrical terms. It can account for mechanics and electro-magnetic phenomena but not for particles, such as the meson or neutrino, nor does it account for quantized phenomena.

While most of the scientific world has accepted the equations of Einstein and the idea of geometric fields, there are those scientists who are seeking the fundamental particle of gravity, the graviton, so that they can dispute the field nature of the universe and return to a particle centered one.

Einstein had concluded that particles could not have velocities greater than that of light. There are a number of scientists who are disagreeing with this idea and who propose that, while particles cannot be accelerated faster than light that there are particles which travel faster than light, tachyons. These persons include: Gerald Feinberg, Columbia University; E. C. G. Sudarshan, Syracuse University; O. M. P. Bilaniuk, Swarthmore College; Sho Tanaka, Kyoto University; M. E. Arons, City College of New York; and J. Dhar, University of Delhi. They have not discovered the particles which they feel must exist.[6]

It should by no means be considered that the matter is cut and dried. The conflict of ideas goes on.

STUDY AND DISCUSSION QUESTIONS

1. What is relativity?
2. What is a frame of reference? What is yours? What was Galileo's? Einstein's? Newton's?
3. Can you think of an example of a value that is in absolutes? Something which is not?
4. Compare Newton's gravity to Einstein's.
5. What were the problems that arose from Newton's universal law of gravitation?
6. What are non-Euclidian geometries?
7. What experiment is famous because it failed? Why?
8. What proofs are available to indicate that Einstein is right?
9. Is Einstein's theories accepted by everyone?

FOOTNOTES

1. Bernard Jaffe, *Michelson and the Speed of Light,* Doubleday & Co., 1960.
2. James A. Coleman, *Relativity for the Layman,* The New American Library, 1954, pp. 74-75.
3. Hermann Bondi, *Relativity and Common Sense,* Doubleday & Co., 1964, p. 160.
4. Coleman, *Op. Cit.,* pp. 83-84.
5. *Ibid,* pp. 89-91.
6. Dietrick E. Thomsen "Particles Through the Looking Glass," *Science News,* February 22, 1969, pp. 196-197. A further implication of tachyons is that they appear before they are created—causality would be destroyed.

For Further Reading:

Bertrand Russell, *The ABC of Relativity,* The New American Library, 1958.

Lincoln Barnett, *The Universe and Dr. Einstein,* The New American Library, 1950.

Hermann Bondi, *Relativity and Common Sense,* Doubleday, 1964.

Clement V. Durell, *Readable Relativity,* G. Bell & Sons (1926), Harper Torchbooks, 1960 reprint.

O. G. Sutton, *Mathematics in Action,* Harper Torchbooks, 1960.

Stephen F. Mason, *A History of the Sciences,* Collier Books, 1962, Ch. 43.

Leopold Infeld, *Albert Einstein,* Charles Scribner Sons, 1951.

Morris H. Shamos, *Great Experiments in Physics,* Holt, Rinehart and Winston, 1959, pp. 315-328. Einstein's 1905 paper.

Albert Einstein, *Essays in Science,* Philosophical Library, 1934.

James A. Coleman, *Relativity for the Layman,* The New American Library, 1954.

Peter Gabriel Bergmann, *Introduction to the Theory of Relativity,* Prentice-Hall, Inc., 1942. This book is recommended only to the student with a mathematical inclination and four or five years of college level mathematics.

For the experiment by Cockcroft and Walton see *Proceedings of the Royal Society,* A, 137: 229-42. (1932) or a reprint in *Classical Scientific Papers,* Mills & Boon Ltd., 1964, pp. 289-303.

XXXV

Relativism

The ideas in the theories of relativity were at first snubbed by just everyone, as any new idea usually is. Einstein presented his special theory in 1905 and the general theory in 1915. As a close friend and colleague remarked in a public lecture a few years ago that Einstein's work was so original that it contained no footnotes which of course meant that it couldn't be considered very seriously because it wasn't scholarly. Even by the late forties there is little mention of relativity in standard physics texts[1] and by the sixties a number of colleges did not teach the general theory in their physics master's programs.

The Nazis tried to prevent the spread of relativity or at least to get it attributed to someone else because Einstein was considered a Jew. He was forced to leave the Kaiser Wilhelm Institute for this reason and later immigrated to this country. The communists were also opposed to relativity because it destroyed the concept of matter and it is a bit difficult to handle dialectical materialism if matter doesn't matter.

Gradually as a generation of classical supporters died out or retired the idea took hold and spread. Possibly too, the explosion of the "A-bomb" did much to popularize Einstein's ideas. Again, the new radio (1921) which revived the lost art of story telling, spread some of the ideas at a popular level.

> I'd been staying the weekend with a friend of mine who lives about fifteen miles out of Bristol.

> There was another man stopping there, too, who lived at Dawlish. Well, on the Monday morning our host drove us into Bristol in time for the Dawlish man to catch his train, which left a good deal earlier than the London one. Of course, if old Einstein had done his job properly, we could have gone by the same train. As it was, I had over half an hour to wait. Talking of Einstein, wouldn't it be almost worth while dying young so as to hear what Euclid says to him when they meet—wherever it is?[2]

Strangely enough the new ideas of Einstein's touched off or revived a number of questions in philosophy. Philosophy, like religion, had been driven back out of areas which it had formerly had before the positive methods and successes of physics until the assumptions and predictions of Einstein. Again the questions of "reality," "mind-body," "extension," "time," "matter," and "space" arose to be dealt with as well as new problems.

> The modern physicist, while he still believes that matter is in some sense atomic, does not believe in empty space. Where there is not matter, there is still *something,* notably light waves. Matter no longer has the lofty status that it acquired in philosophy through the arguments of Parmenides. It is not unchanging substance, but merely a way of grouping events. Some events belong to groups that can be regarded as material things; others, such as light waves, do not. It is the events that are the *stuff* of the world, and each of them is of brief duration. In this respect, modern physics is on the side of Heraclitus as against Parmenides. But it was on the side of Parmenides until Einstein and quantum theory.[3]

Certainly such a view could not but help to reinforce the ideas of Edmund Husserl (1859–1938), the founder of the modern school of phenomenology which tends to place more emphasis on the event and the experience rather than the material existance.

There is also a return to the Pythagorean ideas of the reality of geometry.

> Greek astronomy was geometrical, not dynamic. The ancients thought of the motions of the heavenly bodies as uniform and circular, or compounded of circular. They had not the conception of force. There were spheres which moved as a whole, and on which the various heavenly bodies were fixed. With Newton and gravitation a new point of view, less geometrical, was introduced. It is curious to observe that there is a reversion to the geometrical point of view in Einstein's General Theory of Relativity, which from the conception of force, in the Newtonian sense, has been banished.[4]

In a sense too the destruction of a material universe has led towards the new existential philosophies. Away from the logic and rational sciences into the mystical searches of the East. Man seeks new methods and new meanings. It has been stated that perhaps a thumbnail sketch of the universe would be that matter is condensed energy; energy is condensed space; and space is nothing. The emphasis on nothingness can be seen in the works of Martin Heidegger (1889–   ) and others.

Perhaps the greatest effect of relativity is to cause an examination of the basic assumptions and approaches to knowledge itself as well as the rather unsettling feeling as a new era begins in the unknown and contrary to what was believed to be common sense. The dissolution of the rigid comfortable frame of concepts of the nineteenth century took place in two distinct stages.

> The first was the discovery, through the theory of relativity, that even such fundamental concepts as space and time could be changed and in fact must be changed on account of new experience. This change did not concern the somewhat vague concepts of space and time in natural language; but did concern their precise formulation in the scientific language of Newtonian mechanics, which has erroneously been accepted as final. The second stage was the discussion of the concept of matter enforced by the experimental results concerning the atomic structure. The idea of the reality of matter had probably been the strongest part in that rigid frame of concepts of the nineteenth century, and this idea had at least to be modified in connection with the new experience.[5]

Heisenberg also points to the danger of applying scientific concepts in areas where they do not belong and that perhaps natural language may be more stable in the expansion of knowledge than the rather precise terms of science.

> In trying to distinguish appearance from reality and lay bare the fundamental structure of the universe, science has had to transcend the 'rabble of the senses.' But its highest edifices, Einstein has pointed out, have been 'purchased at the price of emptiness of content.' A theoretical concept is emptied of content to the very degree that it is divorced from sensory experience. For the only world a man can truly know is the world created for him by his senses. If he expunges all the impressions which they translate and memory stores, nothing is left. That is what the philosopher Hegel meant by his cryptic remark: 'Pure Being and Nothing are the same.' A state of existence devoid of association has no meaning. So paradoxically what the scientist and the philosopher call the world of appearance—the world of light and color, of blue skies and green leaves, of sighing wind and murmuring water, the world is designated by the physiology of human sense organs—is the world in which finite man is incarcerated by his essential nature. And what the scientist and the philosopher call the world of reality—the colorless, soundless, impalpable cosmos which lies like an iceberg beneath the plane of man's perceptions—is a skeleton structure of symbols.[6]

Science, more and more, deals with those concepts not within the realm of perception and has aided in the alienation of man. Less and less does the mathematical language give a picture within the sensory range of man.

> The most, apparently, we can say of language is that it indicates relations, and a Symbolist poem does this just as much as a mathematical formula: both suggest imaginary words made up of elements abstracted from our experience of the real world and revealing relations which we acknowledge to be valid within those fields of experience. The only difference between the language of Symbolism and the literary languages to which we are more accustomed is that the former indicates relations which, recently perceived for the first time, cut through or underlie those in terms of which we have been in the habit of thinking; and that it deals with them by means of what amounts, in comparison with conventional language, to a literary shorthand which makes complex ideas more easily manageable. This new language may actually have the effect of revolutionizing our ideas of syntax, as modern philosophy seems to be tending to discard the notion of cause and effect. It is evidently working, like modern scientific theory, towards a totally new conception of reality. This conception, as we find it today in much Symbolist literature, seems, it is true, rather formidably complicated and sometimes even rather mystical; but this complexity may presently give rise to some new and radical simplification, when the new ideas which really lie behind these more and more elaborate attempts to recombine and adapt the old have finally begun to be plain. And the result may be, not, as Valery predicts, an infinite specialization and divergence of the sciences and arts, but their finally falling all into one system.[7]

There has, as yet, been little movement in this direction; but then forty or fifty years is little in the history of an idea. There are moves toward a symbolic language in philosophy to study the validity of ideas and there is a current move to replace words in common language with sounds which may be more expressive, less restrictive and binding.

Scientists, too, have begun to analyze and examine their own approaches and methods. They may wind up in the ditch like the centipede who had been asked to explain how he walked and became so confused that he no longer could. As we have pointed out in an earlier chapter, scientists do not do what they have said they were doing. It is only recently that they have paid attention to what they have been doing.

> . . . science itself is no longer unified, majestic, or what I have called puritanical. Man's shadow falls across the work. It has always been known that scientists differ in temperament, but seldom admitted that they do not follow *the* scientific method, whether by the term is meant a set of rules or a mental, discipline. Long ago the French mathematician Poincaré could distinguish types among his colleagues—the logical and the intuitive. Other scientists have ascribed success to intuition, "feel," and hunch, rather than to method. It was left for a contemporary physicist and philosopher, A. A. Moles, to detail with examples from a wide range of past and recent achievements the bewildering variety of procedures that lead to great results.[8]

Relativity brought much clarification to a number of fields of science. Cosmology, of which we will speak more of later, had one dilemma which depended on the shape of the universe and Einstein provided the theory for a negative curvature which made it possible to test for its validity.

This was not merely an abstract notion of science that only affected science itself or a minor area in philosophy. The shock to all laymen has been great and the trauma extensive.

But relativity is radically new. It forces us to change deeply rooted habits of thought. It requires that we free ourselves from a provincial perspective. It demands that we relinquish convictions so long held that they are synonymous with common sense, that we abandon a picture of the world which seems as natural and as obvious as that the stars are overhead. It may be that in time Einstein's ideas will seem easy; but our generation as [sic] the severe task of being the first to lay the old aside and try the new. Anyone who seeks to understand the world of the twentieth century must make this effort.[9]

[It may be useful] . . . to note a number of labels recently suggested for our era. One author refers to it as the Era of Violence; another prefers the Age of Anxiety, and this along with the Age of Analysis, might lead us to think of the Age of Freud. Others argue that the twentieth century belongs as much to Einstein as to Freud: in other words, the *Age of Relativity, a notion which appeals especially to the cultural relativists who have cast aside absolute values as outmoded.** A professor of philosophy, writing at mid-century, remarks that we live in an "age of unprecedented cultural crisis and confusion where there is little courage and little sanity," which would suggest either the Age of Uncertainty, the Age of Obscurity, or the Age of Fear. What is the common denominator of all this? Fear would seem a pervading characteristic, but it does not adequately express the loss of nerve, the dehumanization, and the vulgarization of culture that some of the aforementioned terms imply. But whatever the word, the climate of opinion is strikingly pessimistic.[10]

What the relativists want is a physical justification for discarding the absolutes of ethical and moral standards. The sociologists poked about in various groups of men and could find no common denominators. From this they concluded that any action is justifiable by man. When they looked at relativity they did not notice that in discarding the absolutes of space, time, matter, causality and order of two events Einstein assumed an absolute value for the speed of light. But facts do not destroy beliefs. "Facts are the enemy of truth."[11] This is amply demonstrated by the Age of Aquarius, or mysticism, into which we are moving.

It is true that Newton assumed time to be an absolute and yet man has always been aware of the partial relativity of time. When we are engaged in a pleasurable activity time passes much faster than when we are doing something that is unpleasant. On those occasions when we are waiting for something to happen, we all know how slowly time seems to drag. In the chronology of our lives we are aware that the amount of time we have lived and the amount of time that an activity or waiting takes must be put into a ratio to see its relative nature. For example, to a man of sixty a minute represents sixty seconds out of the 1.9×10^9 seconds he has been alive whereas a child of six the same minute represents sixty seconds out of 3.2×10^7 seconds he has been alive. This may be easier to see with a month. One month to a child is $1/72$ of his life span whereas to the man it is $1/720$th of his life span. It is the same month but it does not seem the same. The description given by James Hilton in *Goodbye Mr. Chips* of time slipping past for the aged is an excellent illustration.

The idea then is not a new one but Einstein is speaking of a time difference when two frames of reference are in motion relative to each other. This gives rise to what is called the twin paradox. One member stays on the earth while the other goes off into space at a velocity which is a percent of the speed of light, e.g. 161,000 miles/sec. which would be 86% the speed of light. At that velocity each second of his would be the same as two seconds for the brother who stayed on earth. If his trip took ten years then when he returned his brother on earth would be ten years older than he. This idea has seemed contrary to our common sense and we oppose the idea—even if the opposition is a subconscious one. It also gave rise to a verse which has been popularized.

There was a young girl named Miss Bright,
Who could travel much faster than light,
She departed one day,
In a relativistic way,
And came back on the previous night.

This would, of course, not be possible with the assumptions that Einstein makes because as the velocity of light is approached the mass of the body increases and at the speed of light the mass becomes infinite.

Another phenomena explained by relativity which had been predicted by the Lorentz transformation was the decrease in dimensions as velocity approached the speed of light. At speeds of 50%, 90%, and 99% that of light, length is reduced to 86%, 45%, and 14% respectively of their motionless state. This phenomena has not had as much repercussions as some of the other features of relativity except in the field of science fiction and perhaps some poetry.

There was a young man named Fisk
Whose fencing was exceedingly brisk.
So fast was his action,
The Fitz-Gerald contraction
Reduced his rapier to a disk.[12]

It might be fair for the reader to ask that if Einstein's equations are correct then why haven't Newton's equations been discarded? It seems that

Newton's equations hold for the special cases where velocities are low and masses are neither too small nor too large. Since many common problems fall into these realms it has not been necessary to throw Newton's works aside.

In some areas it has been necessary either to modify Newton's equations or to reject them. In the case of high speeds, i.e. near the speed of light, the mass of a moving body becomes considerable. "Such speeds are now common for electrons in hundreds of varieties of radio tubes and for electrons and other subatomic 'particles' in many types of atom-smashers. The theory of all these devices must take into account the relativistic increase in mass."[13] Certainly since the age of steam, it has been customary to add velocities, i.e. if a vehicle is going 10 miles an hour and a passenger is moving down the passageway at 3 miles an hour then his velocity relative to the ground is 13 miles an hour. Velocities have been considered similar to lengths where 5 inches and 2 inches are 7 inches. However, if we add 170,000 miles/second to 170,000 miles/second we get 186,000 miles/second instead of 340,000 miles/second. The relativist answer does not correspond to our usual expectations. Indeed, aside from the assumptions we have made in the past, there is no assurance that the same operations we apply to numbers holds for velocities at all.

Our conceptional views of a number of things has undergone a tremendous transformation and has made our century as traumatic as the one that Copernicus lived in. Cherished ideas have been abandoned and some of their implications strips away the comfortable feeling that nineteenth century man had with his environment. If one event cannot be shown to have occurred before another, as relativity shows, then the idea of cause and effect must be abandoned. The force of gravity, as a force, is gone and with it the deterministic view and the optimism that man can change the world or that the world is subject to understanding. Matter and energy, as separate concepts, have gone leaving us with a language problem and also the realization that the concept of solidity is gone and it is only the repulsion of protons and electrons that keeps us from falling through the earth. The destruction of other absolutes of time, space, length has contributed to a feeling of relativism and uncertainty. However,

> Einstein's achievement is one of the glories of man. Two points about his work are worth making. The first is that his model of the world was not a machine with man outside it as observer and interpreter. The observer is part of the reality he observes; therefore by observation he shapes it.

The second point is that his theory did much more than answer questions. As a living theory it forced new questions upon us. Einstein challenged unchallengeable writs; he would have been the last to claim that his own writs were beyond challenge. He broadened the human mind.[14]

STUDY AND DISCUSSION QUESTIONS

1. Are the complex situations of daily life subject to the same kind of analysis as is used in physics?
2. Why do some speak of the "bankruptcy" of science?
3. Why have some physical paradoxes interested the "man in the street"?
4. Why doesn't it seem possible to add velocities? What are the inherent problems? Why wasn't this discovered earlier?
5. What is meant by "simultaneity"?
6. What kinds of time are there?
7. What is meant by "causality"? Is it necessary or true?
8. What kind of mental devices do you have? Where did they come from?
9. Why can't we use the analyses of previous generations? Is this the message of the "love-in" group? Of the existential?
10. Why had a sharp turn to idealism occurred at the 20th Century?
11. How is the theory of relativity an example of a metaphysical interpretation of science?
12. How does the theory of relativity support Mary Baker Eddy?
13. Why do the communists attack the theory of relativity? Why did the Nazis?
14. How does the theory of relativity refute materialism? Arguments and examples.
15. In what sense is the theory of relativity dogmatic?

FOOTNOTES

1. See e.g. Harvey E. White, *Modern College Physics,* D. Van Nostrand Co., 1948; or Francis Weston Sears and Mark W. Zemansky, *College Physics,* Addison-Wesley, 1947.
2. From a story by A. J. Alan "The Hair" (1928) in an anthology by Dorothy L. Sayers (ed) *The Omnibus of Crime,* Harcourt Brace, 1929, p. 610.
3. Bertrand Russell, *A History of Western Philosophy,* Simon & Schuster, 1945, p. 70.
4. *Ibid,* p. 217.
5. Werner Heisenberg, "Physics and Philosophy" in Arthur P. Mendel (ed) *The Twentieth Century,* The Free Press, 1965, p. 129 ff. Also see Russell, *Op. Cit.,* p. 540.

6. Lincoln Barnett, *The Universe and Dr. Einstein,* The New American Library, 1948, pp. 123-124.

7. Edmund Wilson, *Axel's Castle,* Charles Scribner's Sons, 1931, pp. 296-297. See also Barzan, p. 237 ff. Percy W. Bridgman "Suggestions from Physics" in A. B. Arons and A. M. Bork (eds.) *Science and Ideas,* Prentice Hall, 1964, pp. 28-36.

8. Jacques Barzan, *Science: The Glorious Entertainment,* Harper & Row, 1964, p. 93. For A. A. Moles' study see *La Creation Scientifique,* Geneva, 1957.

9. James R. Newman "Einstein's Great Idea" in Richard Thruelsen and John Kobler (eds.) *Adventures of the Mind,* Alfred A. Knopf, 1958, p. 220.

10. Roger L. Williams, *Modern Europe,* St. Martin's Press, 1964, p. 489.

11. Dale Wasserman, et al, *Man of La Mancha,* Random House, 1966, p. 40.

12. Cited in George Gamow, *1, 2, 3 . . . ∞* , Mentor, 1947, p. 101.

13. Morris Kline, *Mathematics in Western Culture,* Oxford University Press, 1958, p. 447.

14. Newman, *Op. Cit.*

* Italics, mine.

XXXVI

A Nuclear World

There are two principles that have been cornerstones of the structure of modern science. The first of these was put forward in the eighteenth century after weighing the chemicals going into a reaction and those coming out was that matter can be neither created nor destroyed but only changed in form. It was called the law of conservation of mass and gave us a quantified chemistry which could be successful. The second principle came out of the studies of thermodynamics in the nineteenth century and stated that energy can be neither created nor destroyed but only changed from one form to another. This is known as the law of conservation of energy and has been the plague of inventors of perpetual-motion machines. In addition it has been very useful in the design of advanced heat engines as well as certain areas of chemistry and physics.

They were finally fused together in 1905 by Einstein's $E = mc^2$ although they were not changed in general literature and teaching until after the "Atomic" bomb. The equation also meant that if a small amount of mass, say 2.2 lbs (about 1 kilogram), could be completely changed into energy that the amount would be tremendous, on the order of 25 billion kilowatt hours of energy. In 1939 this would have been enough energy to run all the industry in the United States for a period of two months. Compared to the energy produced by coal, the same mass of coal when burned produced 8.5 kilowatt hours of heat energy, the difference stimulated a great deal of speculation among philosophers, physicists, engineers, and science fiction writers. In the 1920's and '30s there were a rash of stories making use of atomic power that appeared in science fiction stories and in daily comic strips. These speculations in the early years had no support from direct experimental evidence.

A strange new property of matter was discovered by Henri Becquerel (1852–1908) while seeking additional phosphorescent materials that had been exposed to radiation with a Crookes' tube. Any newly discovered phenomena, you will recall, always causes a proliferation of persons and material with a given technique. Becquerel found in 1896 that salts of uranium caused a darkening of a photographic film even when the salt had not been exposed to radiation. He therefore supposed that the darkening was caused by some kind of radiation given off by the uranium.[1]

This family of materials was enlarged by Pierre (1859–1906) and Marie Sklodowska Curie (1867–1934) with the discovery in 1898 of two new elements, polonium and radium.[2] Within four years explanation of the nature of radioactivity and the strange rays emanating from radioactive substances was worked out by Ernest Rutherford (1871–1937) and Frederick Soddy (1877–1956).[3]

In addition to blackening photographic plates, radioactive substances may be detected by their property of ionizing gases. Normally air and other gases do not conduct electricity. Only when the atoms are broken into positive negative charges, or when the electrons are stripped from the nucleus does conductivity occur. Becquerel had found that charge leaked from an electroscope due to the air being ionized and ever since the rate of discharge of an electroscope has served as a measure of the intensity of radioactivity. Even today most detection instruments, Geiger-Müller counters, ionization chambers and Wilson cloud chambers, for studying radiation operate on the ionization principle.

The ionization power depends on the type of radiation given off. It was found that the radiation was of three types, which were named after the first three letters in the Greek alphabet in order in which they were discovered. These were (1) alpha, which are actually the nuclei of helium atoms—sort of a sub-microscopic strip show; (2) beta, which are high speed electrons; and (3) gamma, which are electromagnetic radiations similar to X-rays.

Not only did radioactivity show promise as an aid to the discovery of the structure of the atom, it did, as Einstein suggested, point the direction in showing the equivalence of mass and energy. From 1907 through 1917, Rutherford conducted scattering experiments using high speed alpha and beta radiation through matter in order to determine the interior structure. It was somewhat similar to trying to find out about an unseen forest by shooting arrows into it and seeing what came out on the other side. If only a few arrows came through then you would know that the trees in the forest were placed close together but if, on the other hand, a good many arrows came through then it would follow that the trees were not spaced closely together.

Something of the interior structure could also be told from the breakdown or decay of radioactive material. If an atom emits an alpha particle, the alpha particle is a helium nucleus and has an atomic number of two (which also indicates the number of protons) and a mass of four (which is the total number of protons and neutrons), it becomes an atom of a different element with an atomic number lower by two and a mass number lower by four. The emission by a nucleus of beta particle, an electron, increases the atomic number by one and leaves the mass number unaltered. Sometimes in the decay process the changes also occur with the emission of gamma rays. Only those elements with very high atomic and mass numbers are naturally radioactive. There is no way to know when a particular radioactive atom will decay. However, in a large population of atoms the number which will decay in a given time interval can be calculated. The rate at which this decay will occur is called the "half-life" and differs for each radioactive substance.

When decay occurs the new element forms has the same atomic number as another element but differs in its mass. This phenomena was noted by Soddy in 1913 and he called the atoms that differed from each other in mass isotopes.

Before 1919 no one had been able to break down non-radioactive nucleus or in affecting the decay rate of these elements that were naturally radioactive. In that year Rutherford showed that high energy alpha particles could cause an alteration in the nucleus of an ordinary (non-radioactive) nucleus. He managed to change a few atoms of nitrogen into oxygen. This was the realization of one part of the dream of the alchemist to find the philosopher's stone which would enable a person to change base metal to gold and to prolong life or regain youth. In the decade that followed a number of persons did similar experiments and produced other elements artificially.

Polonium naturally gives off very energetic alpha particles and in 1930 Walther Bothe (1891–1957) and H. Becker found that when these particles struck certain light elements, such as beryllium, boron or lithium, a very penetrating radiation was produced.

In 1932 Irene and F. Joliot showed that if this unknown radiation fell on paraffin or any hydrogen-containing compound it gave off high energy protons. The nature of this unknown radiation was revealed the same year by Sir James Chadwick (1891–   ) who found them to be particles of the same mass as protons but not having a charge.[4]

Attempts continued in the '30's to split atoms with high energy particles. Particles such as helium nuclei, protons, and neutrons were given energy in particle accelerators (commonly called "atom smashers") such as the Van de Graaff generator developed in 1931 and the cyclotron developed by Ernest Orlando Lawrence in the same year. The energy required for the splitting was far more than the amount of binding energy released. If we look at the helium nucleus for example we can see the binding energies involved. The nucleus contains two protons and two neutrons. The mass of the proton is 1.00758 and that of a neutron is 1.00893 so that the mass of the component parts is $2 \times 1.00758 + 2 \times 1.00893 = 4.03302$ but the mass of the nucleus itself is 4.00280. The difference, 0.030 mass units approximately, represents the binding energy. This may seem small but the energy required to completely break up a gram of helium would be 190,000 kilowatt hours of energy.

Equations using the binding energies of the atoms were used in astrophysics by persons like Hans Albrecht Bethe (1906–   ). In the 1930's he worked out a theory to explain the heat of the sun by a cycle of nuclear changes involving carbon, hydrogen, nitrogen, and oxygen from which helium was formed as a product and much energy. During the '40's Bethe worked on the Atomic bomb project and in the late forties and early fifties on the theory for the hydrogen bomb.

Recalling that once a new technique or method is developed everyone plays with the new toy it should seem natural that after bombarding the light elements with particles someone should begin the bombardment of heavy elements. Enrico Fermi (1901–1954) began the bombardment of uranium in 1934 with neutrons but what happened was not fully understood by him at that time. Others besides Fermi were also bombarding the heavy elements with a variety of particles. Fermi was forced to flee Fascist Italy.

Lise Meitner (1898–   ) and Otto Hahn (1879–   ) had worked as a team for twenty years and in 1938 were repeating Fermi's experiment. In the process barium seemed to be produced. At about this time, Meitner's ancestors were proven to be non-Aryan and she fled Germany for Stockholm. On the train Dr. Meitner kept thinking of the barium. She did some calculations assuming that krypton was formed then the binding energy released should be 200,000,000 electron volts per atom. This is 5,000,000 times more energy than that produced from the burning of coal. When she got to Stockholm, she prepared a report to be published in a scientific journal and she sent a cable to Copenhagen to Dr. R. Frisch, the son-in-law of Niels Bohr.

Frisch set out to repeat the Hahn-Meitner experiment in Dr. Bohr's physics laboratories and he sent a cable to Bohr who was visiting Einstein and Fermi in this country. The news reached Bohr Tuesday, January 24, 1939. He and Fermi calculated and independently reached the same conclusions as Meitner. They then rushed to Columbia to test the calculations. The results were obtained on Wednesday. On Friday there was a meeting of theoretical physicists at Georgetown University in Washington, D.C. At this meeting the work of Meitner and Hahn was revealed and the confirmation given.

The event was noted in the newspapers without much fanfare. In an article on page 2 sandwiched between "WPA Speeds Work on Grant's Tomb" and "Republican Group Backs Lotteries" *The New York Times* reported the gist of the event and it was commented that it might be 20 to 25 years before the experimental work could be put to practical use.[5] It was also reported in a broadcast over the Columbia Broadcasting System on February 2, 1939.[6] There was great excitment in the scientific community but little elsewhere.

First attempts to interest our government met with little results. Dean Pegram of Columbia arranged a conference between representatives of the Navy Department and Fermi March 1939 and the Navy expressed interest and asked to be kept informed. In July, Szilard and Wigner conferred with Einstein and later the three men discussed the situation with Alexander Sachs. Sachs was encouraged to see President Roosevelt and carried a letter from Einstein to lend weight to his arguments. Einstein's name was a household word and it was felt that it was better to try to start at the top.

The President appointed a committee which met October 21, 1939 and returned a recommendation November 1, 1939 in which both atomic power and an atomic bomb are mentioned as possibilities. By the next meeting of the committee, April 28, 1940, there were two new factors on the scene. It had been discovered that only the isotope 235 of uranium would split with neutrons and secondly that a large section of the Kaiser Wilhelm Institute in Berlin had been set aside for reasearch on uranium. This committee gradually grew into a military project designated as the Manhattan Project.

There was other information about fission, as the splitting process was called, all over the world in the scientific community by June of 1940. Only a few elements seemed to fission—uranium 235, thorium, and protactinium. The last two elements only fissioned if very fast neutrons were used which would require very high energies and hence were not practical to work with. At some neutron speeds uranium 238 could be changed to uranium 239 (plutonium) which was also thought to be fissionable. Fission seemed to produce one to three neutrons in addition to particles in the size range of barium and krypton. With these facts anyone with basic science knowledge can make a bomb—at least a theoretical bomb. Atom bomb secrets were not from science but from the fields of technology and engineering—not knowledge but know-how.

Secrecy came much later. In hindsight it would seem that any educated reading man should have known what was to come. Articles on atomic research came thick and fast as a counting examination of the *New York Times Index* for the years 1930 to 1945 will show. In 1930 there were 14 articles, 1931 had 30 articles, 1932 showed 43 articles—the annual count stays between twenty and forty up to 1942 when it drops to eleven and three the following year. On April 30, 1939 there is an article in which Drs. Bohr and Onsager comment on the damage that an atom explosion could cause. Other interesting articles that year included: May, 1940—"Nier holds U-235 has little current commercial or military value"; "Prof. W. Krasny-Ergen Speeds Up U-235 Production Rate." One might intelligently ask that if U-235 has little military value then why would there be news value in speeding the production rate?

The principle of operation of an atomic bomb or power plant using uranium fission is simple enough. If one neutron causes a fission that produces more than one new neutron, then the number of fissions can increase tremendously with the release of enormous amounts of energy. The problem with neutrons is that other things can happen to them without producing fission: they can escape, they can be captured by uranium without causing fission, or they can be captured by an impurity. If each fission produces more than one fission then a chain reaction occurs—a neutron causes fission and produces three neutrons, the three neutrons cause fission and produce three neutrons each so that there are nine neutrons to cause nine fissions and releases a total of twenty-seven neutrons which continues the reactions.

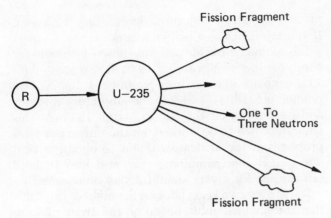

Figure XXXVI-1. Fission.

the size of a baseball. A sphere was found to be the best shape because it has the largest volume with the least surface. Another part of the solution lay in introducing what was called a moderator to reduce the speed of the neutrons since it seemed that the high speed neutrons were the ones captured without causing fission. The substances used would have to be one that was light in weight and yet would not itself absorb neutrons. These include hydrogen, deuterium (heavy hydrogen), beryllium, and carbon (graphite).

In the early phases of the technical work there were considerable problems due to ignorance. Up to this time the use of uranium other than as an item of scientific interest had consisted of pottery glaze and the metal itself had not been prepared in large quantities (in 1941 a few grams existed). They didn't even know the correct melting point.

Another problem was the separation of U-235 from U-238 after the uranium had been purified. Since they are isotopes they differ from each other only in

One of the key engineering problems was what size and shape is necessary to keep a chain reaction going? This size is referred to as the critical mass which was found to be twelve pounds (early work had indicated 200 lbs) and would make a mass about

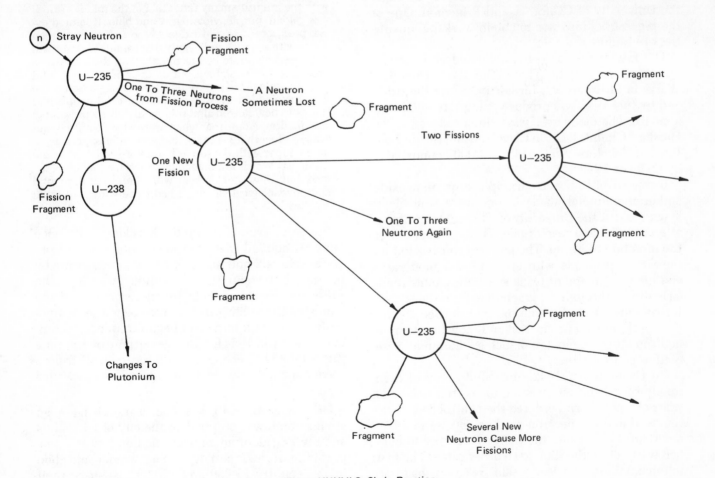

Figure XXXVI-2. Chain Reaction.

weight and cannot be separated by any chemical means. Harold Clayton Urey (1893–   ) who had discovered heavy hydrogen tried the physical method of swinging the atoms in a magnetic field which did cause separation but was too slow to produce separation in quantity. Finally the method used was to change uranium into Uranium hexafluoride, a gas, and to diffuse it through concrete. Chemists had to work out methods even to work with and prepare fluorine as it is extremely poisonous and explosively reactive. In order to separate in quantity a large amount of electric power was needed (one kilogram of plutonium needed 0.5 to 1.5 million kilowatts), a large amount of water for cooling, and a large amount of secrecy. Production sites were set up in two cities which were built—at Hanford, Washington (1943 for plutonium) and Oak Ridge, Tennessee —in a wilderness.

The other alternative was to produce large amounts of plutonium. This work was begun by Fermi in the spring of 1940 (at least some of the theoretical and experimental studies) at Columbia while others worked at the University of Chicago, the University of California, and Princeton. One of the biggest holdups was to obtain workable quantities of uranium.

The raw oxide was being delivered at 15 tons a month by May 1942. The Mallinckrodt Chemical Works in St. Louis was further purifying the oxide and by July 1942 was producing at a rate of 30 tons a month. The metal itself was being produced at the Harshaw Chemical Co. in Cleveland and the du Pont Plant in New Jersey at the rate of 1,000 pounds a day by October 1942.

By the fall of 1942 enough graphite, uranium oxide and uranium metal, about six tons, were available at Stagg Field at the University of Chicago to justify an attempt to build a reactor pile. This was built under the direction of Fermi. The pile was constructed on the lattice principle with graphite as a moderator and lumps of metal or oxide as reacting units regularly spaced through the graphite to form the lattice. Instruments located within the pile showed the intensity of neutron production and movable strips of neutron absorbing material, such as cadmium, were used as control rods.

On December 2, 1942 an excited group of men stood around the pile waiting to see the results of years of work. Fermi ordered the control rods to be removed and the neutron production rose to a tremendous number and the power level rose to one-half watt before the shut down order came. The first self-maintaining nuclear chain reaction had occurred. Today even the building is gone but in its place is a modern sculpture representing the event. If you tap it, it has a hollow sound.

Los Alamos, New Mexico was chosen as the site for bomb testing by November 1942. It was located on a mesa about 30 miles from Santa Fe. J. Robert Oppenheimer (1904–   ) was the laboratory director and he arrived on the site in March 1943. At this laboratory had to be worked out the rather practical problems of the critical size, how to obtain critical size quickly and simultaneously, and how to hold critical size for a very small fraction of a second.

It was over two years later on the night of July 12th that the final assembly began on the atomic bomb. The bomb was mounted on a steel tower 120 feet high. It was detonated at 5:30 A.M. July 16, 1945. There was a brilliant flash—brighter than a thousand stars—the tower was gone and the all too familiar mushroom cloud rising.

> The effects could well be called unprecedented, magnificent, beautiful, stupendous and terrifying. No man-made phenomenon of such tremendous power had ever occurred before. The lighting effects beggared description. The whole country was lighted by a searing light with the intensity many times that of the midday sun. It was golden, purple, violet, gray and blue. It lighted every peak, crevasse and ridge of the nearby mountain range with a clarity and beauty that cannot be described but must be seen to be imagined. It was the beauty the great poets dream about but describe most poorly and inadequately. Thirty seconds after, the explosion came first, the air blast pressing hard against the people and things, to be followed almost immediately by the strong, sustained, awesome roar which warned of doomsday and made us feel that we puny things were blasphemous to dare tamper with the forces heretofore reserved to the Almighty. Words are inadequate tools for the job of acquainting those not present with the physical, mental and phychological effects. It had to be witnessed to be realized.[7]

President Truman was at the Potsdam Conference and was notified that "the baby cried with a lusty voice." He said nothing to Stalin, who was sitting by his side; but Stalin knew it almost as soon as the President. Truman made the decision to use the bomb but the military had decided over a year previously. An air task force had begun training as a unit in the Pacific in 1944 for the sole purpose of dropping the bomb. The technique required take-off from a carrier and a drop at high altitude with a quick turn away.

On August 6, 1945, less than a month later, an atomic bomb was dropped on the city of Hiroshima and the President announced the bomb to the people of the world. It cost every man, woman and child in this country $1,000 and 100,000 Japanese their lives.

STUDY AND DISCUSSION QUESTIONS

1. What have been the cornerstones of modern science?
2. How were they combined and by whom?
3. What were the stages of discovery in atomic energy?
4. When was the potential of atomic energy realized?
5. How was the structure of the atom elaborated? What experiments were crucial and how was the evidence collected?
6. Why didn't Germany develop an atomic bomb?
7. Was the bomb a secret?
8. What were the costs of the atomic bomb?

FOOTNOTES

1. William Francis Magie, *A Source Book in Physics,* Harvard University Press, 1935, pp. 610-613.
2. *Ibid,* pp. 613-616.
3. Stephen Wright, *Classical Scientific Papers,* Mills & Boon Ltd., 1964. Provides accounts of the original papers.
4. *Ibid.*
5. *New York Times,* January 29, 1939.
6. ————, *The Atomic Age Opens,* Pocket Books, 1945 (August), p. 116.
7. Henry DeWolf Smyth, *Atomic Energy for Military Purposes,* Princeton University Press, 1946, p. 254. This is the best single source for information.

XXXVII

Peace or Pieces

"Today the splitting atom is the ingredient of an engine of war—an engine so terrible that it gives the world but two alternatives: *the end of war or the end of humanity.*"[1] Unfortunately it has not meant the end of war—the bomb has fallen into the same category as gas was in WW II in that everyone has it but no one is using it yet.

Before his death in 1931 Edison foresaw that technology would spew forth a terrible weapon—a weapon so terrible that it would make man lay down his arms. Many ages of man have had their weapon of terror. The Pope wanted to outlaw the crossbow for use against Christians. In the fifteenth century Leonardo and Tartaglia shuddered at the thought of what their own military inventions would do. The use of heavy artillery at the siege of Constantinople caused fear and reaction to gunpowder. The first use of Greek fire in the crusades caused dismay and horror. The House of Commons shuddered when Kitchener mowed down the hordes of the Mad Mullah in the Sudan during a battle in which the machine gun was first used. The first use of gas by the Germans at Ypres caused a reaction of horror by the "civilized" nations and by the men who fought in it. Flamethrowers were used in WW I and WW II but some of the particular people opposed the use of napalm in Vietnam. In spite of a thousand years of terrible weapons that scream of man's inhumanity to man there was the hope expressed by many that this time the irrational need of man's violence could be brought under control and end war. The day without war is yet to dawn.

Einstein, when asked of the weaponry of WW III, replied that he did not know about WW III but that after it man would fight with bows and arrows.

For thirty years the threat of a nuclear holocaust has hung over mankind like the sword of Damocles. A whole generation has grown up in the increased tensions of a nuclear age—perhaps it should be called a Benrus age because it's shockproof. Persons cannot live on the raw edge of fear for long and retain their sanity so they learn to live with it without thinking too much about it in much the same way that the Californians have adjusted to living on a geological fault that at any moment might dump them into the Pacific.

The proliferation and improvement of nuclear devices has placed a number of nations in the arms race and the leaders are in a position of overkill so that any war between the major powers, or any accidental firing, will not bring victory to either side but only the promise of retaliation for destruction caused.[2] The military on either side are prepared for heavy losses of the civilian populations and have prepared bomb proof installations for high level politicians and military personnel.

Within a year following WW II tests had been run[3] which showed even with the small Hiroshima type bombs the destruction which could be wrought. There was, at that time, considerable hope that the United Nations would be able to form an effective world government which would be able to control the member nations and others in much the same way that our Federal Government controls the states.

There is still hope that an effective system of nuclear disarmament may be worked out in spite of the failures of disarmaments of the past.[4] The most that can be hoped for is that the bomb will be banned, like gas in WW II, and that "civilization" will survive another war. The hope held by many that the atomic age meant the end of war seems doomed to failure.

Unlike poison gas, which could only be used for the destruction of man, the atomic age offers man a brighter future if he can control himself and survive the new "Four Horsemen of The Apocalypse"—atomic war, and the three "P's" population, poisons (food additives, DDT), and pollution. It may be that the brighter side of the coin may save more lives than those destroyed by the dark side—some might contend that it already has.

Atomic piles which can be used as breeders to produce plutonium for bombs can also be used for the production of heat that can be used in place of coal to make steam, thus reducing air pollution, or it can be used in regions lacking coal and water power. The electricity which can be generated can make possible a new age in the Near East to pump water for irrigation and to provide power for industry. Industry can be free of the bounds imposed by locations of power.

It creates a new problem in the form of thermal pollution because a small rise in temperature in a lake or river can change the ecological balance and destroy life that we wish to maintain. Also, at the present time, fuel for atomic piles is not competitive with fossil fuels.

This tends to be a bar for another application. That is to say in the transportation industry where the only application has been to military craft such as the submarine "Nautilus" which was launched in 1954 and the other related craft since then. No private craft or railways, such as were predicted in the early days of the atomic age, have yet appeared on the scene and it seems unlikely that they will because of the radiation danger.

Radiation is a blessing as well as a danger in a number of useful ways. The chief sources of radiation in the pre-atomic age were radium and X-rays. Today almost any material can be made radioactive by placing it in an atomic pile and indeed this is the chief source of radioactive materials and radioactive isotopes.

In the pre-atomic age radium was used in the treatment of cancer and its use was expensive—radium cost about $25,000 a gram—and limited because only a few grams were in existence. While today cobalt is made radioactive and is used to kill the cancer cells. The cost is reduced and the material is more widely available. A number of cancer victims are alive today because radioactive sources were available to them.

Experiments are underway to use the radioactivity from cobalt to preserve food. Since the inside of most vegetables is sterile treatment of them is somewhat easier. It has already been possible to produce onions and potatoes that do not sprout and whose shelf life is enormously extended. The major problem, still to be solved, is that of taste. Radiation does change the taste in some foods and this has held back development. The time is coming when the refrigerator will become extinct except for making ice cubes and home freezing. It will be possible to store all foods in plastic containers on the shelves and without cooking. One of the major costs of food processing is heat which could be eliminated by radiation also the time for processing would be reduced. The food must be spread out in thin layers so that the radiation can penetrate quickly and passed under and over a cobalt source on a conveyor belt in a pre-packaged form.

One of the fearful aspects of radiation is the increase in the mutation rate. Mutations, sometimes called the production of monsters, are created by a variety of causes chemical and physical which can act on the chromosomes changing the gene makeup. This goes on all the time in nature from cosmic rays and other sources. This increase in mutation rate also has a beneficial side. By the selection of desired mutants it is possible to improve plants and animals to produce those characteristics that are desirable. (Why isn't this selection process used with people?) To look at a specific case let us suppose that there is a disease that affects a particular desirable species of corn. By growing seeds of that species which had been exposed to cobalt radiation a number of mutants would be produced. Most mutants which are produced are called lethal mutants because the particular mutation that they have is not conducive to life, like a white leaf for example. Those mutants that live which also have the desirable characteristic are exposed to the disease. After many tries it may be found that there is one plant produced that is immune to the disease but retains the characteristics we want. That solves that particular problem in an amazing fashion.

The killing of insect pests by chemical means has produced pole-to-pole effects that are undesirable. Radiation is one means to control some insect pests without pollution effects. Male insect eggs are made sterile by radiation. When the eggs hatch the males go forth to mate but the females with whom they mate produce sterile eggs. (Why hasn't this method been used to control human populations? A cobalt source could be put under the counter in welfare offices and each time the adult picked up a check she or he would receive a radiation dose which could be adjusted to produce sterility.)

Industry too has turned to radiation as a cheap means of handling several of its problems.

Radioactive cobalt is used in quality control to detect foreign objects in food products by using a fluoroscope to look inside. Mirrors must be used so that the operator is not exposed to high levels of radiation. This technique also provides a means of examining large metal castings for flaws in an economical way and is far more easy to accomplish than an X-ray machine which would be impossible to position or at least difficult. In these ways, and others, it is possible to produce quality materials at a lower cost.

Industry can also use radioactive sources in various production control techniques. As one example let us look at a rolling mill which produces steel sheets. The metal is forced between two adjustable rollers that squeeze the metal into sheeting. The metal as it comes from under the rollers passes over a radiation source. Above the metal sheet is a detection device which measures the amount of radiation reaching it. This device can be adjusted to relate radiation received to thickness of metal. If the metal sheeting is becoming too thin a signal is sent to the rollers which automatically open wider to make less pressure and hence a thicker sheet; whereas, if the metal is too thick the signal sent increases the roller pressure to make the sheeting thinner. Prior to this a man stood with a vernier caliper and measured the thickness as the sheeting passed through and signaled an operator who adjusted the pressure. Since the sheeting was moving at a high speed this meant considerable sheeting would pass through which was not the correct thickness and hence increased the amount of waste and limited the length of the sheet which could be produced of the same thickness. This application cut costs and increased quality.

Radioactive materials could also be used to measure and meter other bulk materials or items to obtain numbers faster and to further automate bulk processing. Its uses are limited only by the power of man's mind to envision and create.

Other isotopes besides cobalt are used in a variety of ways in a number of fields. Since the isotopes are radioactive there is something different about those particular atoms that can be detected and used to separate those atoms from other atoms of the same kind chemically that are not radioactive. They are then marked atoms. This is a particular property that has found uses in many places.

One problem that is of interest to the scientist, as well as a possible means of solving food problems for large populations, is that of photosynthesis. The green plant changes carbon dioxide, from the air, and water, from the ground, into sugars, plant structures, starches, using sunlight for energy. The question has been of exactly how this is accomplished. By using carbon that has been made radioactive and turning it into carbon dioxide it has been possible to gradually track down a good deal of this process. If we could chemically do the same process to produce only those foods that we wished it would go a long ways toward solving the food problem. This same technique has been applied in areas of chemistry to discover what the atoms are doing when they combine and the steps that they go through.

In medicine the idea of using marked atoms has been applied to chemicals that tend to go and be collected in one particular part of the body. As a specific example iodine, which tends to concentrate in the thyroid, has been made radioactive and used to treat or diagnose disease in that gland. Tumors have also been located and defined with radioactive isotopes. Radioactive isotopes have been used to find blood clots and other circulatory blockage by injecting an isotope of very short half-life and then seeing by using a probe how far the isotope is carried and where it does not go.

Strangely enough even the explosive power of nuclear energy has peaceful uses. The reason being that pound for pound an atomic bomb is less expensive than any other type of explosive.

For some time it has been argued that the United States needs a sea level canal, one without locks, connecting the Atlantic and Pacific oceans because the present one is too small for some large ships and also to prevent the eastern coast from being cut seawise from the west in time of war. It has been calculated that underground explosions, which would not pollute the air with radiation, could produce such a canal in Nicaragua at a considerable savings.

Another problem that nuclear explosions could solve is to provide the developing nations in Africa with harbors. If you examine the coastline of Africa, you will find very few natural harbors anyplace around the whole continent. This factor delayed exploitation and exploration of the "Dark Continent." Today, however, the industrialization of the continent is slowed by the lack of harbors. Ghana, for example, has no harbor and ships are unloaded into small boats that are paddled in across the surf with the loss of considerable cargo. This increases the cost of every import item in Ghana and slows its development and bleeds its economy.

Today, the economy of the first industrial nation of the world, England, is threatened because she is separate from the continent. The plan, which is several hundred years old, has been revived of making a tunnel under the channel to aid in trade with Europe. Import and export items would be cheaper if goods did not have to be taken from trains, put on ships, and then put onto trains. Some boats are used that carry trains but this is still not very effective.

It has also been suggested that our highway system could be improved if there were more gaps in the Rockies. An examination of a highway map will show that there are only certain regions that have highways that go through.

The only application of nuclear explosives so far has been in the field of mining. Any material which is fluid or can be made fluid can be mined by the technique of using nuclear explosives to crush large quantities of ore or to create caverns and vaporize petroleum deposits. The mining of copper sulfate has been done in this manner. After the explosion water is pumped underground and the water dissolves the copper sulfate and it can be pumped to the surface. The production of natural gas is also recent by using underground nuclear explosives to break up the deposits and the heat to produce the gas from low grade oil shale. Low grade oil shale may be the most important source of petroleum and petroleum products in the near future.

As we have seen repeatedly, it is seldom that a device or discovery in technology cannot be put to a useful as well as a harmless purpose. If man acts like a monkey with a machine gun then the results are only too well known to go into. Man must control his inventions and not let his inventions control him.

STUDY AND DISCUSSION QUESTIONS

1. What difference does the manner of a man's death matter?
2. Why did many feel that the atomic bomb meant the end of war?
3. What has happened instead?
4. What are the "good" aspects of the atomic bomb? The "evil"?
5. How can the same properties of the atomic bomb be used for good and evil?
6. Has more good or more evil resulted from the bomb?
7. Should the atomic bomb have been used on Japan?

FOOTNOTES

1. ———, *The Atomic Age Opens,* Pocket Books Inc., August 1945, p. 48. Italics are theirs.
2. Ralph E. Lapp, *Kill and Overkill,* Basic Books, 1962, (especially pp. 23-38).
3. David Bradley, *No Place to Hide,* Little, Brown & Co., 1948.
4. Charles A. Barker (Coordinator), *Problems of World Disarmament,* Houghton Mifflin Co., 1963. Also Morton Grodzins and Eugene Rabinowitch (Eds.) *The Atomic Age,* Basic Books, 1963. Contains original documents e.g. the letter from Einstein to President Roosevelt.

XXXVIII

Cosmology and Terragenesis

Cosmology deals with the origins and make-up of the universe, while cosmogony deals with the origins of the earth. The two terms are frequently confused so that I prefer to use the term terragenesis instead of cosmogony.

Almost all peoples have had some explanation as to the origin of things. To the Egyptians, the earth was flat with Egypt at the center. The sky was supported by four iron pillars and was hung nightly with the lamps of stars by the gods. The sun by day moved along a celestial river and by night through and underground tunnel. To most of the ancients—Babylonian, Egyptian, and Hebrew—the act of creation was explained by one or more gods. In most of these stories there exists a primeval mixture of earth and water which is separated at the time of creation. Only in the Hebrew story is creation accomplished for nothing. The Greek's concept is the bringing of order (cosmos) from chaos but without explaining the origin of the material substance.

It is a blend of the Greek and the Hebrew that is absorbed by the Christian West. The earth is the center of the universe created especially for man. The fixed stars, while beyond the wandering stars (planets), are relatively close and are provided by God to light the heavens at night and to provide a means of navigation for man. The blend of the physical world and the theological are one. It is a unity with each reflecting the totality of the other.

This nice balance is upset beginning in the 16th Century with the Copernican revolution. While opposed by Catholic and Protestant alike, the growth of Neo-Platonism helps to entrench the Copernican view. The earth is dislodged from the center of the cosmos. Kepler and Galileo destroy the perfection of the heavens. Kepler destroys the perfect circular motion of Aristotle while Galileo shows with his telescope that heavenly bodies seem to be made of the same kind of stuff as the earth and to have a physical corruptible nature. Newton applies the capstone by showing that there is a universal law that governs all natural motion. For Kepler, Newton and the other Neo-Platonists, God has become the great mathematician and the essence of God is to be found by an examination of nature to find the great mathematical laws which govern all things.

Newton is one of the pillars that the 18th Century of reason is built upon. And it is during this century that we find the first speculative, rational theories of cosmology and terragenesis. Immanuel Kant (1724–1804), a German philosopher, was the first of these. He had been struck by the idea from the fact that all heavenly bodies seemed to rotate and from the number of nebulae and various spiral forms that could be seen in all parts of the heavens. His theory is somewhat general without too many details but its essence was a cloud of matter which contracted from gravity and began to rotate. As the contraction continued, the rotation increased—much the way an ice-skater does by pulling in her arms to increase her rate of rotation. This rotation flattened the cloud and caused it to form rings. Each of which formed a planet. One obvious fault with this theory is that gravitational contraction cannot cause rotation.

Marquis Pierre Simon de Laplace (1749–1827), a French mathematician, was apparently unaware of Kant's ideas and in 1796 proposed a very similar hypothesis. Laplace overcame the chief objection to Kant's theory by having the cloud already in motion. This theory became known as the Nebular Hypothesis.

The Nebular Hypothesis held until the end of the 19th Century when evidence began to pile up which was inconsistent with the idea. The sun rotated too slowly compared with rate of revolution of the planets. Not all of the members of the solar family rotate in the same direction—the moon of Neptune, three of the moons of Jupiter, and one of Uranus. The planet Uranus does not rotate in the same plane as the other planets. Furthermore, a diffused gas, instead of being drawn together as suggested, would tend to expand infinitely.

These and other objections gradually worked for the destruction of the Nebular Hypothesis. In the first four decades most of the theories had the earth and the rest of the solar system having its origin from the sun. The material was either swirled off the sun in some manner or it was caused by a near collision with a passing star.

About 1900 two men, Thomas Chrowder Chamberlin (1834–1928) and Forest Ray Moulton (1872–1952), from the University of Chicago, proposed what came to be known as the Planetesimal Hypothesis. Chamberlin was a geologist and Moulton was an astronomer. The theory that they elaborated thus had to agree with facts from both the fields. Essentially the theory had a star pass close enough to the sun to draw tremendous tides from the surfaces of the star and the sun. These tides exploded into spurts of matter and was given a whirling motion by the star as it passed. All of the matter given off cooled rapidly giving rise to what are called planetesimals. The larger planetesimals collected the small ones. This occurred rapidly at first and then more slowly. It is a process still going on as the meteorites constantly demonstrate. This theory overcame some of the problems presented by the Nebular Hypothesis but was still not adequate. Mathematical analysis of the dynamics involved using a single passing star showed that not enough motion would have been given. Most geologists also objected to the theory as they felt that the earth must have been molten in its early stages.

Sir James Hopewood Jeans (1877–1946) and Sir Harold Jeffreys (1891–   ), in 1918 proposed a theory to overcome the latter objection. In their theory, called the Tidal Hypothesis, they proposed that the star which passed the sun was very much larger and that when it passed the tide raised on the sun was drawn off in a long filament which broke into drops of planet size. These large drops would be given a rotational motion by the star as it left our sun. While the large drops would cool quickly, there would still be time for the material to separate. Iron would sink to the core, and the lightest material, various light silicates, would form the crust. This theory and others like it were current until the 1940's, although there were a number of problems that these theories did not adequately handle.

A theory of turbulent flow has been developed in the past thirty years. It is largely founded on the mathematical work of Sir Geoffrey Ingram Taylor (1886–   ) in England, Werner Heisenberg (1902–   ) in Germany, Andrei Kikolaevich Kilmogorov (1903–   ) in Russia, and Theodore von Karman (1881–1963) in the United States. By the application of this theory to a mass of cosmic gas, Baron Carl Friedrick von Weizsacker (1912–   ) in Germany and Gerard Peter Kuiper (1905–   ) in the United States were able to revive the Nebular Hypothesis during the 1940's.

The extreme isolation of the stars in the immensity of space casts strong doubts that two stars could pass close enough to cause the tidal effects to create planets. The more detailed calculations made possible by the mathematical theory shows that there are certain critical distances at which rings of material might be collected together. The Weizsacker theory does explain several observable phenomena. The best conditions for the collection of this material would be about midway in the formative cloud or about the location of the orbit of Jupiter—and Jupiter is the largest planet. The distribution of the cloud would explain the sizes of the planets. The flattened disc which would occur in the formative stages would account for all the planets being in the same plane. All the planets except two are within 5° of this plane about the sun. Mercury and Plato are the two exceptions.

There were a number of theories based on modifications of Weizsacker's theory. The most comprehensive of these is the Dust Cloud Hypothesis proposed in 1948 by Fred Lawrence Wipple (1906–   ) of Harvard.

The Dust Cloud Hypothesis begins with the facts that there is more matter in space between the stars than there is in the stars themselves; and that there are collections of dust, known as nebulae, throughout space. Whipple took a suggestion of Lyman Spitzer (1914–   ) that the pressure of light might be enough to drive these dust particles together as the beginning basis of his theory. Once some of these dust particles had been collected they would begin to exert a gravitational influence on other dust particles and would grow in mass. The movement of various parts of the dust cloud under the influence of light and gravity would create turbulences in the cloud. These turbulences would create a spiral motion towards the more dense part of the cloud which would provide a center. As the gas in the largest mass continued to accumulate there would be an increase in temperature due to the force of gravity pulling the particles toward the center. When this temperature reached about 100,000° C. a reaction would begin among the hydrogen atoms producing helium and a vast amount of energy. This body would be the sun. The other smaller bodies that had collected but which had not reached enough mass to create the high temperature would become the planets.

Theories of this nature would account for the planets traveling in the same plane and direction but due to the nature of the turbulences would still allow an explanation for the bodies which do not rotate in the same directions.

HOW THE UNIVERSE HAS GROWN

To the ancients the stars were relatively close and friendly but then they were not able to measure anything but relative distances. In 1672 Cassini of Paris and Richter of Caynne joined forces to measure the distance to Mars using parallax. Then using Kepler's Laws they determined that the distance from the earth to the sun was 87 million miles. (Actual c. 93 million.) While after Copernicus it was known that the stars were very distant, because they showed no measurable parallax, it was not known how far.

The first reliable value for the parallax of a star was made by Friedrich Whilhelm Bessel (1784–1846) on December of 1838. He made his observations on star number 61 in the constellation of the Swan (Cygnus). He found that the amount of apparent shift during a year was 0.32 second of arc.

Hold your finger at a distance of one foot in front of your nose. Close one eye, then alternate eyes and your finger will seem to move back and forth compared to distant objects. If your finger were held twenty feet from your eyes it would appear to move about 1°. For it to appear to move 1 min. your finger would have to be 1,200 ft. away and for it to appear to move 1 sec. it would need to be 12 miles away. A disarming thought.

Great distances in astronomy are measured in: astronomical units = 9.3×10^7 miles; light years = 6×10^{12} miles; or parsecs = 1.9×10^{13} miles. An astronomical unit is the radius of the earth's orbit. i.e., the distance of the earth to the sun. A light year is the distance that light will travel in one year. (the speed of light is c. 186,000 miles/sec.) A parsec is that distance that a star must be from the earth to show a parallax of one second. (parsec from *par*-allax and *sec*-ond).

The method of parallax is only useful up to a distance of 300 light years. What was needed was another method for determining distance. It had been known for some time that any radiation followed the law of inverse squares. i.e., the intensity is reduced by the square of the distance. Light is a form of radiant energy so that the law would also apply here. If we know the intensity of a light at a given distance then at 10 times the distance, the intensity would be reduced to 1/100th of what it had been. But the catch in this problem was to find out how bright a star really was. Stars are a number of different types and their brightness is dependent on what type they are and how far away they are.

In 1917, Henrietta Swan Leavitt (1868–1921) at Harvard and Harlow Shapley (1885–   ) at Mount Wilson Observatory, found a kind of star, called a Cepheid variable because the first star of this type was in the constellation Cepheus, whose real brightness was related to the time or period of the change in brightness. This new tool gave a means of measuring the universe.

The star nearest to us is the sun. It is about 93,000,000 miles away and light from it gets to us in about 8 minutes. The next nearest star is about 4 light years away. i.e., it takes light 4 years, traveling at 186,000 miles/second, to reach us. Stars are not spread out uniformly in space but seem to be grouped in what are called galaxies. Our own galaxy, The Milky Way (what we call the Milky Way is our inside view of the edge), was determined by the above method to be 100,000 light years long and about 20,000 light years thick. It is believed to be shaped like two saucers, one inverted over the other.

Our nearest neighbor outside our galaxy are the Magellanic Clouds. They are about 80,000 light years from the earth, about 70,000 light years from our galaxy. Some of the most distantly observed galaxies are billions of light years away. Man, the solar system, even his own galaxy shrinks to the insignificant beside such distances. Not even a grain of sand on the beach of the universe.

In addition to the large distances in space and between stars, it was found that the stars are traveling at high speeds. Our sun, which is located about 27,000 light years from the center of our galaxy, is traveling around the galactic center at about 180 miles/second. (about 10,800 miles/hour.) The galaxies themselves seem to be moving. Our own galaxy is moving at about 186 miles/second while some galaxies more than a billion light years away are moving at 6,000 miles/second.

How do we know the speeds with which the stars travel? Our knowledge is based on a phenomena similar to that existing for sound that all of you know about. In sound the phenomena is called the "Doppler Effect." When a moving observer passes a fixed source of sound or a moving source of sound passes a fixed observer, the pitch or wave length of the sound changes. You may have observed this at the races or films of races that when the racer passes the observer or camera that the pitch of the engine changes. Edwin Powell Hubble (1889–1953) in 1920, observed that the same phenomena occurs with light. That a light source moving away from the observer will cause a shift in spectral lines toward the red end of the spectrum. The amount of the shift can be used to calculate the velocity of the star. Between

1928 and 1936, Hubble and Milton Humanson (1891–   ) carried out studies of the motion of galaxies.

The energy source for the stars and sun was a great puzzle to the astronomers of the 19th century. No chemical reaction would fill the bill, not even the burning of coal or hydrogen. The sun would burn out in a mere thousand of years. Besides that the composition of the sun was known from spectrographic studies to be over 90% hydrogen so that the amount of material with which the hydrogen could react was limited. Another suggestion that the heat of the sun was due to friction of the atoms rubbing together due to the immense force of gravity would not stand investigation. Lord Kelvin made the calculations that showed that the sun could not be older than 40 million years on that basis and geological evidence alone showed that this time was too short.

About the turn of the century with the discovery of radioactivity and the formulation by Einstein of the mass-energy relationship, astronomers became convinced that the energy of the stars involved some kind of nuclear reaction. Working from pressure and temperature data, as well as the knowledge of the kinds of atoms present, astrophysicists were able to work out the energy reactions on the stars and in the process to come up with the idea of making a hydrogen bomb.

The temperature of the stars is known from the intensity of certain colors in the spectrum. The method is somewhat similar to the method used by a steelworker in determining how hot the steel is from the color of the metal. In some cases this can also be done with very sensitive pyrometers. The total energy of the sun can be calculated by constructing an imaginary sphere with a radius of 93 million miles. Then making allowances for heat loss in the atmosphere the heat striking a particular area can be measured. This quantity of heat is then multiplied by the total area of the sphere. For our sun this amounts to 4×10^{26} joules/sec. Using Einstein's equation we can calculate the amount of mass lost per second as 5 million tons. The mass of the sun is large enough for this reaction to go on for billions of years.

Hans Bethe (1906–   ) suggested several equations by which the sun might produce its energy by the reaction of the nuclei of hydrogen atoms together to produce helium. This is sometimes called the proton-proton reaction. Bethe, Oppenheimer and Teller were members of a group who first suggested that it might be possible to make a hydrogen bomb. There were other reactions of a nuclear type that occur on stars depending on the conditions of temperature and pressure within a particular star.

One of the newer tools of astronomy is the radiotelescope which was developed after World War II. The principle of radioastronomy was discovered in 1931, by a Bell telephone worker who was working on a communication problem. He reported it in a trade magazine and it went unnoticed. During the war interference with radar suggested that somehow the Germans had found a way to bug the system and a group of physicists were put on the problem. They discovered what the Bell man had—that stars, including the sun, produce radio waves. It was the sun, not the Germans, that was "bugging" the system. The advantage of radioastronomy is that while clouds in space, the nebulae, obscure the vision using optical telescopes this is not true for the radiotelescopes. It is, therefore, possible to map portions of the sky that were never possible before. Much can be learned about the structure of the galaxy using this tool.[1]

As new facts are found and as mathematical systems are developed and analyzed, our concept of the universe changes. Questions such as: did the universe have a beginning and will it end? and what shape is the universe? have often been raised and are still with us.

From a strictly logic standpoint there are only four alternatives to the first question: the universe (1) had a beginning and will have an end, (2) had a beginning but will not end, (3) had no beginning but will end, and (4) had no beginning and will not end.

The first alternative as described in Genesis and Revelation was firmly held until the 19th Century. As suggested in the Bible, the end should be one of fire. With the discovery of the second law of thermodynamics it appeared logical that the end should be one of gradually losing heat until all the universe is near absolute zero. This is the so-called heat-death theory of the universe. The second and third alternatives, while logically possible, have never had much support. Today, there are two strong theories that have the most feasibility based on our current knowledge.

The first and the one that has the greatest current support is the "Big-Bang" theory postulated by George Gamow (1904–   ) and others. According to this theory the universe began from a single point in space with an explosion of matter. All the elements were formed within milli-seconds of the first moment and the matter formed is still going away from the center. Calculation based on the velocities of the galaxies in our apparently expanding universe and backtracking these bodies to the point where they would have been a single point coincides with the age of the universe as calculated by aging methods. (The age of our galaxy e.g., seems to be about 20 billions of years.)[2] Walter Baade's (1893–1961) work in 1953, in which he found two kinds of Cepheid variables, changed Hubble's work and tended to support an expanding universe.

The other contending theory is one proposed by Hermann Bondi (1919–   ), Fred Hoyle (1915–   ), and Thomas Gold, (1920–   ) in 1948, called the "steady state" theory. Sometimes it is called the universe according to Hoyle. According to this theory the universe had no beginning and will have no end. The universe then is always in the process either of contraction or expansion. The formation of the elements is explained in terms of being created in the older stars of the system as part of their energy building cycle. When each star reached its novae or exploding state, the elemental materials are dispersed to take part in the building of a new system —new matter is constantly coming from energy and new energy is constantly coming from matter in a never ending cycle. The principle objections to the "steady state" have been from two sources: the shape of the universe, which to support this theory should be spherical, and newly discovered objects in space called quasars. Quasars or quasi-stellar objects seem to be large bodies—larger than stars but smaller than galaxies—not within galaxies. They seem to have greater light than stars and are powerful radio sources. Their nature is not fully understood, but certain theoretical considerations pose problems for the "steady state" theory.[3]

The universe was considered to be Euclidian—i.e., parallel lines do not cross, a straight line is the shortest distance between two points—until it was found that there are other geometries—i.e., non-Euclidian. A Euclidian universe would be planar, infinite, and would not have any curvature. By the 1920's it became apparent from the work of Einstein that the universe was not Euclidian. This left two possibilities: (1) that the universe was spherical, finite and with a positive curvature or (2) that the universe was hyperbolic, infinite and with a negative curvature—a so-called saddle-shaped universe. The evidence collected seems to support the idea that the universe is saddle-shaped and hence would go against the idea of a "steady-state" system.

There are, of course, a number of other theories which we have not considered here. The trailing gas theory by Schmidt and Levin which is supported by Urey's chemical evidence; a modification of Gamow's theory by Goldhaber which proposes an anti-matter universe; John Wheeler's theory which makes the universe and all matter merely aspects of geometry.[4]

STUDY AND DISCUSSION QUESTIONS

1. What is the basis of all the early origins of the earth?
2. When do theories begin to be developed about the origin of the earth?
3. What is the nebular hypothesis and when does it originate?
4. What are the theories of the late 19th and early 20th century based on? Why are they unsatisfactory?
5. What are the differences between the big band and the steady state theories? Which seems most reasonable? Why?
6. What are the various measuring sticks of the universe and their units?
7. How do stars get their energy?

FOOTNOTES
1. See *Science* (April, 1968), pp. 20-42.
2. See *Science* (March, 1964), pp. 1281-1286.
3. See *Science* (December, 1966), pp. 1281-1287. Also see Ben Bova, *In Quest of Quasars* (Cromwell Collier Press, 1969).
4. *Princeton Alumni Bulletin,* (Sept. 21, 1962), pp. 11-16.

Suggested Reading: Cosmology and Cosmogony

Boschke, F. L., *Creation Still Goes On,* McGraw-Hill, 1964.

Struve, Otto, *The Universe,* MIT Press, 1962.

Munitz, Milton K., *Theories of the Universe,* Free Press, 1957.

Gamow, George, *The Creation of the Universe,* Viking Press, 1961.

Bondi, Herman, *The Universe At Large,* Anchor Books, 1960.

Bondi, Herman, *et al, Rival Theories of Cosmology,* Oxford, 1960.

Bondi, Herman. *Cosmology,* Cambridge, 1961.

Whipple, Fred L. *Earth, Moon and Planets,* Harvard, 1963.

Hoyle, Fred. *The Nature of the Universe,* Harper and Row, 1960.

Moore, Patrick. *The Planets,* W. W. Norton and Company, 1962.

"The Man on the Mountain," *Time* Magazine, Vol 87 (March 11, 1966), pp. 80-84.

"The Birth and Life of the Universe," New York *Times* (June 12, 1966), p. 26.

Coleman, James A. *Modern Theories of the Universe.* New American Library, 1963.

Vancoleurs, Gerard de. *Discovery of the Universe.* Macmillan Company, 1957.

Hoyle, Fred, *Galaxies, Nuclei and Quasars.* Harper and Row, 1965.

Suggested Readings: Stars and Recent Stellar Studies.

Bok, Bart J., *The Astronomic's Universe,* Cambridge, 1958.

Durse, W. and Dieckooss, W. *The Stars,* University of Michigan Press, 1957.

Page, Thorton, *et al. Stars and Galaxies,* Prentice-Hall, 1962.

Aitken, Robert G. *The Binary Stars,* Dover Press, 1935.

————. Frontiers in Astronomy (Readings from *Scientific American*), W. H. Freeman and Company, 1970.

Bova, Ben. *In Quest of Quasars.* Cromwell Collier Press, 1969.

Abell, George. *Exploration of the Universe.* Holt, Rinehart and Winston, 1964.

Kahn F. D. and H. P. Palmer, *Quasars.* Harvard University Press, 1967.

Lovell, Sir Bernard. *Our Present Knowledge of the Universe.* Harvard University Press, 1967.

XXXIX

We've Taken the -d out of God

The various forms of expression—philosophy, art, science, literature, poetry, drama, music—of a people often act as a mirror, reflecting an image. Look then and see your culture as it is reflected. Do you see beauty? Or perhaps it is the beast.

All good and all evil can be found in the culture of man. Nothing other than man contains the good and the evil. A pair of hands may create immortal beauty or they may strangle and destroy. A knife—in the hands of a surgeon may cut out a growth and prolong life; or in other hands, may stab and carve death. Never try to blame things as evil. Things are only as good or evil as the hands that hold them.

Look into the mirror. See the irrational, the incommunicable, the discordance, the jarring, the shocking age that is now. It is an age of the rootless. Always on the go—faster and faster—but with less meaning as to the destination—less reason to go. It is an age of the relative—absolutes are gone. It is an age without idealism—two generations living, disillusioned by two major conflicts, and all growing up and living under the threat of imminent destruction. The mirror once bright with the idea of progress and the reform of man, is now tarnished. It is an age where the intellectual has rejected the burden of reform and has ceased to attempt communication of his message to the many.

In this section, we want to examine particularly the state of current philosophy, music, drama and art; and try to determine the extent to which science and technology have played a role in shaping the present portrayed in each.

In philosophy there are two major strands today: the logical empiricist and the existential (Zen Buddhism frequently is adapted). The logical empiricist is objective and deals with problems using systems of logic. To him, validity is sensory based and is subject to laws of logic. On the other hand, the existentialist is subjective and deals with problems using feelings and mysticism. For him God is dead and there is no purpose in the traditional sense. There is no order in the universe. There is only chaos and the irrational. He rejects all that has no meaning for him. This includes all traditions and social customs.

"Existentialist thinking was conditioned by three nineteenth century situations: (1) the situation in philosophy, following Kant and Hegel and the advance of the sciences; (2) the situation for Christianity after the Enlightenment; and (3) the situation of the person lost in the masses of a progressive society, one among many, isolated and organized."[1]

Under Newton's world system, there is order and a divine plan that extends from atoms to the stars and lasts from the 17th century to the 20th. There is predictability. Newton's world system crumbles before Einstein, and unit predictability falls before Heisenberg and the advent of statistical approach. The idea that the world of man can be improved by man—an optimism in progress—begins to fade as some men improve their material world but also improve the means of destruction.

The process of dehumanization is evident in modern science ... in the sense that science reveals phases of natural life which are not connected with the natural milieu which man is habituated. Physics has revealed sounds that we cannot hear and colours we cannot see. And the technical results of modern physics place man in a new and untried sphere, a non-humanized, cosmic milieu. Physics takes pride in its completely ex-centric attitude toward man. The breath-taking achievements of modern technics are connected with the great discoveries of modern physics. That modern technics are dehumanizing man and turning him into a mere technical function is clear to everyone ...[2]

... the chief cosmic force which is now at work to change the whole face of the earth and dehumanize and depersonalize man is ... technics [the summation of modern scientific progress, expecially technical and mechanical progress] ... Man has become a slave to his own marvellous invention, the machine. ... Technics is man's latest and greatest love. At a time when he has ceased to believe in miracles, man still believes in the miracle of technics. Dehumanization is, first of all, the mechanization of human life, turning man into a machine. The power of the machine shatters the integrity of the human image. ... The machine was intended to liberate man from slavery to nature, to lighten his burden of labour, but instead of that it has enslaved man anew and cursed him with unemployment.[3]

Through mass production and automation the individual has lost a sense of identity and the feeling of pride in craftsmanship. The machines can do it better—can do it faster. Man becomes enslaved by the factory bell and whistle and the tyranny of the clock.

Science, in pushing back the boundary of the unknown, help the Biblical critics in the destruction of God. The Christian-ethical tradition begins to disappear. Misuse of the ideas of relativity destroy the absolutes of ethical behavior. "Well, it may not be right for you, but it's all right for me—values are only relative."

In the mid-nineteenth century was celebrated the opening of the Great Exhibition in London which attested in material form their belief in the idea of Progress. This belief continued to grow by leaps and bounds. By the early 1900's there existed men who had seen the last stagecoaches and the first railway trains directed by the new telegraph. From these same telegraph wires they had seen the massive growth of the penny press. They saw the beginnings of a technology which would not only process breakfast foods but ideas as well. They saw electric lights moved into the homes and the first movements of organized labor. Their sons, the last to be breast fed in large numbers and to wear sailor suits, had the excitement of seeing the first men fly.

> They had already witnessed the advent of the motorcar, that major propellant of the twentieth-century momentum. They had seen the first moving pictures, the greatest cultural force of the era. They heard the first radio signals, cheeping like a fledgling lost in outer space.
>
> No heirs of the ages had been granted a more triumphant sense of the human potential. Science was about to set men free and make them happy.[4]

The belief that Progress as produced by technics would solve all man's problems and would continue a bright new golden age was shattered not only by the beastial war that should not have been but by the very inhumanity of man to man as civilians were involved in large numbers. There had been no major world conflagration since 1815, and the near century of peace had lulled men into believing that reason in politics was reigning. Man learned that the very technique which could produce the telegraph, and the airplane could also produce the tank, machine gun, and poison gas.

> The war was a revelation of the evil, the hatred and jealousy which had been accumulating in mankind ... The War revealed the personality of our civilization. It mobilized action, but evil action, rather than good. "Everything for the war." ... It cheapened human life, it taught man to take no thought for human life and personality, to consider them as means and instruments in the hands of the fatality of history.... The war educated a generation of believers in force. The demons of hatred and murder then released continue their activity.[5]

Following the disillusionment of World War I there were attempts to revive the idea of material progress as chemistry came into its own with synthesis of many products unknown in nature and as the sulfa drugs gave a new meaning to Ehrlich's dream of chemotherapy and man's dream to prolong life.

Like the war that should not have been came the Great Depression to bring home the point that man could go hungry and starve in a world filled with plenty. Progress was becoming again a tarnished mirror in which the dreams of man reflected a horror —a nightmare.

The nightmare was accentuated by World War II which occurred in an age which had the promise of great material benefits but seemed only to come when the economy was pumped by war. World War II eliminated the depression but left behind the grim pictures of Auschwitz, Belsen, Hiroshima and Nagasaki.

The "bomb" leaves stress with the feeling that the individual can do nothing to avert the cataclysm that is to come—a philosophy of doom.

Critics of existentialism have called it "A philosophy which has abandoned a rational universe, which insists that the world cannot be thought, must be preoccupied with the super-rational or with the irrational, and absurd."[6]

In drama, the theatre of the absurd illustrates many of the same tendencies that are found in existential philosophy. Frequently drama reflects the current philosophical position and sometimes philosophers use the theatre as a vehicle to communicate, or attempt to communicate, their thoughts.

In the following selection there is a loss of identity shown, the meaninglessness of it all, and the irrational. Although I am sure that all of you have experienced the same sensation that the playwright is trying to convey when you are at a family gathering or listening to your parents talk about those that they know well but that you do not know. The next time this occurs, you might amuse yourself by using the same name for all the people they are talking about and you may see the full meaning of this segment of a play.

Mr. Smith: Bobby and Bobby are like their parents. Bobby Watson's uncle, old Bobby Watson, is a very rich man and very fond of the boy. He might very well pay for Bobby's education.

Mrs. Smith: That would be proper. And Bobby Watson's aunt, old Bobby Watson, might very well, in her turn, pay for the education of Bobby Watson, Bobby

Watson's daughter. That way Bobby, Bobby Watson's mother could remarry. Has she anyone in mind?
Mr. Smith: Yes, a cousin of Bobby Watson's.
Mrs. Smith: Who? Bobby Watson?
Mr. Smith: Which Bobby Watson do you mean?
Mrs. Smith: Why, Bobby Watson, the son of old Bobby Watson, the late Bobby Watson's other uncle.
Mr. Smith: No, it's not that one, it's someone else. It's Bobby Watson, the son of old Bobby Watson, the late Bobby Watson's aunt.
Mrs. Smith: Are you referring to Bobby Watson, the commercial traveler?
Mr. Smith: All the Bobby Watson's are commercial travelers.
Mrs. Smith: What a difficult trade! However, they do well at it.
Mr. Smith: Yes, when there's not competition.[7]

In all of the arts there is the feeling of the creation of something new. Of using new techniques, new media, new form, and a rejection of the traditional and the accepted. The feeling of the irrational—of chaos—is to be found in them all. In painting and sculpture, I am sure that all of you have seen examples of where the artist is expressing himself—getting his subjective feelings into a medium—but not communicating his feeling or ideas to you. In this type of expressionist's art there is the danger or at least the occurrence of those whose art is a fraud. It may be a case of the Emperor's New Clothes or a case of good taste—misplaced.

To the extremist in avant-garde music—all sounds are music. Very literally all sounds—the breaking of glass, a child's toy, a jackhammer, traffic noise—are music. And here again the emphasis is not order but chaos—no particular timing, no particular beat. Rather, noise arranged to suit the feel of the performer-composer. In this type of expression, which I even hesitate to dignify by the name of music, each performer, must of necessity, be his own composer. The same performer would not make the same performance twice. Composition becomes a matter of chance arrangement of sounds with strong discords and a lack of any determination—the indeterminacy principle at work in the arts.

According to some of the avant-garde, there is today an increasing synthesis and overlapping of the arts. This can be illustrated in the case of the poet who reads to the accompaniment of the bongo drums or other noise-making devices. Or in the case of some of the newer motion pictures where there is a blend of noise, color, poetry, and changing forms—this can be seen. The latter is outside of the experience of most of you since it is not the usual fare provided for entertainment. In general, it does not communicate to a very wide audience.

The mirror has not provided a very total picture of philosophy and the arts, nor a very wide one in terms of total population. It is, however, important because the relatively small per cent of the world's population who are doers, the activists, of the rather passive mass. It is a rather elite group communicating amongst itself.

Some facts of the future may be dimly seen with an analysis and comparision of the sixteenth and seventeenth centuries to the developments in the twentieth century.

A means of disseminating information more quickly and to more people became possible with the development of the printing press about 1456, which corresponds to the radio (1920) and television (1950). The dates are when they became generally available. These were means of broadening man's mental horizon.

His physical horizon was broadened in the past by the circumnavigation of the globe in 1522, and by setting foot on the moon in 1969. In both cases there is a strong feeling of new worlds conquered and of the unknown breached. The wealth of the new world came from the first and the wealth of technological innovation from the second.

Man's ideas of himself, of reality, and his philosophy is forced into an unsettling change. In the past the change was wrought by the Copernican revolution (1543) reinforced by Galileo and others in the seventeenth century. In the twentieth century the ideas from Einstein, Heisenberg, and Planck which have been reinforced by the physical observations of the universe have made similar changes in man and his ideas.

René Descartes, in the seventeenth century, attempted to give man a new meaning after he had been displaced from the center of the universe by the Copernican revolution. By eliminating everything but doubt itself, he showed that the action of the mind proved reality of the individual. "I think, therefore I am." The self becomes the center of worlds with absolute freedom in God. There is a certainty which is reinforced by Newtonian physics. With the crumbling of Newtonain physics in the twentieth century and the destruction of certainty by Heisenberg and Einstein man is again left without a position in the universe and without knowledge that he is.

In the works of Jean Paul Sartre (1905–   ) there is an attempt to again reestablish man. The self becomes the center of worlds and Descartes' statement becomes "I am therefore I think" with absolute freedom in man. Mechanical things are pushed around but man is no-thing. The fact of his own existence is enough and is creative in that it makes others exist.

From the seventeenth century came the methods of science and the growth of materialism. What will come of the revolutions of thought after the twentieth century? There are indicators on all fronts that there is a growing rejection of the methods of science to solve all problems. Politically, the government is withdrawing financial support from a number of projects. Socially, there are an even larger number of indicators. There has been a steady growth in the magical, mystical and occult in the last twenty years. More is spent on astrology and its commercial aspects than ever before. There are a number of shows on television which emphasize this direction—"Dark Shadows," "The Munsters," "The Adamm's Family," "Bewitched," "I Dream of Jeannie," and "Twilight Zone" for example. These are mystical or magical and are contrary to the materialistic trend of a generation ago. In science fiction, the general direction of the past thirty years has been the world destroyed or the aftermath of the world destroyed with a return to a simpler life or a movement toward a mental upsurgence. This trend is reflected in the number of oriental mystical revivals such as Zen Buddhism. This antimaterialist trend is seen in the numbers of young persons who have flocked to the Peace Corps and who have formed collectives with a rejection of material society.

Perhaps we are at a new watershed as great as that of the sixteenth and seventeenth centuries. It is hard to say what direction we are going but there are indications that it could be the frontier of a mental rather than a physical world and if man could control his mind then our greatest age is in the future. Certainly science will have a role to play in dealing with the material aspects of the future world and its roots will be in science. Scholasticism did not die out instantly with the rise of Cartesianism and Newton, nor will science die out but it will play a smaller role than it has.

STUDY AND DISCUSSION QUESTIONS

1. In what ways do the arts and sciences act as a cultural mirror? Cite examples.

2. What is characteristic of our age according to the author? Do you agree?
3. What characterizes the existentialist?
4. In what ways does science dehumanize man?
5. What is the difference between technology and technics?
6. What was the significance of the Great Exhibition?
7. What were the effects of World War I and World War II on the idea of progress? Why?
8. What is the significance of the part of the play quoted from Ionesco's "The Bald Soprano"?
9. How do all of the arts seem to say the same thing?
10. What parallelisms does the author see between the sixteenth and seventeenth centuries and the twentieth century? Do you agree?
11. What is the role of science in the future?
12. What hope for mankind may lie in the future?

FOOTNOTES

1. H. J. Blackham, *Reality, Man and Existence.* Bantam Books, 1965, p. 2.
2. Nicolas Berdyaev, *The Fate of Man in the Modern World.* University of Michigan Press, 1935, pp. 36-37.
3. *Ibid.,* pp. 80-81.
4. Reginald Pound, *The Lost Generation of 1914.* Coward-McCann, Inc., 1965, p. 14.
5. Berdyaev, *op. cit.* pp. 13-14.
6. Blackham, *op. cit.* p. 14.
7. Eugene Ionesco, "The Bald Soprano," in *Four Plays* by Eugene Ionesco. Translated by Donald M. Allen, Grove Press, 1958, p. 13.

Suggested Readings:

William Barrett, *Irrational Man,* Doubleday Anchor, 1958, pp. 1-7; 23-41; and 42-65.

E. H. Gombrich, *The Story of Act,* Phaiden, 1966, pp. 455-469.

Alvin Toffler "Future Shock" *Playboy* (Feb., 1970).

Suggested Readings: On Existentialism

Jacques Maritain, *Existence and the Existent,* Pantheon Books, 1948.

Jean Paul Sartre, *Being and Nothingness,* Citadel Press, 1956.

Paul Zillich, *The Courage To Be,* Yale University Press, 1952.

May Rollo (ed.), *Existential Psychology,* Random House, 1960.

DATE DUE		
APR 25 '86		
MAY 14 '90		
FE 5 '94		
MAY 2 7 1997		